Springer Biographies

The books published in the Springer Biographies tell of the life and work of scholars, innovators, and pioneers in all fields of learning and throughout the ages. Prominent scientists and philosophers will feature, but so too will lesser known personalities whose significant contributions deserve greater recognition and whose remarkable life stories will stir and motivate readers. Authored by historians and other academic writers, the volumes describe and analyse the main achievements of their subjects in manner accessible to nonspecialists, interweaving these with salient aspects of the protagonists' personal lives. Autobiographies and memoirs also fall into the scope of the series.

William D. Brewer

Kurt Gödel

The Genius of Metamathematics

 Springer

William D. Brewer
Fachbereich Physik
Freie Universität Berlin
Berlin, n.a., Germany

ISSN 2365-0613   ISSN 2365-0621   (electronic)
Springer Biographies
ISBN 978-3-031-11311-6   ISBN 978-3-031-11309-3   (eBook)
https://doi.org/10.1007/978-3-031-11309-3

Cover photos—Front cover: portrait of Kurt Gödel, Vienna 1935, cropped. Photographer unknown,
licensed by picturealliance/IMAGNO/Wiener Stadt- und Landesbibliothek. Back cover: Gödel's birth house
in Brno. Private, AMZ 2021. Below: Gödel portrait, Princeton 1951, cropped from a photo by Alan
Richards, courtesy of the IAS Princeton archives, from the Shelby White and Leon Levy Archives Center.

This Springer imprint is published by the registered company Springer Nature Switzerland AG
The registered company address is: Gewerbestrasse 11, 6330 Cham, Switzerland

Preface

Yet Another Gödel Biography!?

Kurt (Friedrich) Gödel, one of the most remarkable scholars of the twentieth century, was virtually unknown to the general public during his lifetime (1906–1978), although he enjoyed considerable renown among mathematicians, especially those interested in mathematical logic and fundamentals of mathematics, and among philosophers of science, computer scientists, and theoretical physicists. That situation changed the year after his death with the publication of Douglas Hofstadter's first book[1], the famous *Gödel, Escher, Bach—An Eternal Golden Braid* (*GEB*). It was neither Hofstadter's intention to write a biography of Gödel nor of J. S. Bach, nor of M. C. Escher, but rather to explain his ideas about *strange loops*, his term for self-referential cyclic processes, and to demonstrate them using examples from Bach's music, Escher's paintings and drawings, and Gödel's exposition of the incompleteness theorems, which underlies the latter's scientific reputation to a large extent. Hofstadter also later wrote a biographical article on Gödel for *Time* magazine,[2] which included him among the "100 most influential persons of the 20th Century".

After another 25 years, Gödel had become something of a cult figure, in particular in the years 2006–2008, near the 100th anniversary of his birth and the 30th year after his death, when newspapers and magazines devoted space to articles on his life and works (often with the general theme "Genius &

Madness…"), and objects such as T-shirts and baseball caps bearing his signature or stylized images of his face were offered for sale. That "Gödel craze" has in the meantime subsided for the most part, although the recent publication of some popular biographies is reviving it to some extent (see Tieszen 2017 (Item 43 in Appendix B) and Budiansky 2021 (Item 48 in Appendix B).

Three years after Gödel's passing, Adele, his wife of many years, also died, leaving his papers and correspondence to be archived at the IAS (and later at the Firestone Library) in Princeton. Several mathematicians and historians of science then recognized the need to catalogue and transcribe the items in his legacy, leading to the publication of his *Collected Works*[3] in the following years. In addition, biographical memoirs appeared in the publications (later archived on the Internet) of various scholarly institutions and academies to which he had belonged,[4] and one of his last close associates, Hao Wang, wrote a biography entitled *Reflections on Kurt Gödel.*[5] The first complete scientific biography, *Logical Dilemmas: The Life and Work of Kurt Gödel*, was later written by John W. Dawson,[6] mathematician and historian of mathematics, who was one of the editors of the *Collected Works*. John Dawson also published a brief summary of his biography in *Scientific American* in 1999[7]. Dawson's work set the standard for later biographies, but it has been criticized[8] because Dawson did not know Gödel personally. Hao Wang, who did have close personal contacts to Gödel, wrote two other accounts of Gödel's (and his own) intellectual journey from the fundamentals of mathematics to philosophy. The first of them, *From Mathematics to Philosophy*, was published in 1974, years before Gödel's death, and was in part the result of interactions between Wang and Gödel; Gödel later said that it best summarized his views on philosophy.[9] Wang's later book,[10] *A Logical Journey*, was published posthumously in 1997.

Several other biographies and memoirs have been written subsequently, in particular during the period surrounding the anniversaries of Gödel's birth and death. They in the main emphasized a single aspect of his work (either mathematical logic and set theory, computability, philosophy, or general relativity), and were written from a professional standpoint (by mathematicians, philosophers, computer scientists, or historians of science or philosophy). Two more books, the popular works mentioned above, have also appeared more recently.[11]

A list of (biographical) publications on the life and work of Kurt Gödel is given at the end of this book as Appendix B. Interested readers who want to explore Gödel's mathematical, physical, and philosophical works in greater depth should consult the specialized books listed there. Edited editions of his unpublished essays and notebooks have also appeared in 2016, 2019, and 2020 (Items 16 and 17 in Appendix A). That Appendix summarizes Gödel's own publications, which during his lifetime were not numerous, but nevertheless momentous. Appendix A also includes posthumously published works. [A bibliography of Gödel's published works by John Dawson[12] (1983) appeared in advance of the *Collected Works*].

So why, then, is yet another biographical work on Gödel—this book—of interest? As Gödel's most prominent biographer, John W. Dawson, has pointed out in an article describing the writing of his own Gödel biography[13], scientific biographers approach their subjects from many different avenues, depending on their personal and professional viewpoints, and new ones can often add to the overall picture of the subject in distinctive ways. This is exemplified by the numerous published biographies of Albert Einstein and Richard P. Feynman, to mention two prominent cases. Furthermore, the availability and accessibility of the earlier Gödel biographies for an informed general audience may be limited and will decrease over time. A new biography within the Springer Biographies series, aimed at the audience of that series and widely available as both print-on-demand and e-book, and written from the point of view of a physicist, thus seems very appropriate at present. We emphasize that this book does *not* pretend to serve as a textbook, nor as a work on mathematical logic; there are many such books in the literature, written by competent experts. It is also not a popularization. Rather, it is a detailed introduction to Kurt Gödel's life and work, covering his mathematical-logical, physical, and philosophical contributions in an informal manner, and meant for a general but informed audience.

It has often been noted that Gödel's life and work can be divided into three distinct periods. The first comprises his childhood and youth, which he spent in the city of his birth, *Brünn/Brno*, during his childhood a provincial capital in the Austro-Hungarian Empire, now the second-largest city in the Czech Republic, located in southern Moravia. He finished secondary school in 1924, at 18 years of age, and soon afterwards went to Vienna to begin his higher education. Vienna remained his principal place of residence for over 15 years, and this constitutes the second period in his life. During his first 6 years in Vienna, he completed his undergraduate and graduate studies, obtaining his doctorate in mathematics in 1930.

After writing his doctoral thesis in 1929 (the degree was conferred in early 1930), Gödel continued with his important contributions to mathematical logic, with which he was able to obtain the *Habilitation* in 1933, an advanced degree beyond the doctorate, at that time a prerequisite in German-speaking countries for an academic career as a university teacher and researcher. During the decade of the 1930s, he traveled often, spending all together around 2 years abroad, mostly in the USA, and there mostly at the Institute for Advanced Study (IAS) in Princeton. He also suffered several breakdowns, which necessitated stays in various sanatoria. During this period, the rise of fascism in several European countries changed the political landscape drastically, and after Austria was occupied by Nazi Germany in 1938 (an action known as the '*Anschluss*', the annexation), those political changes began to impinge on Gödel's life in Vienna. He married his long-term lady friend Adele (Nimbursky) Porkert in 1938, and in early 1940, when World War II had already paralyzed most of Western Europe, the two of them left Austria and entered upon a long and complicated journey to Princeton—via Lithuania, Russia (and the Trans-Siberian Railway), by ship across the Pacific, and continuing by transcontinental railway from San Francisco to New York and then, finally, to Princeton.

Gödel was eagerly awaited in Princeton, where he became a member—initially on an annual basis—of the IAS, somewhat later a permanent member, and finally professor of mathematics at the IAS (this last only in 1953). This marks his third period, which covered almost exactly 38 years, from early 1940 until his passing in January 1978. [Another division of the periods of Gödel's life counts the first one—his youth and education—to include the years 1906–1929, while the second, the period of active research into mathematical logic and set theory after taking his doctorate, comprises 1930–1943, and the final period, when he was concerned mainly with physics and philosophy, already in Princeton, extends from 1944–1978. See Wang (1987)].

During this third period, his interest and productivity in mathematics diminished, but he made important contributions to theoretical physics (undoubtedly stimulated in some indirect way by his friendship with Albert Einstein, also a professor at the IAS, which continued until Einstein's death in 1955). His main interest after about 1944, however, was in philosophy, although much of his work in that field remained unpublished during Gödel's own lifetime (see Appendix A, and also Items 7, 19, and 41 in Appendix B).

While there can be no doubt that Gödel's publications on mathematical logic, in particular on the completeness [Gödel (1930)] and the incompleteness theorems [Gödel (1931)], were his most important works, for which he would be remembered (at least within mathematical circles) even if he had done no more work in any other area, they will not form the central topic of this book. There have been many attempts to explain them to a more general audience (i.e., everyone who is not a professional logician or metamathematician, or perhaps a philosopher of science)—see e.g. Rosser (1939), Findlay

(1942), Kleene (1952), Mostowski (1952)—but the first generally accessible explication appears to be the article written by Ernest Nagel and James R. Newman[14] and published in *Scientific American*, in June 1956. Gödel himself was impressed by that article and mentioned it in a letter to his mother (cf. Wang 1997). (In his letter dated August 24, 1956, he complains that the magazine sent a photographer, who took about 50 photos, and then picked the very worst one for the article). It was the precursor to a book by the same authors[15] which expanded on the theme, first published in 1958.

Their book has been reprinted several times, long after its authors' passing. For the new edition published in 2001, Douglas Hofstadter wrote an introduction (and edited the text to some extent). In his Introduction—and later in a Foreword to Karl Sigmund's book[16] on the *Wiener Kreis*, he tells the fascinating story of how the book by Nagel and Newman (which he picked up 'by chance' in a bookstore in 1959, at age 14) became the wellspring of his interest in 'strange loops' and in a real sense the origin of his own book, *GEB*. 42 years later, he had come full circle by assisting in the publication of a new edition of Nagel and Newman, making it accessible to a wider audience. The Nagel & Newman book had a similar effect on the young Gregory Chaitin (cf. Chap. 12).

Together with Hofstadter's own books [*GEB* and Hofstadter (2007)], as well as Nagel & Newman (1958) (Item 1 in Appendix B), and also Franzén (2005), Item 26; Hintikka (1999), Item 16; and Smith (2020) and (2021), Items 32 and 49 there—these works provide a rich source of information and explanation on Gödel's famous theorems for readers who are not necessarily professional mathematicians. Readable, brief but authoritative summaries of Gödel's contributions to various fields are given by the articles in the *Gödel Centennial* issue of the *Notices of the American Mathematical Society* (April 2006); see [Davis (2006)] and the website linked there.

In addition, there are various books on other particular aspects of Gödel's work; for example, on *computability*, see the book edited by Copeland, Posy, and Shagrir (2013), Item 39 in Appendix B; or for Gödel's excursion into *relativity theory and cosmology*, cf. '*A World without Time…*', by Palle Yourgrau (2005), Item 24 in Appendix B.

In Chaps. 6, 7 and 8 of the present book, the historical and scientific background of Gödel's important early works on metamathematics will be explored in some detail, setting the stage for an appreciation of his contributions. There, his doctoral and *Habilitation* theses, which contain his most significant work on mathematical logic, are also presented in some depth, keeping in mind that readers who wish to penetrate still further into the details of this work should also consult the more specialized literature mentioned above.

In Chap. 9, we take a closer look at Kurt Gödel's health throughout his life, both mental and physical. This was an important issue for him during most of his life and an aspect that must be considered if we are to understand his personality and career. The next four chapters concern aspects of his life and work in Vienna after 1931, and up to his emigration to the USA, together with Adele in 1940. The final 7 chapters in this book consider Gödel's path in his 'third period', roughly the second half of his life, in Princeton, where he found a refuge that allowed him to work peacefully, apart from the psychiatric problems which plagued him increasingly as he grew older.

A few remarks on notation and nomenclature are in order: References to articles and books in the literature list ('References') at the end of this book are given in the form '[Name (year)]', where 'Name' is that of the author(s) or editor(s). They are sometimes mentioned directly in the text, but in the main are referred to using *backnotes*, denoted by a raised Arabic numeral in the text, and collected at the end of each chapter. These are numbered in sequence, beginning with '1' in each chapter. Those publications which are of biographical character are in addition collected in Appendix B for quick reference. Supplementary information directly pertaining to items in the text is also contained in some of the backnotes, replacing *footnotes* (which are not appropriate for the e-book version). A summary list of Gödel's own works, published and non-published, is provided in Appendix A, as mentioned above. A more complete and detailed listing, with commentary, can be found in the *Collected Works* [Feferman, Dawson et al. (1986–2014)], and in Dawson (1983) and (1984b). A third Appendix (C) contains a summary of publications on the 'Gödel Universe' by a Brazilian group of theoretical physicists.

Finally, one of those ominous 'notes to the reader': Gödel's insightful and innovative works in mathematical logic, set theory, computability, cosmology, and philosophy are nearly all path-breaking and will, as has often been mentioned in his praise, live on in the annals of scholarship and science. They are also of course not always easy to understand, and I have tried to walk the thin line along explanations which, while not claiming to be rigorous, still contain all the essentials—between genuine mathematically and physically precise and complete expositions, on the one hand; and superficial, perhaps misleading simplifications on the other. I may have strayed from that thin line on some occasions, and my advice to readers is to pass over the more dense material lightly, on first reading. Those who want more can return to it after it has 'settled in', or else go to the scholarly literature which is cited generously in this book. And those who merely want to 'get the flavor' of Gödel's life and works can leave it after the first pass, or simply skip over those 'dense passages' altogether. But it's worth giving them a try!

Notes

1. Hofstadter (1979).
2. Hofstadter (1999).
3. Feferman, Dawson et al. (1986–2014); cf. also Item 5, Appendix B.
4. See 'Gödel' on the websites of the FRS (London); IAS (Princeton); NAS (Washington DC); AMS (Providence/RI); Stanford Encyclopedia of Philosophy; cf. also Item 4, Appendix B.
5. Wang (1987).
6. Dawson (1997); see the CRC Edition (2005).
7. Dawson (1999).
8. Schimanovich (2005) (critique of Dawson 1997), pp. 368 ff.
9. Quoted in Wang (1997), p. 7.
10. Wang (1997).
11. Tieszen (2017) and Budiansky (2021).
12. Dawson (1983) and (1984b).
13. Dawson (2006).
14. Nagel & Newman (1956).
15. Nagel & Newman (1958).
16. Sigmund (2015–2017).

Berlin, Germany   William D. Brewer
April 2022

The original version of the book was revised: Some typographical errors in the chapters have been corrected. The correction to the book is available at https://doi.org/10.1007/978-3-031-11309-3_21

Acknowledgments

Heartfelt thanks are due to my wife, Anna Maria Zippelius (AMZ), for her patience and support during the writing of this book, and for her photography in Brno and Vienna. I am grateful for the cooperation with members of the editorial and production staff at Springer/Nature, in particular Angela Lahee, Ashok Arumairaj, and Gowtham Chakravarthy. And, of course, the book would not have been possible without the scholarly effort of many people who have catalogued, transcribed, and collated the written legacy of Kurt Gödel, which forms the basis for essentially all Gödel scholarship; notable are in particular John W. Dawson and his wife Cheryl, and many others in Princeton and Vienna who have made this material accessible to the interested scholarly public. Furthermore, thanks are due to the many people and institutions who have made drawings, photos and texts available for reproduction or quotation in this book. They are named where the material is cited and attributed, and supplementary information is often given in backnotes. I am grateful for their generosity. WDB, May 2, 2022.

Contents

About the Author

William D. Brewer was born in Boise, ID/USA in 1943. After a BA in Chemistry/Mathematics at the University of Oregon, he obtained his Ph.D. in Physical Chemistry ('*Weak-Interactions Studies by Nuclear Orientation*') at UC Berkeley, working in the group of David A. Shirley.

Following postdoctoral work at the *Freie Universität Berlin*/Germany and the *Laboratoire de Physique des Solides* in Orsay/France, he completed his *Habilitation* at the FU Berlin in 1975 and was University Professor there from 1977 until his retirement in 2008. He has worked in nuclear and solid-state physics by low-temperature nuclear orientation, magnetism on surfaces and in thin films, valence determinations and magnetism by XAS and XMCD with synchrotron radiation. Visiting Scholar at Stanford and IBM Yorktown

Heights/USA, KUL Leuven/Belgium, USP and CBPF/Brazil. He has translated over 20 scientific textbooks and monographs (German to English), translates for the Einstein Papers Project (Pasadena/Jerusalem/Princeton), and is co-author, with Alfredo Tiomno Tolmasquim, of the book '*Jayme Tiomno— A Life for Science, a Life for Brazil*' (2020) in the Springer *Scientific Biographies* series.

1

Prologue

The city of *Brünn* is located between the two old imperial capitals, Prague and Vienna. It is in fact the only urban area of any size between them, lying about 130 km (ca. 80 mi.) north of Vienna and 185 km to the southeast of Prague. The modern city is known by its Czech name, *Brno*. It is the second-largest city in the Czech Republic, after Prague, and the Czech-Austrian border runs in an east–west direction about halfway between Brünn/Brno[1] and Vienna. Before the collapse of the Habsburg monarchy in 1918 and the subsequent founding of the 1st Czechoslovakian Republic, Brünn was a provincial capital in the Austro-Hungarian Empire (it was the capital of the Province of Moravia).

The fabled Austro-Hungarian Empire had a number of epithets—it was called the 'Habsburg monarchy' after its ruling dynasty; the '*Donaumonarchie*' ('Danube monarchy'), after the river that flows through its heartland; the '*Vielvölkerstaat*' ('Nation of many peoples'), given the multiple ethnicities and cultures that were united within its boundaries; and the '*k.-und-k.-Reich*' (from the German adjectives '*kaiserlich*' and '*königlich*', imperial and royal, since its rulers were Emperors of Austria and Kings of Hungary). As Dawson[2] points out, this latter form was shortened to the less ceremonial 'k.-k.' or 'K.K.' after the bonds to Hungary were loosened in 1867; the second 'k' then stood for the Kingdom of Bohemia. Even shortly before World War I, the Empire stretched from Dubrovnik (now in Croatia) in the south to near Dresden in the north, and from the Lake of Constance in the west almost to

© The Author(s), under exclusive license to Springer Nature
Switzerland AG 2022, corrected publication 2023
W. D. Brewer, *Kurt Gödel*, Springer Biographies,
https://doi.org/10.1007/978-3-031-11309-3_1

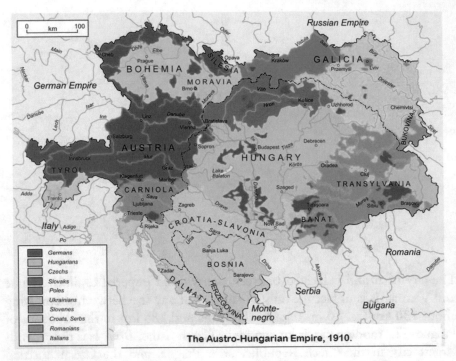

The Austro-Hungarian Empire, 1910.

Fig. 1.1 Map of the Austro-Hungarian Empire in 1910, with colored areas showing the predominant language and culture in each region. Brünn/Brno is a red spot in Moravia, indicating its predominantly German-speaking population at the time, surrounded by a blue sea of Czech speakers. *Source* WikiCommons, public domain[3]

the Black Sea in the east (see the map in Fig. 1.1, showing the extent and the ethnic groups of the Empire in 1910).

Historically, the Austro-Hungarian Empire was the direct descendant of the Holy Roman Empire, which dated back to Charles the First (Charlemagne), crowned 'Emperor of the Romans' in 800 A.D. by Pope Leo III in the old St. Peter's Cathedral in Rome. The initiative for the coronation came from the Pope himself, who had political motives: He wanted to establish a counterweight to the Empress Irene of Constantinople, who was considered to be the heiress to the Byzantine or Eastern Roman Empire, and also to guarantee his own protection by the powerful King of the Franks, which Charlemagne himself considered to be his principal role. Charlemagne made little use of the imperial title, returning to *Aachen/Aix-la-Chapelle*, which served as the capital of his realm. When he died in early 814, he was buried in the cathedral church at Aachen (today, the westernmost city in Germany). His immediate successors retained the title but also made little use of it. The realm of the Franks at that time encompassed most of modern Germany and France, as well as the northern part of Italy.

Charlemagne's dynasty had come to an end by the early tenth century, and his line was succeeded by Otto I, who founded his own dynasty, becoming King of the Germans in 936 A.D. He was crowned as Holy Roman Emperor by Pope John XII in 962. He revived the title and consolidated the territory of the Empire, which had lost some lands in the intervening period; at the time of his death in 973, it encompassed most of modern Germany, the Netherlands and northeastern France, and most of Italy; and it had expanded eastwards to include Bohemia and Moravia. Its eastern boundary bordered on Hungary, from which the Magyars had attempted to invade the Empire in the middle of the tenth century, until they were repulsed by Otto in 955 A.D.

The Empire, often referred to by its Latin initials, S.I.R. ('*Sacrum Imperium Romanum*'), was not at all a traditional kingdom. During most of its history, it was composed of a number of semi-independent regions: individual duchies or principalities, whose local rulers swore loyalty to the Emperor. Some of them, the Prince-Electors (German '*Kurfürsten*'), had the privilege of electing the new Emperor when that became necessary (initially, as 'King of the Romans'; only after being crowned by the Pope could the elected King reign as Holy Roman Emperor[4]). That restriction was however relaxed after the early sixteenth century. The empire did not become a separate nation, but remained for much of its history a sort of superstructure or umbrella organization, uniting a region of independent states (in many ways resembling the modern European Union).

In its early period, the Empire had no real capital; the monarch and his court traveled every few months to a different provincial capital, where they resided in a 'palace' (German *Pfalz*) built for that purpose (more like a longhouse than a modern palace). The word '*Pfalz*' has survived in modern German as the designation of certain geographical regions (e.g., the State of *Rheinland-Pfalz* in the southwest of modern Germany, and the regions *Vorderpfalz* in the west and *Oberpfalz* in the east, in Bavaria); it is usually translated as 'Palatinate' in English. Beginning in the fifteenth century, the suffix '*deutscher Nationen*' ('of German nations') was often added to the name of the Empire, emphasizing the predominance of Germanic languages and cultures within it; that became part of its official title in 1512. A map of the Empire around that time, showing the 'Circles' or groups of member states, is shown in Fig. 1.2.

The center of gravity of the Empire gradually moved southwards and eastwards, and it acquired something like capital cities, where the imperial court

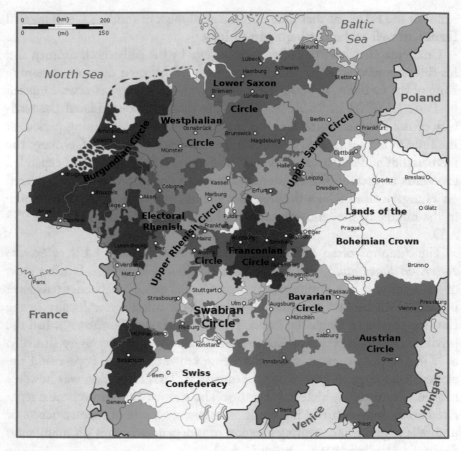

Fig. 1.2 Map of the Holy Roman Empire in 1560, with colored areas showing the 'Circles' or Territories, major regions containing one or more member states with a common ethnic, cultural or political heritage. *Source* Wiki Commons, CC A-SA license[5]

and the emperor resided most of the time. Various dynasties provided the candidates for election as emperor in various periods, beginning with the *Carolingian* period (heirs of Charlemagne), followed by the *Ottonian*, the *Salier*, the *Hohenstaufen*, the *Welf*, the *Luxembourger*, the *Wittelsbacher*, and the *Habsburger* periods. The last two alternated as ruling families during the final 500 years of the Empire, from 1314 until its dissolution in 1806. The Wittelsbacher monarchs traditionally chose Munich as their residence city, while the Habsburger emperors resided in Vienna or Prague (Prague beginning in 1345, for two periods totaling nearly 120 years; Vienna for three periods beginning in 1438 and totaling about 330 years).

The political and integrative power of the empire declined gradually throughout the eighteenth century, and in 1806, its Emperor Franz II, threatened by Napoléon's military successes and wishing to prevent him from usurping the title of Holy Roman Emperor, officially dissolved the Empire, having taken the title *Emperor of Austria* in 1804. After Napoléon's defeat, the Empire was reorganized as the Austro-Hungarian (*'k.-und-k.'*) Empire by the Congress of Vienna (1814–1815). The new Empire survived for just over 100 years, until its last Emperor, the young Karl I, was forced to abdicate in 1918, following the defeat of the Central Powers in World War I, and its territories were reorganized as individual nations, mainly republics or constitutional monarchies. Many of the inhabitants of those new countries later came to regret the passing of the old Empire, whose relative tolerance and freedom for local ethnic groups had been replaced by still more authoritarian regimes which were anything but tolerant.

But now, with this historical background, let us take a step back and focus on Brünn in the mid-nineteenth century, where our story begins. 1848 was a 'revolutionary year', when nationalist and democratic movements in many parts of Europe rose up against the old regimes, in part with a certain success, and attempted to found republics which would be based on the democratic principles propounded by the Enlightenment a few generations earlier. None of these putative revolutions was entirely successful, although they did achieve local regime changes in some places (notably in France, where the "bourgeois king" *Louis Phillippe* was forced by the February revolution to abdicate, and was replaced by Napoléon III, later Emperor of the *Second Empire*).

In Vienna, there were two uprisings, one in March, which obtained some concessions from the monarchy, and a second in October that forced the emperor (Ferdinand I) to flee the city, which was later bombarded. After its recapture, the leaders of the uprising were executed. A more serious uprising took place in Hungary at the same time. In the end, order was restored, and Ferdinand I abdicated in favor of his nephew, Franz Joseph, who became the longest-reigning Emperor in Austrian history after acceding to the throne in December 1848.

In that eventful year, a son was born in Brünn to the family of Josef Gödel and his wife Aloisia Zedniček: their fourth child, who was named *Josef Bernard* Gödel. Josef Sr. was the son of Carl Gödel, also of Brünn. The family worked in the leather trade, and they were reputed to have lived for some time in Bohemia or Moravia. Josef Bernard had three sisters and two older brothers; later, we shall hear more about the oldest brother, Alois Richard (born 1841), and the middle brother, August Josef (born 1846), as well as his sister Anna (born 1854).

Sometime around 1872/73, Josef Bernhard, now grown, married Aloisia Keimel, presumably in Vienna, where her family lived. Apart from the fact that her first name was the same as Josef's mother's name, her family name 'Keimel' is not uncommon in Germany and the Netherlands, and she or her forebears probably came to Vienna from further west. The young family lived in Brünn, like many other Gödels, including Josef Bernhard's siblings Anna and August. Their first child, Rudolf August Gödel, was born on February 28th, 1874. He was followed by two daughters, Maria and Hermine, and another son, Bertold, born in rapid succession in 1875, 1876, and 1877. (One source however states that Rudolf was born in Vienna (cf. backnote [2], Chap. 1, Endnote [5]).) The precise circumstances of Rudolf's childhood and youth are somewhat mysterious. His family moved to Vienna sometime before 1880, when Rudolf was six; but he remained in Brünn, in the care of his Aunt Anna. Just why this arrangement was made is not clear; Dawson[6] suggests that it was due to the early death of his father, Josef Bernhard, who died "shortly after" Rudolf's birth. But this contradicts the family tree, where Josef's death year is listed as 1894, when he would have been 46 and Rudolf, at 20, was already grown. Perhaps the care of four young children, so closely spaced in age, was simply too much for their mother Aloisia, and she wanted to be nearer her own family in Vienna. Josef Bernhard's early death is rumored to have been due to suicide. Dawson[1] has made some remarks about the origin of the family name *Gödel*. It is presumably the diminutive form of Göde, Gode or Gote, old German words derived from the Germanic *Gott* (God), and meaning 'Godfather'; thus Godel or Gödel may refer to 'Godmother'.[7]

Considering the geographical distribution of those names in 1890,[8] one finds them mainly in the north, near the Baltic Sea, in the former province of Pomerania, between the modern Polish cities of *Szczecin* and *Gdansk* (formerly the German cities *Stettin* and *Danzig*), and also in a region just west of Berlin. Some few incidences are found in Saxony and in southern Poland. In modern times, the incidence of Gödel is highest in the south-west of Germany, in Rheinland-Pfalz. Thus, it seems likely that the family may have originated further to the north, moving southward in steps, with a longer stay in Bohemia or Moravia, now parts of the Czech Republic. In Pomerania, the land has been occupied and fought over by Germanic, Slavic and Nordic peoples for centuries, and the name *Göde* may well be derived from a Slavic or Scandinavian root rather than the Germanic. This is suggested for the given name '*Gode*', for example, believed to be derived from Old Norse *góði*, meaning 'good'. Another possibility is the Old Slavic root '*goditi*', 'to be thankful'.[9] Josef Bernhard's mother's family name, *Zedniček*, is of Czech origin, and her family probably originated in the Czech-speaking regions of the Austro-Hungarian Empire, i.e., Bohemia or Moravia.

In any case, Rudolf August spent most of his childhood in Brünn, and he apparently was not unhappy with that arrangement. He attended elementary school there, and started in a *Gymnasium* (preparatory school for academic studies) when he was around 10, but did not do well, so that at age 12 he transferred to a trade school which prepared its pupils to work in the textile industry as weavers. This was an apt choice, since Brünn was a center of the textile industry, and young Rudolf apparently took to the trade quickly.

The map in Fig. 1.3, dated 1903, shows the central, older part of Brünn at that time. The street names were mostly German, and many of the buildings from the later nineteenth century—and still older—remain standing today. After the founding of the Czechoslovakian Republic in 1918, whose population was in the main Czech or Slovakian—apart from German-speaking groups in the northwest and the north (*Sudetenland*, *Schlesien*) and in some cities (the *linguistic islands*), as well as some Hungarian (Magyar) speakers in the southeast and Ukrainian (Ruthenian) and Polish speakers in the northeast—its official languages became Czech and Slovak. Accordingly, the German street names in Brünn, now *Brno*, were replaced by Czech names. The old house numbers were often retained and can still be seen on some buildings. The streets mentioned on the map now have the following names[10]: *Große Bäckergasse* (Great Baker St.) = *Pekařská*; *Elisabethgasse* = *Opletalova*; *Heinrich-Gomperzgasse/ Zwetschkengasse*[11] = *Bezručova*; *Spielberggasse* = *Pellicova*; and *Strassengasse/Wawragasse*[12] = *Hybešova*. The buildings indicated are all still standing and in good condition. Of particular interest is the *Spielberggasse*, a small street which runs along the base of a hill, topped by a much older structure (the '*Spielberg* Castle'). The castle was later a fortress and a prison, and it is now a museum.

In that street, Rudolf August later built a house for his own family, at *Spielberggasse 8a*. The Gödels' villa, a charming structure, has many gables and shows some *art nouveau* influence. Today, it houses a private art gallery called '*galerie 8a*'. Its present-day appearance is shown in Fig. 2.7.

Rudolf August was by all accounts a practically-minded young man, who was not interested in academic pursuits but showed an aptitude for applied problem solving, logic and organization, of which he made good use in his adult life. We will leave him here, in Brünn at age 12 in the year 1886, having set the stage for his later life, and rejoin him in Chap. 2, around 15 years later as a young adult, still living and working in Brünn.

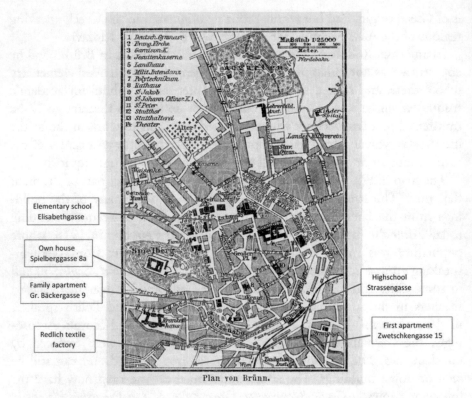

Plan von Brünn.

Fig. 1.3 Historic map of Brünn from 1903, showing the central portion of the city. The arrows point to places that were important to Kurt Gödel or his family in the early years of the twentieth century. *Source*[13] *discusmedia*

Notes

1. We will use its German name *Brünn* in this book, since that was the city's name during Kurt Gödel's early youth there.

   Often, cities and geographical features in Central Europe have both Germanic and Slavic names, and they are frequently used in a rather arbitrary manner in other languages (e.g., the river which runs through Prague is called *Vltava* in Czech, but *die*/the *Moldau* in German and English).
2. Dawson (1997), CRC Edition (2005), p. 12.
3. Wiki Commons, public domain. Reused from: https://commons.wikimedia.org/wiki/File:Austria_Hungary_ethnic.svg
4. See for example Heer (1967).
5. Wiki Commons, Creative Commons Attribution Share-Alike 3.0 license. Reused from: https://commons.wikimedia.org/w/index.php?search=map+holy+roman+empire+1560&title=Special:MediaSearch&go=Go&type=image.

6. Dawson, *ibid.*, p. 3.
7. Apparently, 'Göd' refers to 'godfather' and 'Godl' to 'godmother' in the Viennese dialect. See Schimanovich & Weibel (1997), Note [16] there.
8. From https://nvk.genealogy.net/map/1890:Goedel,1890:Göde, consulted on Jan. 10, 2021.
9. From https://www.namensforschung.net/dfd/woerterbuch/liste/, consulted on Jan. 10, 2021.
10. Street names in Brünn: https://encyklopedie.brna.cz/home-mmb/?acc=profil_ulice&load=1418
11. The word '*Zwetschken*' denotes a kind of plum, and the street may have been near an orchard when it was opened in 1844. In 1896, it was renamed for a local philanthropist, Heinrich Gomperz, who coincidentally had the same name as the Viennese philosopher who was one of Kurt Gödel's teachers in the 1920s. But the name *Zwetschkengasse* was still used on city maps as late as 1910. See the website linked in backnote [10].
12. The older street *Strassengasse* (opened 1782) was renamed '*Wawragasse*' in 1898, after a surgeon of Czech origin in Brünn, Dr. Jindřich Blažej Vávra (1831–1887; German *Wawra*); but the older name was still used on maps up to WW I. It was given its modern name *Hybešova* in 1946.
13. Map: public domain, from *discusmedia*, the 1900 Map Collection; reused from: https://www.discusmedia.com/maps/czech_city_maps/6080/

2

La Belle Époque in Brünn

La Belle Époque—'the Beautiful Epoch'—refers to the roughly 30 years beginning around 1885 and continuing until the outbreak of World War I in 1914. The term arose in Paris but was soon applied to other places throughout Western Europe. It was a time of relative peace and prosperity, corresponding roughly to the 'Gilded Age' in the USA, although it began somewhat later. *Relative*, since there was still great income inequality and poverty; but a solid middle class had arisen in many countries and it had money to spend. Science and technology were producing new achievements almost daily, and the second Industrial Revolution was in full swing, replacing the steam and waterpower that had predominated in the earlier nineteenth century with cleaner and more flexible electrical energy—produced to be sure at central steam- or water-powered plants—and establishing modern communications (telephones, later radiotelegraphy). The arts and music flourished, and towards the end of the nineteenth century a new artistic and architectural style, the 'New Art' (*art nouveau*, or, in German, *Jugendstil*) became popular, filling whole quarters in some cities with the romantically ornate houses and furniture that it favored.

Brünn, the capital of the Austro-Hungarian province of Moravia, was particularly prosperous, a center of the textile industry, sometimes known as 'the Manchester of Moravia' for that reason. Fortunes were made with its spinning, weaving and dyeing works. Like many other cities in the Austro-Hungarian Empire, it had three dominant ethnic groups: the Germanic, the Slavic, and the Jewish. In many places—typical examples are of course

W. D. Brewer, *Kurt Gödel*, Springer Biographies, https://doi.org/10.1007/978-3-031-11309-3_2

Prague, but also *Chernovitz (Chernivitsi)* in the *Bukovina* province and *Lemberg* (*Lvóv/Lvív*) in *Galicia*, today both in the Ukraine—those three cultures interacted to produce a lively and productive creative scene, especially in the arts, literature and theater. The Jewish residents were mainly *Ashkenazi*, whose families had lived in previous times in German-speaking regions, primarily in the Holy Roman Empire, and they usually spoke German as their "public" language, sometimes Yiddish at home. Together with the Germanic residents, they made up over 60% of Brünn's population, while the remainder spoke Slavic languages, mainly Czech. Outside the city, in the surrounding villages and countryside, the population was almost entirely Czech-speaking, as can be seen on the map in Fig. 1.1. By the late nineteenth century, there was a strong Czech nationalist movement, whose proponents were however not necessarily in favor of dissolving the Empire, but rather of obtaining increased political and cultural independence, as had been granted to the Hungarians in the 1860s.

When we left young Rudolf August Gödel at the end of the previous chapter, he was 12 years old, a pupil at a trade school, preparing to work in the textile industry. After he finished school (with honors), he obtained work at the textile factory of *Friedrich Redlich*, where he did well, advancing up through the ranks to become director of the factory, and later a partner in its ownership. At the turn of the century, he still lived with his Aunt Anna in a large and imposing building on the *Große Bäckergasse* (see Fig. 2.1, and also the map in Fig. 1.3). Their building was at No. 9 on that street in Brünn, and its address is now 3–5 *Pekařská* in the modern Czech city of Brno. It was built around an inner courtyard and had common galleries (*Laubgänge*) for the residents on each floor, as was the custom in many multi-family dwellings built in the later nineteenth century in Central Europe.

It also housed the *Handschuh* family. Their *pater familias*, Gustav Handschuh, likewise worked in the textile industry. He had brought his family to Brünn from the Rhineland in what is now southwestern Germany some decades earlier. Gustav Handschuh worked at another textile firm, *Schöller*, where he had a middle-management position. The name *Handschuh* means 'glove', and it is most likely derived from the profession of some forebears of the family, glove-makers or glovers (the latter a not-uncommon name in English, also). However, not much is known about the earlier history of Gustav Handschuh, nor of his wife Rosita Bartl.

Their family had 5 children, two boys and three girls, including a daughter Marianne, who was about 5 years younger than Rudolf Gödel (she was born on August 31st, 1879). The two of them had essentially grown up together, although they never attended the same schools. Marianne was educated at the

Fig. 2.1 The building at *Große Bäckergasse* 9 (now *3–5 Pekařská*), where Kurt Gödel was born. At the left is the northwest entrance, today leading to a pharmacy. That corner of the building is topped by a small tower. At the right is a view from the inner courtyard, looking up at the '*Laubgänge*' (glassed-in corridors running around the inside of the building, serving all the apartments on each floor). Photos: AMZ, 2021

French *Lycée* (*Lyceum*) and was more 'cultivated' than Rudolf. Indeed, their interests and personalities were apparently rather different, as was even their religion—Rudolf's family were traditionally Old Catholic, while hers were Lutheran-Evangelical; however, neither family was seriously religious.

In spite of their different backgrounds and personalities, they were attracted to each other and became engaged to be married. Their marriage was also eminently practical and suitable, given the customs and mores of the time and place: Rudolf would provide the material security for the new family, and work hard to support them, while Marianne would design and organize their household and provide an appropriate personal, cultural and educational environment for their children, and coordinate the family's social life. This was a common arrangement at the time, and both of them could count themselves lucky to have found such an appropriate partner. Marianne, at nearly 22, was at the time definitely of 'marriageable age' and was no doubt happy to be able to move into her own home. Their older son, Rudolf (Jr.), wrote over 80 years later in a memoir of the family (quoted by Wang (1987) and by Dawson (1997)) that their parents' marriage "was not a marriage of love" but was harmonious and provided for a contented family life; and they were certainly materially well off. The various memoirs of their family life hardly mention servants, but given their economic status and the customs of the time, they probably had a cook and a maid, perhaps also a nanny when the children were small, who would most likely have come from the Czech population outside the city. Rudolf Gödel's memoir (1987) does mention a governess, who would have been German-speaking, and they later also had a chauffeur.

It is interesting to compare the family life of the Gödels in Brünn during the later *Belle Époque* to that of the Wittgensteins in Vienna.[1] The youngest son of the latter family, Ludwig, was born nearly a generation before Kurt Gödel, in 1889. His and Gödel's intellectual paths crossed more than once, although they never knew each other personally, and Gödel much later specifically denied any influence of Wittgenstein's philosophy on his own. Ludwig Wittgenstein's father Karl, like Gödel's father Rudolf, was a self-made man, spectacularly successful as an industrialist, proprietor of the largest steel works in Austria-Hungary and incalculably rich by the turn of the century. With his wife Leopoldine (*née* Kalmus), he had 9 children, five sons and four daughters. Their forebears were Jewish, but they were assimilated and were practicing Catholics by the mid-nineteenth century. In spite of their much greater fortune—they lived in a palace in Vienna, and participated actively in the artistic, musical and cultural life of the city—the division of roles between Karl and Leopoldine Wittgenstein was probably quite similar to that between Rudolf and Marianne Gödel. We will return to the indirect influence of Ludwig Wittgenstein on the philosophy and work of Kurt Gödel in later chapters.

The wedding of Rudolf August Gödel and Marianne Handschuh took place in April 1901, when Rudolf was 27 and Marianne was approaching 22. Rudolf was by then already established at the Redlich factory and was financially secure. After their wedding, the young couple moved from the family apartment in the building on the *Bäckergasse* (Fig. 2.1) to their own first apartment at *Heinrich—Gomperzgasse* 15 (or *Zwetschkengasse*; now *Bezručova*).[2] The latter building is shown in Fig. 2.2.

The house on the *Zwetschkengasse* is a somewhat more modest building, in a quiet street outside the center of town, but there, they had their own apartment! The following year, on February 7th, 1902, their first son, Rudolf (Jr.) was born.

Sometime not long after Rudolf's birth, they moved back to the *Bäckergasse*, to an apartment adjacent to Marianne's parents and Rudolf's Aunt Anna. Very probably the advantages of having their own apartment were

Fig. 2.2 The house at *Zwetschkengasse* 15: *Left*, the façade; *right* the entrance door. Photos: AMZ, 2021

outweighed by those of having nearby relatives once they had a small child to care for.

There, on April 28th, 1906, their second son *Kurt Friedrich* was born (Fig. 2.3). They certainly would not have dared to imagine at that time that he would make their family's name famous across the world a century later! As we can see, there was nothing particularly remarkable in the families of Kurt Gödel's parents, nor in his childhood surroundings, that might have presaged his later achievements in mathematics, physics, and philosophy.

An interesting insight into the lives of some of Kurt Gödel's paternal relatives is given by the contribution of a long-time resident of Brünn of nearly the same generation as Kurt Gödel. *Dora Müller*, born in 1920, arrived with her family in Brünn at the age of 10, just when the Rudolf August Gödel family had effectively departed that city. But she soon became acquainted with her schoolmate, Kurt's second cousin *Pauli*, the granddaughter of *Alois Richard Gödel* (born 1841; he was the older brother of Josef Bernhard and thus Rudolf August's paternal uncle). Their branch of the family had remained in Brünn and worked in the leather trade, which the Gödels had practiced for generations. Dora Müller and her schoolmate Pauli Gödel became good friends and remained so throughout their lives, and long after their school days they became in-laws when Dora's daughter married Pauli's son, so that Dora became a "relative by marriage" of the Gödel family.

Fig. 2.3 Kurt Gödel and his brother Rudolf (Jr.), around 1907, when Kurt was about 14 months old. Photo[3]: Gödel archive, IAS/Firestone Library, Princeton

In a talk[4] given at the '*Brno Kurt Gödel Days*', a conference held in Brünn/Brno in 2006 in honor of Kurt Gödel's 100th birthday, Dora Müller recounted some of what she had learned of the family from Pauli and her brother Adolf, and from older members of their family. She also corresponded with Kurt's brother Rudolf in this regard. Her interest in that other branch of the Gödel family was aroused by reading Douglas Hofstadter's book *Gödel, Escher, Bach* (which was published in German in 1985).

Dora learned the story of Alois Richard Gödel, the oldest son of Josef Gödel and Aloisia Zedniček, and thus Josef Bernhard Gödel's oldest brother. Alois, following the family tradition, was apprenticed to a leather cutter (presumably in Brünn), and after several years of learning that trade, he followed the tradition for young men who had completed their apprenticeship in some handcraft and went '*auf die Walz*', that is he traveled for a year throughout Central Europe, wearing the traditional apprentice's clothing, alone or in the company of other apprentices, living off the land and the generosity of colleagues. When he returned in 1865, he set up shop in Brünn, near the *altes Rathaus* (the old City Hall, No. 8 on the map in Fig. 1.3) in the *Rathausgasse* (*Radnická*) street.

Fig. 2.4 *Left*: the entrance to the Old City Hall (*altes Rathaus*) in Brno. The passage leads right through the building to the street behind (*Schwertgasse/Mečová*). Photo[5]: Wiki Commons, used under CC A-SA 4.0 license. *Right*: the *Rathaus* tower as seen from the inner courtyard. Photo: AMZ, 2021

The old *Rathaus* has long since been supplanted by a newer structure, but it is still preserved as a historic building, with an impressive Neogothic entrance leading to a passage which traverses the whole block, emerging behind the *Rathaus* in the *Schwertgasse* (*Mečová*) (see Fig. 2.4). In that small street, Alois' wife Karoline (*née* Baumann) operated a café with the charming name *Zur Gödelmutter* ('At Mother Gödel's'), making a pun on the meaning of the name. She opened in the early morning to serve the workers and market sellers from the nearby *Zelný trh*, the '*Kohlmarkt*' or Vegetable Market, where fruit and vegetables for the whole city were (and still are) sold. According to Dora Müller's informants, it was Karoline's earnings from her café which kept her family supplied with food and clothing, since Alois was not very diligent at his leather business, being more interested in preparing historical drawings and descriptions.

The passage through the *Altes Rathaus* contains a stuffed crocodile ('the Brno dragon'), hanging from the ceiling, and a large wagon wheel mounted

Fig. 2.5 The family of Rudolf August and Marianne Gödel, around 1910. Kurt, center, was about 4 years old, and Rudolf (Jr.) was about 8. Photo[7]: from the Gödel archive, IAS/Firestone Library, Princeton

on the wall. They are the symbols of Brno today, and their stories are given on the city's website (in several languages).[6]

A second story recounted by the 86 year-old Dora at the conference was that of the third brother, August Josef Gödel, born 1846, another great uncle of Kurt Gödel. Defying the family tradition of leatherwork, he became a schoolteacher, whose unconventional ways made him a favorite of his youthful pupils, if not of the school administrators. His significance in our story, and the punchline of Dora Müller's talk, is the fact that the public records of Brünn list August Josef, supposedly a "leather dealer", as a witness to the marriage of Rudolf Gödel and Marianne Handschuh on April 22nd, 1901.

We should note that most of what we know about the Gödel family (Fig. 2.5) during the first 25 years of the twentieth century is due to the memoirs and letters of their older son Rudolf Jr., written many years later, and to the correspondence between Kurt and his mother Marianne in the years following World War II, when he was living in the USA and she in Vienna, beginning in 1946 and continuing up to her death 20 years later.[8] Kurt himself supplied a certain amount of information on a questionnaire which was sent to him by Burke D. Grandjean, a statistician interested in data collection via personal questioning, and therefore called 'the Grandjean questionnaire'.

He sent it to Kurt Gödel in 1974, and Gödel wrote out answers by the following year, but—in typical Kurt Gödel fashion—never returned it to Grandjean. It was found in Gödel's papers after his death, and it is included in its entirety in Wang's (1987) Gödel biography, and is also quoted by Dawson (1997) and other biographers. We will quote from it in later chapters where appropriate, but readers interested in seeing all of it in detail should consult Wang (1987). Wang also quotes from a letter sent to him by Kurt's older brother Rudolf in response to an inquiry, which predates Rudolf's later family memoir [the latter is included in Weingartner & Schmetterer (1987) and in Köhler et al. (2002)]; we denote that letter here simply as RG (1985).[9]

In any case, the most important thing for young Kurt in his first years of life was to have a stable, loving family environment that provided for both his physical and for his emotional and developmental needs. That was clearly the case. As noted in RG (1985), "*It was a harmonious family life; I got along with my brother very well, as did both of us with our parents. We played mainly with each other and had few friends; quiet games—a construction set, a model railroad, in the war of course also tin soldiers.*" This remark concerns mainly the period of Kurt's school days, after he began elementary school in 1912. There is little mention of his early childhood, except that "*he was a happy child, but however around age 5 he had an anxiety neurosis, which later completely vanished.*" Rudolf also reported that during this time, at age 4 or 5, Kurt was unusually attached to their mother, Marianne, and would cry whenever she left their apartment; this is unusual for a child of that age. This was probably considered by Kurt's parents to be a harmless 'phase' which passed without lasting effects, and clearly that is what his brother also thought (many years later, after Rudolf had practiced medicine for some 40 years). It was perhaps indeed more significant, and we will return to this question in Chap. 9.

It is clear from Rudolf's description that the Gödel boys spent most of their free time at home, even after they were both in school. During Kurt's first 6 years, he was surrounded by adults and his considerably older brother. His maternal grandparents and his aunt (Pauline, Marianne's older sister) lived next door, as did his paternal Great Aunt Anna. Their house in the *Bäckergasse* had an interior courtyard and ample space inside where the boys could play (compare Figs. 2.1 and 2.6). The street in front was busy and not suitable for young children. The little street beside the house, leading back down to the next major street below (*Husova*), is now called '*ulička Kurta Gödela*' ('Kurt Gödel's alley').

When young Kurt was 7, soon after he began elementary school, his family moved from the older apartment on the *Bäckergasse* to a newly-built villa

Fig. 2.6 *Left*: The north façade of the *Bäckergasse* house, seen along the modern street *Pekařská*. *Right*: The street sign on the west side of the house. Photos: AMZ, 2021

not far away (cf. the map in Fig. 1.3), on the southern slope of the *Spielberg*, a small mountain in the center of Brünn, at *Spielberggasse* (*Pellicova*) 8a (Fig. 2.7). This short-distance move changed their living situation considerably. The house itself had upstairs apartments intended for Aunt Anna and for Marianne's sister Pauline, who however did not occupy them right away. Downstairs was ample room for Rudolf August's family; but most importantly, the plot of land on which the house stood was large and included a parklike area behind the house which extended some distance up the hill. This meant that the Gödel boys had a 'private space' where they could play undisturbed.

Kurt's brother Rudolf however reports that Kurt did not make much use of their large garden. He was more interested in quiet, indoor pursuits, in particular intellectual pursuits, and did not spend much time in outdoor games or gardening, in contrast to Rudolf himself.

Young Kurt, according to his brother's reports, was a rather quiet child, not at all outgoing, who stuck firmly with his own opinions once he had formed them, and was very concerned with having a structured environment, both physically and mentally. He, like many children, went through a period of asking incessant questions—it was important to him to understand the world in which he found himself. The family called him "(*der kleine*) *Herr Warum*" (loosely translated, 'Little Mr. Why') because of his many questions. It would seem that his earlier childhood was accompanied by a certain anxiety

Fig. 2.7 *Left*: The Gödel villa seen from *Pellicova* Street (formerly *Spielberggasse*). Note the old-fashioned lightning rods on the roof, and the fanlight in the gable. *Right*: The villa viewed from above and behind, looking down from the Spielberg Park. Photos: AMZ, 2021

and a longing for security. Wang (1987, 1997) writes repeatedly of his 'need for security and precision'. In short, he was a thoughtful, sensitive child who had an innate desire to understand the world, and a deep-seated need for *predictability*.

What, then, was the intellectual and cultural environment into which Kurt Gödel was born? Brünn had a long history and a certain intellectual tradition, although, unlike Prague or Vienna, it did not have an old and venerable university (only a (German) Technical College. Later, in 1919, after the founding of the 1st Czechoslovakian Republic, a Czech university—the Masaryk University—was founded, and it is now the second-largest university in the modern Czech Republic, after the Charles University in Prague. The latter is one of the oldest universities in Europe, founded in 1348 by the Emperor Karl IV).

Brünn was in fact an enclave, a *Sprachinsel* (a 'linguistic island'). Dawson[10] and Goldstein[11] refer to its German-speaking inhabitants as '*Sudeten* Germans', but that is a modern term which has sometimes been applied to *all* of the German-speaking citizens of Bohemia and Moravia after 1918, and especially in 1939–45—incorrectly. The name '*Sudeten*' originally referred to a mountain range at the northern borders of Bohemia. The adjoining region had been inhabited by both Germanic and Slavic peoples for many centuries, beginning well before the Carolingian period.

More appropriate terms are the older appellations '*deutschböhmisch*' and '*deutschmährisch*' (Bohemian Germans and Moravian Germans), which apply generally to all the German-speaking peoples in those provinces. There were several other *Sprachinseln* besides Brünn: *Budweis* and *Iglau* in Bohemia, and *Olmütz, Schönhengstgau* and *Wischau*, to the east and northeast of Brünn in Moravia.

When the Hapsburg Empire broke apart in 1918, many of the German-speaking people in the '*Sudetenland*'—and also in nearby Austrian/Moravian *Schlesien* (Silesia), that part of the Province of Silesia which had remained under Austrian rule after the Silesian Wars with Prussia, 1740–63—tried unsuccessfully to obtain independence from the newly-founded Czechoslovakian Republic (and to join Germany or Austria). The German-speaking citizens of Brünn (including the major portion of its Jewish inhabitants), around 2/3 of its population in the census of 1906, accepted the new country, if at times reluctantly, as did the populace of the other *Sprachinseln* in Bohemia and Moravia. They were culturally, politically and linguistically distinct from the *Sudetenland-Schlesien* Germans; and, being isolated by land from larger German-speaking regions in Germany and Austria, had little hope of obtaining citizenship in those countries. The map in Fig. 1.1 clearly shows the distribution of German-speaking citizens in Bohemia and Moravia: the *Sudeten* and *Silesian* Germans are the red bands to the northwest and the north of Bohemia and Moravia, while the *Sprachinseln* are distinctly seen as red spots against the blue (Czech) background.

One can find long lists of prominent natives and citizens of Brünn/Brno in a variety of books and Internet sites, and just a few are given here to illustrate the variety and quality of the talent brought forth by that city up to the beginning of World War I:

Theodor Gomperz was born in Brünn in 1832. After studies there and in Vienna, he became a *Privatdozent* and later Professor for Classical Philology at the University of Vienna. He resigned from his professorship in order to work on his most important book, '*Griechische Denker*' (*Greek Thinkers*, 1893), receiving many academic honors for his works. His son *Heinrich*, born in Vienna in 1873, obtained his doctorate under Ernst Mach in 1896, and became Professor of Philosophy at the University of Vienna in 1920. He was one of Kurt Gödel's teachers there, and a member of the *Wiener Kreis* ('Vienna Circle'; see Chap. 4). Heinrich was discharged from the faculty in 1934 during the brief fascist period in Austria ('*Austrofacism*'), and he emigrated in 1935 to the USA, where he taught at the University of Southern California until his death in 1942. [Coincidentally, his name was the same as that of a rich philanthropist in Brünn, not a close relation and a generation older, after whom the *Zwetchkengasse*, where the Rudolf August Gödel family had their first apartment, was renamed in 1896 (cf. Chap. 1)].

Johann Gregor Mendel, the Augustinian friar who discovered the basic facts of inheritance and effectively founded the science of genetics, was born in Austrian Silesia in 1822. He entered the Altbrünn Cloister in Brünn at age 21 and spent the rest of his life there, except for two years while he studied physics in Vienna, and it was in Brünn that he carried out his research on genetics using pea plants. He also performed research into meteorology, and later became the Abbot of his cloister, a post that he held until his death in 1884. Had he lived longer, he would certainly have received a Nobel prize for his pioneering work on genetics and inheritance.

Ernst Josef Mach, who became a famous physicist and from 1895 professor for the History and Philosophy of Science at the University of Vienna, and whose writings influenced both Albert Einstein and Kurt Gödel, was born in 1838 at the *Chirlitz* Estate, a suburb of Brünn, and educated at home until he was 14. He then attended the *Gymnasium* in *Kremsier* (now *Kroměříž*), about 50 km east of Brünn, and later studied physics at the University of Vienna, obtaining his doctorate in 1860. He occupied various academic positions in the Habsburg realm, carrying out research in optics, acoustics, the physics of sensory perceptions, physiology and philosophy, and developed Mach's Principle, a non-Newtonian interpretation of inertia. He was Dean and Rector at the German University (*Karlsuniversität*) in Prague before he became a professor in Vienna. He adopted what some have termed a phenomenological[12] philosophy of science.[13] After his retirement from the University in 1901, he served in the Austrian parliament, and he died in Bavaria in 1916.

Robert Musil, an Austrian engineer, philosopher and writer, best known for his epic novel '*Der Mann ohne Eigenschaften*' (*The Man without Qualities*), was born in Klagenfurt in 1880 but spent the later part of his childhood and youth in Brünn, where his father was a professor at the German Technical College. He received his engineering degree from the college, studying literature and philosophy on his own, and later obtained his doctorate in philosophy from the University of Berlin. His later life was devoted to writing, in Berlin and Vienna, and he was nominated for a Nobel prize in literature. He died in Swiss exile in 1942.

Adolf Loos, an influential Austrian architect, was born in 1870 in Brünn and graduated from a technical school there at age 19. After spending three years in the USA, he returned to Vienna and began studying and working as an architect. He adopted a utilitarian building style with little ornamentation, and wrote a number of controversial essays on architectural theory. Several of his buildings have become protected monuments in Vienna, and he is considered to be one of the principal founders of Viennese Modernism. He died near Vienna in 1933.

The Czech composer *Leoš Janáček* was born in *Hukvaldy*, Moravia, about 125 km northeast of Brünn, in 1854. From age 11, he was given instruction in piano and choral singing at the St. Thomas Abbey in Brünn. He later studied at the Prague Organ School and taught music in Brünn before studying in Leipzig and Vienna. Back in Brünn, he became an important composer, first with adaptations of Moravian folk music, later with a series of operas. He died in *Ostrava*, Moravian Silesia, in 1928.

Viktor Kaplan was an Austrian engineer, born in the *Steiermark* in 1876. After school in Vienna and studies at the Technical College (now *Technische Universität*) there, he served in the Austrian (*k.-und-k.*) military and worked as a construction engineer until obtaining a position at the German Technical College in Brünn in 1903, doing research in civil engineering for the Chair occupied by Alfred Musil, the father of Robert Musil. He remained in Brünn for 30 years, becoming full professor, and invented—among many other things—the *Kaplan turbine*, a type of water turbine which is still in widespread use for power generation at sites with a low head of water but high flow rates. He died in 1934 in Unterach, central Austria.

Erich Wolfgang Korngold, an Austrian/American musician and composer, was born in Brünn in 1897 and was recognized as an exceptional musical talent early in life. He went to Vienna at around age 11, where he was promoted as a musical prodigy and had considerable success with his early compositions, in particular his opera '*Die tote Stadt*' (at age 23). In 1934, he went to Hollywood at the invitation of theater and film director Max Reinhard, and finally settled there permanently after the *Anschluss* in 1938, receiving two Oscars for his film music before 1940. His 'serious' music was less well received in his later years, but it has been rediscovered in the twenty-first century. He died in Los Angeles in 1957.

This brief list gives an impression of the cultural and intellectual level and diversity in Brünn around 1900. Kurt Gödel later maintained that he felt like a citizen of the Austro-Hungarian realm, and never was at home in the 'new' Brno after 1918, although he finished his schooling there. His family maintained ties to the city until the end of World War II, but he never went back for a longer time after moving to Vienna in 1924. In the next chapter, we will follow him during his school years, 1912–1924, which were an important formative period for his later life and achievements.

Notes

1. Eilenberger (2018), Geier (2017).
2. See backnote [11] in the previous chapter.
3. Image from Gödel archive, IAS/Firestone Library, Princeton; public domain. Reused from: https://longstreet.typepad.com/thesciencebookstore/2015/03/great_babies.html
4. Müller (2007).
5. Image from Wiki Commons, uploaded by Radler59, Sept. 3, 2017. Used under Creative Commons Attribution-Share. Alike 4.0 International License; reused from: https://commons.wikimedia.org/wiki/File:Brno_Old_Town_Hall-02.jpg
6. See https://www.visitbrno.cz/en/brno-wheel/23/ and https://www.visitbrno.cz/en/brno-old-town-hall/22/.
7. Image from Gödel archive, IAS/Firestone Library, Princeton. Reused from: https://www.privatdozent.co/p/kurt-godels-brilliant-madness?s=r.
8. See backnote [19], Chap. 3.
9. See Wang (1987), p. 12.
10. Dawson (1997), CRC Edition (2005), p. 7.
11. Goldstein (2005), Kindle edition, p. 53.
12. See Fisette 2012 on '*Phenomenology and Phenomenalism, Ernst Mach and… Husserl*'.
13. A better term for Mach's philosophical position is perhaps '*Descriptivism*'.

3

School Days. A New Nation

Elementary School

In September 1912, young Kurt Gödel entered his first year of school—the beginning of an educational process which would take him to the highest level, the *Habilitation*, just 20 years later. He was sent to a semiprivate elementary school, the *'Evangelische Privat-Volks- und Bürgerschule'* ('Evangelical Private Elementary and Citizens' School'), located at *Elisabethgasse* (*Opletalova*) 6, about 600 m (6 city blocks) to the northeast of the family apartment (cf. Fig. 1.3). The school building is still intact and nearly unchanged (compare the modern photo in Fig. 3.1). Today, it houses the 'Evangelical Academy and Higher Vocational School of Social Law', operated by the Czech Evangelical (Lutheran) Church. When little Kurt Gödel entered the school in 1912, it was governed by the *'Mährischer Ausgleich'*, the 'Moravian Compensation', a set of laws passed by the Habsburg monarchy in 1905 to alleviate the tensions between the German and the Czech ethnic groups in Moravia. The four laws in the *Ausgleich* regulated the ratio of German and Czech members in the provincial parliament; the election process; the official languages of the province; and the school system in Moravia (separation into German-speaking and Czech-speaking schools).[1]

The *Ausgleich* was the culmination of a series of laws beginning in the eighteenth century which regulated the founding and operation of schools in Austria-Hungary. Originally, the schools were established exclusively by the Catholic Church, but a series of liberalizations had made it possible later for

© The Author(s), under exclusive license to Springer Nature Switzerland AG 2022, corrected publication 2023
W. D. Brewer, *Kurt Gödel*, Springer Biographies,
https://doi.org/10.1007/978-3-031-11309-3_3

Fig. 3.1 *Left*: The school building at *Elisabethgasse* (*Opletalova*) 6, where Kurt Gödel attended grade school. *Right*: The entrance to the building. It now houses an Evangelical (Lutheran) Academy. Photos: AMZ, 2021

private Protestant schools to be set up. The elementary school in the *Elisa-bethgasse* was a combination of the earlier *Volksschule*, an elementary school usually offering five grades, and the *Bürgerschule*, a middle school offering grades 6–8. Completing 8 grades was the legal minimum schooling expected of all citizens according to the 'Population Education Act of 1868'.[2] This school would have offered instruction only in German, and its pupils would have come from the German-speaking majority (ethnic German and Jewish populations) of the city. Other elementary and middle schools were available for the Slavic population. Kurt Gödel evidently identified strongly with the Austrian culture and the German language, and always claimed to speak no Czech (although it has been reported that he spoke a Slavic language to an immigrant in Princeton many years later[3]). Given his talent for languages and his immersion from his earliest years in a society in which Czech was an important (official) language, it seems unlikely that he really knew no Czech, even if it was not his 'preferred' language.

In practice, the Gödel family were not religious, although both of their sons were baptized in the Evangelical Lutheran church, of which mother Marianne and her family were nominal members. Father Rudolf August and his family were nominal members of the Old Catholic church. The boys were sent to the private Evangelical elementary school, perhaps simply because it was located conveniently close to the family apartment (and later, to their villa in the *Spielberggasse*).

As we have seen, the child Kurt was shy and inclined to introspection, and had little experience with other children of his own age up to his 6th birthday. Nevertheless, he adapted well to school and received high marks during his entire school career.[4] His brother Rudolf ('Rudi'), 4 years older, had probably left elementary school and moved on to the Gymnasium the same year that Kurt began school. Most likely a nanny or a governess made sure that Kurt arrived at his school on time and picked him up after the school day was over, at least in his first school years. The distance from his school to the family dwellings was not great, but it involved crossing the large *Elisabeth Str.* (*Husova*), which even in those calmer days might have been a danger for a young child alone.

Kurt himself (in the Grandjean questionnaire) reports that he was not religious in his early years. As we will see, he later became, in his own words, 'a believer', although he was never an active member of any church congregation. Much later, he devised an 'ontological proof' of the existence of God, using the methods that he had developed for formal-logical proofs,[5] and it was found to be (computationally) correct in 2013[6]; we shall return to this topic in Chap. 18.

Kurt played with his brother much of the time, and up to the age of 7 or 8 he was active outdoors; but when he was 8, he suffered an attack of rheumatic fever, probably caused by a streptococcus infection, and missed some months of school as a result. In those pre-antibiotic days, the only treatment was rest and recuperation. He recovered from the infection, apparently without lasting health problems, but later worried that he might have suffered heart damage after reading a description of the infection in some medical books; and he maintained those worries throughout the rest of his life, even though they were contradicted by various doctors. His brother reported later[7] that this episode was "the beginning of his lifelong hypochondria" and his obsessive preoccupation with his health and diet in later life, and of his mistrust in the diagnoses and therapeutic suggestions of doctors. This is an example of his tendency to form and maintain extremely strong opinions, resistant to any attempt at correction or moderation, also mentioned by his brother.

In any case, after his illness, Kurt was no longer interested in outdoor activities and spent his spare time reading and studying. There is no record of friendships with his classmates during his elementary-school career. Around the age of 10, he did develop an interest in playing chess, and no doubt found partners with whom to play, certainly including his brother. By this time, he had finished 4th grade and was preparing to move on to the *Gymnasium*, the next step in his education. There, he found a suitable chess opponent in the person of the school's chess master, Adolf Hochwald.[8] Dawson[9] attributes his interest in chess to World War I, which by Gödel's 10th year had been raging for two years. It apparently aroused the interests of the Gödel brothers in war and strategy games. Rudolf also notes that Kurt was very upset and chagrined on those rare occasions when he lost a game of chess.

Their mother Marianne had received some musical education and could play the piano well, which she often did for her sons. Her taste ran to 'light music', operettas and dance music, which were popular in Viennese society around the turn of the century, and Kurt adopted that preference.[10] He never learned to play an instrument himself, nor did he show any interest in more 'serious' music. This is in contrast to many other people with mathematical/theoretical talent, who often are also musically talented and use music as a form of relaxation or meditation—for example Albert Einstein, Gödel's much later elder friend. Einstein was a competent violinist and often performed in public as well as in private amateur groups, and he greatly admired the music of Bach, Mozart, and Beethoven.

After Kurt Gödel finished his fourth year at the Evangelical School in July 1916, he had the option of continuing there for another year, and then attending the adjacent *Bürgerschule* ('Citizens' school', a middle school) for three more years, after which he would have completed the legal requirement of 8 years of schooling. Since he was clearly destined for higher education, however, he transferred in the fall of 1916 to a *Realgymnasium*, which would allow him to pass the *Matura*—high-school leaving examination—after a further 8 years of schooling, the prerequisite for entering a university.

High School

Gödel entered the next stage of his education at the '*k.-k. Staatsrealgymnasium mit deutscher Unterrichtssprache*, located at *Strassengasse/Wawrastr.* 15 (now *Hybešova*; about 5 blocks southeast of the family villa, somewhat closer than his elementary school; cf. Fig. 1.3 and Backnote 12 in Chap. 1). The rather cumbersome name of his high school was the result of many decades of changing educational policy in the *k.-k.* Empire, as well as attempts to placate nationalist groups by a just distribution of 'official' languages. The school building (Fig. 3.2) still exists and has been competently restored. It currently houses the *Centre Hybešova*, belonging to the Brno School System, which comprises the Preschool Department for the Brno city region, the

Department of Elementary Schooling for the Brno province region, and the Department for Secondary and Higher Vocational School Students for the Brno province region.

Dawson[11] describes the history of academic secondary schools in Austria-Hungary in a footnote. Rudolf and Kurt Gödel were fortunate to have entered high school soon after the latest school reform in 1908, which combined the *Realschulen* and the *Gymnasien* into the '*Realgymnasium*' schools, where both purely academic subjects (ancient languages and culture, history, philosophy) formerly limited to the *Gymnasien*, and the *realia* (modern science, modern languages, social sciences), considered to be more 'practical' and previously taught at the *Realschulen*, were combined.[12] This gave them a wider choice of subjects from which to choose, and a more complete 'school culture'. The Austro-Hungarian school system had been oriented towards producing effective civil servants during the first half of the nineteenth century and earlier, and it was only after the educational reforms stimulated by the revolutionary protests in 1848 that something resembling the modern secondary-school and university system was put in place. Already in 1862, there was a first attempt to combine the *Gymnasien*, preparatory schools for university studies, with the *Realschulen*, which gave a practical education as a preparation for non-academic professions taught at technical colleges. That first move to establish *Realgymnasien* was defeated

Fig. 3.2 *Left*: A modern view of the building at *Strassengasse/Wawragasse* (*Hybešova*) 15, formerly the *k.-k. Staats-Realgymnasium mit deutscher Unterrichtssprache* ('*State Real-Gymnasium with German as instructional language*'). *Right*: The entrance to the building. It now houses part of the education administration for the Brno region. Photos: AMZ, 2021

by the conservative educational establishment a few years later, and only after another forty years, with the reform of 1908, were the two types of secondary school put on an equal footing and combined.

Dawson[13] also gives a rather complete description of the *Realgymnasium* attended by Gödel, and of his experiences there. His entering class had over 90 members (among the more than 400 pupils in the entire school). They were in the majority Catholic (55%); 40% were Jewish, and there were only a few Protestants. Nearly all were German native speakers, although there were some few with the Czech mother tongue. After four years, his class was reduced to only 36 members, many having left the school after completing their required 8 years of school education. As in elementary school, Gödel had few friends at the *Gymnasium*. One of those few, Harry Klepetař (1906–1994) later answered an inquiry from John Dawson.[14] He had shared a desk with Gödel during their entire 8 years at the school, and knew him as well as anyone at that point. Klepetař was of Czech origin; he later studied law at the Charles University in Prague, and still later became a journalist, writing in German. In his letter to Dawson, he confirms that Kurt Gödel was very introverted but brilliant, with a wide range of interests. Klepetař had a much more positive impression of their school than Gödel, who wrote to his mother much later that "*it* [was] *worthy of little praise*".

Wang[15] points out that "*Gödel was far above the maximum demand of the schools*" and was thus able to master the material in his courses and achieve the highest marks, while learning what interested him 'on the side'; so that for example by the time of his graduation in 1924, he had mastered much of the material of the usual university courses in mathematics, in addition to high school mathematics.

Mathematics and Philosophy

At first, Kurt put his energies into learning languages—primarily Latin. When he was 14, his interest turned to mathematics,[16] and later included philosophy. His brother Rudolf (1987) writes that Kurt had the reputation of never having made a mistake in Latin throughout his school career (an unlikely claim, which however reflects Kurt's desire for precision and his language competence; and it is also typical of the reputation that a brilliant pupil can acquire in high school). Gödel's interest in philosophy beginning around age 16 was probably stimulated by philosophy instruction in his high school, which led him to read some of Kant's writings, an experience that he cited much later in the Grandjean questionnaire as having aroused his interest in philosophy. (Two years of instruction in philosophy were usual in Austrian secondary schools after the reforms beginning in 1848). According to Wang,[17] Gödel mentions a calculus textbook as having awakened his interest in mathematics two years earlier. Dawson[18] also quotes his mention of the calculus text, but he suggests a different origin for Gödel's interest in mathematics and science: reading about Goethe's *Farbenlehre* (color theory) and its clash with the Newtonian theory during a trip with his family to *Marienbad* in the summer of 1921. Kurt later recalled that occasion in letters to his mother. This was confirmed by Rudolf Gödel in an interview given in 1986 for the television documentary described in Schimanovich & Weibel (1997). Kurt's letter[19] of August 26th, 1946, to his mother is worth quoting in this connection:

> *...Das Buch 'Goethe' von Chamberlain, von dem Du schriebst, bringt eine Menge von Jugenderinnerungen für mich. Ich las es (sonderbarer Weise genau jetzt vor genau 25 Jahren) in Marienbad u. sehe noch heute die merkwürdig lilafarbigen Blumen vor mir, mit denen damals alles übersät war. Es ist unglaublich, wie sich einem so etwas einprägen kann. Ich glaube, ich schrieb Dir schon 1941 aus dem Mountain Ash Inn, dass ich dort dieselben Blumen wiederfand und wie eigentümlich mich das berührte. Dieses Goethe-Buch ist auch der Anfang meiner Beschäftigung mit der Goethe'schen Farbenlehre u. seinem Streit mit Newton gewesen u. hat dadurch indirekt auch zu meiner Berufswahl beigetragen. So spinnen sich durch's Leben merkwürdige Fäden, die man erst entdeckt, wenn man älter wird...*

[English translation by the present author]: ...The book 'Goethe' by Chamberlain that you wrote about brings back a flood of youthful memories for me. I read it (strangely enough, exactly 25 years ago from just now) in Marienbad, & today I still can see the remarkable lilac-colored flowers that grew all over there. It is unbelievable how such a thing can leave such a lasting impression. I believe that I wrote to you in 1941 from the Mountain Ash Inn [in Brooklin, Maine, where Kurt and Adele were vacationing in the summer of 1941] that I had found the same flowers there and how strangely that touched me. That Goethe book [was] also the beginning of my interest in Goethe's color theory and the controversy with Newton [his theory of light] & thus it contributed indirectly to my choice of profession. Bizarre threads spin themselves through our lives, and we discover them only as we grow older...

Rudolf's version as given in his interview for the 1986 television documentary[20] was the following:

After the War [WWI], we traveled several times with my brother to Marienbad, and I recall that one time, we read together the Goethe biography by Chamberlain. And he [Kurt] was especially interested in Goethe's color theory, and came back to it several times. That is one of the reasons for his interest in the natural sciences. In any case, he preferred Newton's analysis of the color spectrum over that of Goethe.

It is fascinating to note the parallel between Kurt Gödel's discovery of science through a book that he read, more or less by chance, at age 15, and the parallel discovery 38 years later by Douglas Hofstadter, at age 14 and also through a book that he found by chance, of Gödel's logic—which launched him on a career in science and mathematics and was a major incentive for his later book *GEB*, that in turn made Gödel famous (compare the preface to the present book). Bizarre threads indeed!

What Kurt did take from his school years was his proficiency in using a type of shorthand writing, the *Gabelsberger Schnellschrift*, which he later employed for note-taking and manuscripts—a serious obstacle for the scholars who later undertook the editing and transcription of his written legacy, as this form of shorthand was long since obsolete when they began their work.

Near the end of his school days, Kurt had his first romantic adventure. He became amorously interested in the daughter of friends of his family—she was however about 10 years older than he. She was said to be 'an eccentric beauty',[21] and was still unmarried at 28. Gödel himself might have been considered 'eccentric' already at age 18, and this may have made her even more attractive to him. His parents quickly put an end to the affair, which they considered unseemly.

In spite of his principal interests in mathematics and philosophy, Kurt resolved to major in theoretical physics at the university after finishing his high school education. However, the physics instruction that he received in high school seems to have been oriented mainly toward practical aspects such as units of measure and simple experiments,[22] with some treatment of the associated mathematical descriptions, and he apparently did not find it inspiring. Just why he nevertheless initially chose physics as his major subject remains something of a mystery, although his experience with the book on Goethe may have had a strong influence, dealing as it did with a topic in physics.

World War I and the End
of the Austro-Hungarian Empire

The period of the Gödel brothers' high-school education (Rudolf: 1912–1920; Kurt: 1916–1924) coincided with earth-shaking events in Europe, which affected Austria-Hungary in the most dramatic ways. In August 1914, Europe plunged into World War I—which many people had been expecting for at least a decade, but in general had underestimated. Austria-Hungary in fact precipitated the war after the assassination of its heir to the throne, the Archduke Ferdinand, by a Serbian nationalist in Sarajevo. The Habsburg monarchy prepared to punish Serbia, but was opposed by Russia, which supported the Pan-Slavic movement and considered Serbia to be an ally. Russia in turn was backed by its allies Great Britain and France (the *Entente cordiale*), while Austria was joined by Germany to form the *Central Powers*.

The German military expected that they could easily win the war within a few months. They were very wrong, and WWI in fact dragged on for more than 4 years, resulting in incalculable death and destruction throughout much of Europe (and elsewhere), and marking the end of the previous era of relative peace and prosperity. When the Central Powers were finally defeated in late 1918, their (imperial) governments collapsed and were replaced by republics. The Russian Empire had already collapsed in 1917, and Russia was ravaged by a civil war. The effects on Austria-Hungary were particularly drastic, since the Empire was subsequently divided up into relatively small nations corresponding to the ethnic boundaries within the '*Vielvölkerstaat*' (Fig. 1.1), leaving Austria as a comparatively small country in Central Europe, an 'Alpine Republic', and combining Bohemia and Moravia, as well as Slovakia, parts of Silesia, and parts of Hungary and the (later) Ukraine into the 1st Czechoslovakian Republic.

The new Czechoslovakian Republic had in fact considerable advantages: Germany and Austria were held responsible for the war by the winning allies, and they were militarily disciplined (their armies were essentially disbanded) and economically punished (through reparation payments). Austria had lost its *Hinterland*, and with it much of its industry, natural resources, and agricultural lands, and it fell on hard times in the years 1919—1924, as did Germany. The Czechoslovakian Republic had 'inherited' many of those resources, kept its army active, and had intact industries. Its idealistic promises to become a new 'land of many nations' were not kept in practice, and it used military force to suppress dissent among some of its new citizens (including the *Sudeten-* and *Schlesien*-Germans).

By the end of 1918, Brünn had become the Czech city of *Brno*, and almost nothing was the same as before the war. Nevertheless, the Gödel family remained practically unscathed, and even their economic standing survived the political changes and the postwar inflation comparatively intact, although father Rudolf August Gödel apparently lost a considerable sum to war bonds which were never repaid after the defeat of the Central Powers. But by the mid-1920s, he had recovered financially. There was no reason for a successful businessman like Rudolf August to leave Brünn, where he had long-standing roots and continued economic success. For his sons, the situation was different.

Both Rudolf and Kurt were fortunate to be young enough that they were not involved in the war as combatants, and in fact their family escaped the deprivations which affected the civilian population in many other places. Kurt, in answering the Grandjean questionnaire, states that the war had '*little effect on his family*', not even economically. In retrospect, this is rather surprising, given the havoc wrought by the war and its consequences in much of Europe.

The immediate result of the war for the Gödel family was that they became citizens of the Czechoslovakian Republic, as the Austro-Hungarian Empire into which they had been born no longer existed. Czechoslovakia had now become a 'nation of many languages', dominated by Czech and Slovakian but also including a considerable German-speaking minority, as well as Hungarian (Magyar), Ukrainian (Ruthenian), and even Polish ethnicities, as we saw in Chap. 1. Austria, in contrast, had only German as its official language, and was now a small, relatively poor country.

In June of 1924, Kurt (see Fig. 3.3) passed his *Matura*, the school-leaving examination that qualified him to begin university studies. His brother Rudolf had been studying medicine at the University of Vienna since 1920, and there was no doubt that Kurt would also go there. Rudolf had a hard life in his first years in Vienna, and he was dependent on support—not only financial—from his family in Brünn, given the hardships of life in immediate postwar Vienna. But things had normalized there to a considerable extent by the time that Kurt was also ready to move to Vienna.

Kurt could have studied at the German branch of the Charles University in Prague (the *Karlsuniversität*), a venerable and respected institution, like his school friend Harry Klepetař; but he apparently never seriously considered that possibility. Going to Vienna was eminently practical, since he could live with his brother Rudolf, who already had a working knowledge of life in the city and at the University, and it also fit with his cultural and political preferences—he still considered himself to be an Austrian (and indeed he

Fig. 3.3 The young Kurt Gödel, photographed in the mid-1920s. This picture has been reproduced in many books and articles about Gödel. Dawson (1997) refers to it as a 'school portrait' (p. 19) and dates it to around 1922. Schimanovich & Weibel (1997) denote it as 'Kurt Gödel at the age of 20' (p. 30), i.e., 1926. Wiki Commons puts it in the interval 1924–27. Photo: from Wiki Commons[23]

again acquired Austrian citizenship, after renouncing his citizenship of the Czechoslovakian Republic, some years later in 1929). We know that he began his studies as a student of theoretical physics, and in the next chapter we shall see how his intellectual life evolved in the following years, a critical period for his later role in mathematical logic.

Notes

1. Fasora (ed.) 2006; Závodnik 2008.
2. Cvrcek & Zajicek 2019.
3. Dawson 1997, CRC Edition (2005), Footnote 6, p. 15.
4. *Ibid.*, pp. 15, 16.

5. Fuhrmann 2005: https://www.information-philosophie.de/?a=1&t=4435& n=2&y=1&c=50.
6. Gödel (1941). See also https://www.spiegel.de/wissenschaft/mensch/formel-von-kurt-goedel-mathematiker-bestaetigen-gottesbeweis-a-920455.html.
7. Rudolf Gödel (1987).
8. Dawson (1997), CRC Edition (2005), p. 14.
9. *Ibid.*, p. 12.
10. Schimanovich & Weibel (1997), p. 27.
11. Dawson (1997), CRC Edition (2005), Footnote 2, p. 39.
12. Stachel (1999),
13. Dawson (1997), CRC Edition (2005), pp. 12–19.
14. Letter to Dawson dated December 30th, 1983, quoted in Dawson (1997), CRC Edition (2005).
15. Wang (1987), p. 14.
16. *Ibid.*, p. 21.
17. *Ibid.*, pp. 16 and 19, 'the Grandjean questionnaire'.
18. Dawson (1997), CRC Edition (2005); see Endnote [42] and Footnote 8, p. 18.
19. Kurt Gödel's postwar letters to his mother Marianne (Gödel 1946–66) are available online from the Vienna Library, as facsimiles of the (German) originals, at https://www.digital.wienbibliothek.at/wbr/nav/classification/255 9756.
20. Schimanovich & Wiebel 1997, p. 27. Translation by the present author.
21. Hintikka (1999).
22. Dawson (1997), CRC Edition (2005), p. 17.
23. Image from Wiki Commons: from a family album of the Gödel family, photographer unknown; public domain; reused from: https://commons.wik imedia.org/wiki/File:1925_kurt_g%C3%B6del_(cropped).png.

4

Student Life. *Moving to Vienna*

In October 1924, a few months after completing his *Matura* in Brünn, Kurt Gödel went to Vienna to begin his university studies. He initially lived in an apartment at *Florianigasse* 42 in Vienna's 8th District, the *Josefstadt* (see Figs. 4.1 and 4.2), with his brother Rudolf. They lived there for 2-½ years, until April 1927. It was located some distance behind the *Rathaus*, around 800 m (½ mile) west of the University.

Vienna in 1924 was an extraordinary place. Only a few years before, it had been the Imperial Capital of a far-flung empire covering most of Central Europe and including a dozen ethnicities, each with its own language and culture. Now, since late 1918, it was still a capital city, but only of a small 'Alpine Republic', comparatively homogeneous, where everyone spoke German. The city itself however had not lost its cosmopolitan flair overnight. It had been a center of intellectual and cultural activity for many years, and the focal point of an empire that dated back more than a millennium. The disastrous 1st World War had diminished its worldly glory considerably, but it remained a magnet for creative people from many places and in many fields—from music and the arts to science and philosophy, the humanities, and the social sciences. The brief interlude 'between the wars' saw a blossoming of intellectual and creative activity in Vienna, which came to an end with the rise of fascism in Europe, the annexation (*Anschluss*) of Austria by Nazi Germany in 1938, and another catastrophic World War, which left it an occupied city in ruins by 1945. But when young Kurt Gödel arrived in 1924, all that lay in the unforeseeable future, and going there must have been the realization of a

© The Author(s), under exclusive license to Springer Nature
Switzerland AG 2022, corrected publication 2023
W. D. Brewer, *Kurt Gödel*, Springer Biographies,
https://doi.org/10.1007/978-3-031-11309-3_4

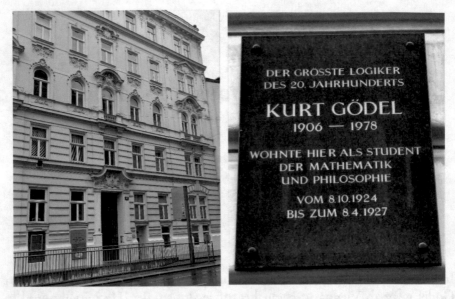

Fig. 4.1 *Left*: The building at *Florianigasse* 42 in Wien-8 (*Josephstadt*), where Kurt and Rudolf Gödel lived from October 1924 until April 1927. *Right*: The memorial plaque placed by the Vienna *Kurt Gödel Society* and the city. Its text reads: 'The greatest logician of the twentieth century, *Kurt Gödel*, lived here as a student of mathematics and philosophy (dates)'. *Photos* AMZ, 2021

dream for him—he was now at the center of what he considered his culture and people (although he was not in any sense a racist or a cultural chauvinist, as we can see from many of his remarks and even more so from his lifestyle and manner of dealing with others).

The University

The University of Vienna, where Kurt matriculated in October 1924, was a venerable institution, only a few years younger than the Charles University[1] in Prague (it had been founded just 17 years after the latter, by the Archduke of Austria, Rudolf IV). Both universities were modeled on the University of Paris, and both were allied with Jesuit academies in the mid-sixteenth century; but the Charles University became Protestant a century later, while Vienna continued under Catholic (Jesuit) influence for two centuries. The Jesuits were disciplined and strict, and there was little question of academic freedom under their control. Around 1750, the Empress Maria Theresia, and later her son Joseph II, reduced the role of the Church in the universities (and in general in the educational system of the Empire), but not with the goal

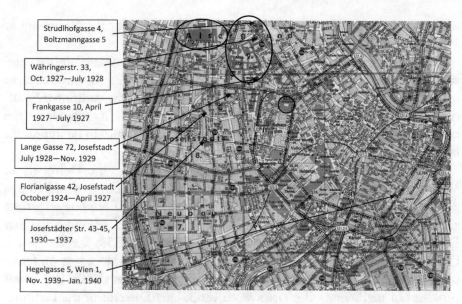

Strudlhofgasse 4, Boltzmanngasse 5	
Währingerstr. 33, Oct. 1927—July 1928	
Frankgasse 10, April 1927—July 1927	
Lange Gasse 72, Josefstadt July 1928—Nov. 1929	
Florianigasse 42, Josefstadt October 1924—April 1927	
Josefstädter Str. 43-45, 1930—1937	
Hegelgasse 5, Wien 1, Nov. 1939—Jan. 1940	

Fig. 4.2 Vienna, map of the city center. The arrows indicate the locations and dates where Kurt Gödel lived, and other places of interest to him. Not shown is the dwelling at *Himmelstr.* 43 in *Wien-Döbling* (19th District), about 5 km to the north-west of the University, where Gödel lived with Adele from 1937–1939. The ellipses show the University Medical Faculty (*upper left*) and other faculties (*to its right*). The original Main Building of the University is just north of the *Rathausplatz* (small circle). *Map* from vidiani.com[3]

of increasing academic freedom and excellence at the universities—rather, to ensure state control and the establishment of a 'talent foundry' for the education of future civil servants.[2]

After the reign of Joseph II ended in 1790, the French Revolution struck fear in European monarchs, including his successors Leopold I and Franz II(I),[4] and they set up a system of police spies who watched over the universities in particular, thought to be breeding grounds for subversives. This type of authoritarian regime suppresses creativity and provides little room for research or academic freedom, and indeed the emperors of the period up to 1848 often seemed to consider the university faculty as dispensable, a potential asset for economizing in order to finance reforms in other parts of the educational system. This policy continued under Emperor Ferdinand I, who reigned from 1835 until 1848, when he was forced to abdicate by the revolutionary uprisings of that year. In fact, one of the factors fueling those uprisings in Vienna was discontent with the state of the university and its lack of academic freedom. His successor, Franz Josef I, instituted widespread reforms over the next 20 years, leading to a renaissance at the University of Vienna as well as in many other places.

Physics, Mathematics and Philosophy were all well-established and successful departments at the University of Vienna by 1924 (in those days, as institutes within the Philosophical Faculty). The other three traditional Faculties—Theology, Law, and Medicine—were essentially training schools for the corresponding professions, although research had also begun to be important by the early twentieth century.

The Physics Institute (now the *Fakultät für Physik*) at the University of Vienna had considerable renown in the later nineteenth century, and counts among its earlier professors and researchers a number of famous scientists: *Christian Doppler* (1803–1853), of Doppler Effect fame, who was the first Director of the newly-founded *Physikalisches Institut* in 1850; *Josef Loschmidt* (1821–1895), known for early work on the atomic hypothesis, was professor of physical chemistry from 1872 until 1891; *Josef Stefan* (1835–1893) was professor of mathematical physics from 1863, later Director of the Institute, Dean of the Faculty, and Rector of the University (1876/77); he was known for his thermodynamics research (Stefan-Boltzmann law). *Ernst Mach* (1838–1916) had been Dean and Rector at the Charles University in Prague before becoming professor of philosophy (for the history and theory of science) in Vienna from 1895–1901. *Ludwig Boltzmann*, famous for his development of statistical mechanics, was professor of mathematics and theoretical physics in Vienna from 1894–1900 and again from 1902–1906. Those scholars had established a tradition and left a legacy of excellent research at the Physics Institute in Vienna, which was interrupted by World War I but had regained its momentum by the time of Gödel's arrival in 1924.

The generation immediately preceding Gödel included *Stefan Meyer* (1872–1949), who was an assistant to Boltzmann and, from 1909, professor of physics in Vienna. He became a pioneer in radioactivity and was Acting Director of the Institute for Radium Research (and a mentor of *Lise Meitner* (1878–1968), the second woman to receive a doctorate in physics in Vienna). *Victor Franz Hess* (1883–1964) was assistant in the Physics Institute from 1906–1920, and later at the neighboring Radium Institute. He was awarded the Nobel Prize in physics in 1936 for his discovery of cosmic rays. *Erwin Schrödinger* (1887–1961), one of the founders of quantum mechanics, also a Nobel prizewinner (in 1933), received his doctorate in Vienna and remained there as assistant until 1920. *Hans Thirring* (1888–1976) also received his doctorate in Vienna, became professor of physics there and was head of the

Physics Institute when Gödel was a student. He, together with the mathematician *Josef Lense* (1890–1985), discovered the 'frame-dragging' effect in General Relativity (Lense-Thirring Effect) in 1918.

There was thus no shortage of talent in physics in Vienna when Kurt Gödel began his studies there. The receipts for the books that Gödel borrowed from the library or ordered from bookstores (which he scrupulously kept in envelopes that were found in his legacy many years later) show that he read mainly physics literature during his first three semesters. But sometime during his second year (1925/26), he began to shift his emphasis toward mathematics (with philosophy as a 'minor subject'). In fact, at that time there were no requirements for students at the University of Vienna to declare major and minor subjects or to complete prescribed schedules regarding coursework; they simply had to pass the examinations at the end of their studies in their chosen field. The courses that Gödel attended have, however, been largely discovered by later research using University records.[5] Gödel later told Hao Wang[6] that it was his "desire for precision" that moved him to abandon physics in favor of mathematics. He had indeed since his earliest years been eager to find precise answers to many questions about the world and the phenomena that we encounter in it, both physical and mental. Physics, as an empirical science, can offer no such certainty; its 'laws' may be modified or rejected whenever new experimental or observational evidence turns up, or a new theory offers a more convincing description of the experimental results.

Mathematics, in contrast, deals with laws which (ideally) can be rigorously proven, and which may be supposed to represent 'truth' on a higher (Platonic or Kantian) level. It also seems likely that the effect of lectures attended by Gödel and given by charismatic professors, especially the lecture series on *'Einführung in die Zahlentheorie'* (*'Introduction to Number Theory'*) taught by mathematician *Philipp Furtwängler* in 1925/26, and that on the *History of European Philosophy* taught by philosopher *Heinrich Gomperz*, also in 1925/26, inspired him to change his emphasis from physics to mathematics and philosophy. Indeed, his later career encompassed significant contributions to all three of those fields (although certainly *most* significantly to fundamentals of mathematics).

The website of the *Fakultät für Mathematik*[7] at the University of Vienna describes the period from the end of World War I to the later 1930s as a *'Blütezeit'* ('heyday', 'zenith') of that department, listing its important professors as *Franz Mertens*, later *Wilhelm Wirtinger*, *Philipp Furtwängler*, and *Hans Hahn*, with the students and docents *Kurt Gödel*, *Eduard Helly*, *Witold Hurewicz*, *Walther Mayer*, *Karl Menger*, *Johann Radon*, *Kurt Reidemeister*, *Otto Schreier*, *Gabor Szegö*, *Alfred Tauber*, *Olga Taußky(-Todd)*, *Heinrich*

Tietze and *Leopold Vietoris*. Professor *Philipp Furtwängler* played an important role in Gödel's 'conversion' to mathematics, and he must have been an impressive figure to many of his students. He was a cousin of the famous orchestra conductor Wilhelm Furtwängler,[8] and both had formidable 'lion heads' (Fig. 4.3). **Philipp Furtwängler** was born in 1869 into a musical family in Lower Saxony, western Germany. He studied in Göttingen and wrote his doctoral dissertation there under Felix Klein on a topic in number theory. He served as assistant in Potsdam and professor in Bonn and Aachen before becoming professor of mathematics in Vienna in 1912, where he remained until his retirement in 1938. Due to an illness, he was paralyzed from the neck down, and gave his lectures from a wheelchair, speaking from memory while an assistant wrote proofs and equations on the blackboard. He has been compared with Steven Hawking in terms of his presence as a lecturer.

Gödel later mentioned that Furtwängler's lectures were "the best he ever heard in mathematics".[10] Unlike his fellow student Olga Taußky-Todd, however, who specialized in number theory and wrote her doctoral thesis under Furtwängler, Gödel chose *Hans Hahn* to be his mentor. Hahn was interested in the fundamentals of mathematics and mathematical logic, and that was apparently more attractive to the young Kurt Gödel than the more pragmatic number theory, which deals with mathematical objects rather than metamathematical systems.

Fig. 4.3 Philipp Furtwängler lecturing, probably in the early 1930s. *Photo* Wiki Commons, public domain[9]

Fig. 4.4 Hans Hahn, ca. 1905. *Photo* by Theodor Bauer, from Wiki Commons, public domain[11]

Hans Hahn (Fig. 4.4) was born in Vienna in 1879. He began his university studies at the University of Vienna in 1898, initially majoring in law; but he transferred to mathematics and continued his studies in Strasbourg and Munich, returning to Vienna in 1901. His dissertation (1902) was on the '*Theory of the 2nd Variation of Simple Integrals*'. He was for some years an assistant at the Technical College in Vienna (now *Technische Universität Wien*), and he completed his *Habilitation* at the University of Vienna in 1905.

The 'First Vienna Circle'

A group of young academics, all working in Vienna in the first decade of the twentieth century, including the physicist Philipp Frank, the mathematician Hans Hahn, and the philosopher and economist Otto Neurath, all admirers of Ernst Mach and Ludwig Boltzmann, began meeting regularly as a 'coffeehouse circle', a frequent phenomenon in Vienna, where a wide variety of coffeehouses in every district and for every taste in décor, clientele and *ambience* were available as comfortable meeting places.

Frank and Hahn were *Privatdozenten* at the University of Vienna, Neurath a teacher of economics at the Vienna Commercial Academy. They discussed modern philosophy and science and their social implications in this 'first Vienna Circle' ('*der erste Wiener Kreis*'),[12] beginning in 1907 and continuing for about five years. They were active in publishing, translating and reviewing literature on science and philosophy, and they followed the work of Frege and of Russell on the fundamentals of modern logic, and of Hilbert on an axiomatic basis for the sciences. Both Hahn and Frank spent some semesters in Göttingen, where Hilbert worked. They also published contributions and gave lectures to the *Philosophische Gesellschaft* (Philosophical Society) at the University of Vienna, led at that time by Alois Höfler.

In 1912, owing to his early interest in and engagement with Einstein's theory of Special Relativity, Frank was offered a position as professor at the German branch of the Charles University in Prague, the Chair which had been vacated by Einstein himself in 1912 after only a year. Frank remained there until 1938. Hahn obtained the offer of a professorship in *Chernowitz*, at the northeast boundary of the *k.-k.* Empire, and he went there in 1909. After being wounded in the War, he moved to Bonn in 1916, returning to a full professorship in Vienna in 1921. Neurath, who had obtained his doctorate in Berlin, spent much of WW I working in a ministry in Vienna, but received his *Habilitation* in Heidelberg in 1917, was a museum director in Leipzig, then moved to Munich, where he became involved in the political upheavals following 1918. He was arrested and served some time in prison; after his release, he returned to Vienna. He was the most 'political' of the Vienna Circle's active members, and later wrote a 'manifesto' for the circle which caused several members to leave it, including Gödel.

Hahn, back in Vienna, was able to convince other faculty members of the *Philosophische Fakultät* to support the appointment of **Moritz Schlick** as professor in 1922—Schlick had been a student of Max Planck and a friend of Albert Einstein in Berlin—to the Chair previously occupied by Ernst Mach and Ludwig Boltzmann. Schlick (Fig. 4.5) had written his dissertation in 1904 on '*The Reflection of Light by an Inhomogeneous Layer*' under Planck's guidance, and then returned to philosophy, his first interest. He obtained his *Habilitation* at Rostock (1910) with a thesis on *Truth in modern logic*, and published several papers on Special and General Relativity. When he arrived in Vienna, Schlick was encouraged by Hahn to begin a new 'Vienna Circle', discussing topics on the philosophy of science and meeting bi-weekly

Fig. 4.5 Moritz Schlick, photographed in 1930. *Photo* by Theodor Bauer, from Wiki Media[13]

in a lecture room at *Boltzmanngasse 5*, a building belonging to the Physics Institute (Fig. 5.3).

The Vienna Circle (*Wiener Kreis*)

The Circle, initially referred to as the '*Schlickzirkel*' (the Schlick Circle), began meeting in 1924, and it soon attracted other participants, so that by 1926 it included *Rudolf Carnap, Herbert Feigl, Otto Neurath* and *Friedrich Waismann*. Hahn brought his student Kurt Gödel in 1926; *Olga Taußky* and later *Karl Menger*, the latter a few years older than Gödel, also began attending at Hahn's invitation.

Menger, in his book '*Reminiscences of the Vienna Circle and the Mathematical Colloquium*',[14] gives a vivid picture of the many 'Circles' which existed in Vienna in the 1920s, some political/philosophical, some psychological, some purely philosophical (including one led by Heinrich Gomperz on the history of philosophy), some on religion and some on phenomenology. Menger was invited by Hahn to join the 'Schlick Circle' in the fall of 1927, soon after he had taken up his professorship of geometry at the Mathematics Institute. He first met Gödel as a student in his one-semester course on dimension theory in the fall of 1927, of whom he says, "*He was a slim, unusually quiet young man. I do not recall speaking with him at that time*".[15] Soon after, Menger again met Kurt Gödel at a meeting of the 'Schlick Circle', which Gödel had been attending since 1926 at the invitation of his mentor Hahn.

Neither Menger nor Gödel was dedicated to positivism—the philosophical school of the Circle—nor to Wittgenstein's philosophy, which dominated the Circle in the years 1925–27. But they apparently enjoyed the discussions and attended regularly, although Gödel in particular seldom made a comment, and then only when it concerned a topic on which he was well informed. One day, after long discussions of Wittgenstein's *Tractatus*, during which both Menger and Gödel had not spoken, Menger remarked as they were leaving the seminar room together, '*Today, we out-Wittgensteined the Wittgensteinians. We remained silent!*'. He was referring to the final sentence of Wittgenstein's *Tractatus*, "*Wovon man nicht reden kann, darüber muss man schweigen.*" ("*Concerning what cannot be spoken of, one must remain silent*"). And Gödel replied, "*The more I think about language, the less I can comprehend how people can ever understand each other at all*".[16]

Olga Taußky (-Todd) was just a few months younger than Kurt Gödel, but she began her studies of mathematics at the University of Vienna a year later, in the fall of 1925. She met Gödel at a seminar series given by Moritz Schlick[17] on Bertrand Russell's book, *Introduction to Mathematical Philosophy*. After 1926, she also attended the 'Schlick Circle', and of Gödel's relationship to that institution, later famous as the 'Vienna Circle', she wrote[18]:

> Although Gödel's home was in the mathematics seminar, although he was to become a student of Hahn's, although he was not a member of the Vienna Circle (Wiener Kreis) *[at that time, 1925]*, he was nevertheless an offspring of the Vienna Circle to which Hahn, Menger, Carnap, Waismann, and others belonged. The Vienna Circle was a group created by Schlick and concerned, if I understand it correctly, with the development of a language for science and mathematics.

After he left theoretical physics in mid-1926, Gödel concentrated his studies on mathematics and philosophy, beginning seriously in the autumn of 1926. Major motivations for his move were certainly the lectures of Furtwängler (on number theory) and of Gomperz (on the history of philosophy) in 1925/26, but, as he wrote to Wang in the 1970s, he was also motivated by his search for 'precision' (a recurring theme in their correspondence), which he felt was lacking in physics. His attendance at the meetings of the Vienna

Circle, which also began in the autumn of 1926, was certainly an additional contributing factor, although he was not at all in accord with the philosophical direction which predominated in the Circle. Schlick was very impressed with the philosophy of Ludwig Wittgenstein, as expressed in the latter's *Tractatus* (published 1921/22, first in German, then in English, with a preface by Russell). The Circle devoted the academic year 1926/27 to a (re-)reading of the *Tractatus*, which Gödel accepted passively without developing much enthusiasm.

Nevertheless, he continued to regularly attend the meetings of the Circle, and he contributed to it in ways which other participants remembered as positive, when he had something relevant to say. Otherwise, he remained in the background. After 1927, Wittgenstein's philosophy played a lesser role in the discussions of the Circle. In 1929, a brochure[19] (a sort of manifesto, proclaiming the aims of the Circle), entitled '*Wissenschaftliche Weltauffassung: der Wiener Kreis*' ('The Scientific Conception of the World: The Vienna Circle') was written by Neurath with input from Carnap, ostensibly in honor of Schlick. However, neither Schlick nor Hahn were pleased by its contents, even though Hahn signed it formally as an author; and several other people, including Menger and Gödel, effectively left the Circle as a result, feeling that it no longer represented their philosophical positions. In the meantime, Menger had founded his *Mathematical Colloquium*, and Gödel was taking an active part in it, becoming an editor of its regularly-published proceedings and contributing often to them with his own original work.

Just what *was* the philosophical position of the Vienna Circle? **Rudolf Carnap** (1891–1970) later played a major role in publicizing (and perhaps determining) it—although he was not a 'founding member' of the Circle. He was born in Ronsdorf, near Düsseldorf, Germany, and graduated from a *Gymnasium* in Jena in 1910. He studied at the University of Jena, obtaining his first degree in physics in 1914, having spent some time at the University of Freiburg, and he attended Frege's lectures on mathematical logic while in Jena. He served in the German army during World War I, but he was released in 1917 to begin graduate work at the University of Berlin. He wrote his doctoral thesis back at Jena, on an axiomatic theory of space and time. It was rejected by the physics department as 'too philosophical', so that he wrote a second version, relating it to Kant's work, which was accepted by the philosophy department in 1921. He stayed on at Jena, writing on *logical positivism* (not yet called that), and met Hans Reichenbach, the founder of *logical empiricism* from Berlin, at a conference in 1923. Reichenbach introduced him to Schlick, who offered Carnap a position as docent in Vienna.

Carnap (Fig. 4.6) arrived in Vienna in the spring of 1926, and immediately began attending the Circle. It was he who suggested that the Circle devote itself to Wittgenstein's *Tractatus* in the ensuing year. That was Gödel's (by his own statement, his *only*) exposure to Wittgenstein's philosophy. Gödel and Carnap began having discussions on mathematical logic around 1927, and some of them were recorded in Carnap's diaries.[20] Carnap became the spokesman of '*logical positivism*', as the philosophy of the Circle came to be called (for the first time in a 1931 publication by Schlick's former students Herbert Feigl and Albert Blumberg[21]).

Carnap remained in Vienna for some years, and obtained his *Habilitation* in 1928 with a thesis entitled '*Der Logische Aufbau der Welt* ' ('The Logical Structure of the World'). It became his first major publication.[23] He went in 1931 as a professor to the German University (*Karlsuniversität*) in Prague (with the support of physicist Philipp Frank, one of the founders of the 'First Vienna Circle'), but he continued to participate in the meetings of the Circle, traveling to Vienna on a bi-weekly basis, until the increasingly fascist environment forced him to emigrate to the USA in 1935. He became an influential figure in American philosophy. He was a faculty member in Chicago and at Harvard, then on the staff of the IAS in Princeton, and finally a professor of philosophy at UCLA. He died in Santa Monica/CA in September 1970. The *Wiener Kreis* effectively came to an end in the mid-1930s, after Hahn's early

Fig. 4.6 Rudolf Carnap, around 1938. *Photo* Courtesy of the University of Chicago Photographic Archive[22]

death in 1934, Schlick's murder by a deranged former student in 1936, and the emigration of several of its other members in 1933–35.

Logical positivism (later practically merged into '*logical empiricism*', Reichenbach's Berlin version of a similar (but distinct) philosophy, and now collectively termed '*neopositivism*'; cf. backnote [16]) is based on the premise that meaning, or knowledge, can be obtained only from verifiable observations (empiricism) or from logical proofs. It strived to avoid inaccuracies due to the imprecision of language (thus its attraction to Wittgenstein, who however was never a logical positivist nor himself a member of the Vienna Circle). Like a number of other philosophical programs in the first third of the twentieth century (e.g., Heidegger, Wittgenstein), it aimed at redefining philosophy as a whole, intending to introduce a 'scientific philosophy' based on empirical principles similar to those underlying classical physics. While Gödel shared some of its ideas, and he was attracted to fundamentals of mathematics through his early attendance at the meetings of the Circle, he was himself a Platonist (beginning in about 1925, before he began to attend the Circle), and never really accepted logical positivism.

Gödel's 'Turn' to Metamathematics

During the period 1926–28, Kurt Gödel continued taking courses, now predominantly in mathematics, and pursued his own studies. (During that period, he attended a course on differential geometry, of which he much later made good use when he became interested in Einstein's General Relativity. He had already taken a course given by Thirring on General Relativity while he was studying physics in 1924–26. We will take up this theme—Gödel's later excursion into theoretical physics—in Chap. 15).

Gödel's interest shifted after 1927 towards fundamentals of mathematics and mathematical logic, probably as a result of several factors: (i) the influence of the Circle, and especially a seminar on mathematical logic given by Rudolf Carnap in 1928/29; (ii) his study (beginning in the late summer of 1928)[24] of the *Principia Mathematica* (PM) by Bertrand Russell and Alfred North Whitehead,[25] a three-volume work on fundamentals of mathematics (in modern terms: metamathematics), published in 1910–13; and of the book '*Grundzüge der theoretischen Logik*' (Principles of Mathematical Logic)[26] by Hilbert and Ackermann (HA), which appeared in late 1928; (iii) his choice of Hahn as mentor for his doctoral work (Furtwängler had impressed him with his lecture course, but he was a traditional number theorist and not interested in fundamentals. Hahn was not a logician, but he had wide interests, had recognized Gödel's brilliance, observing him at meetings of the Circle and coffeehouse sessions afterwards, and he was younger and more flexible). And—as noted by various authors— (iv) the influence of two lectures given in Vienna by *L.E.J. Brouwer* in March 1928. Gödel most probably attended both of them. We quote from the summary of a talk by Teun Kotsier at a symposium[27] on Brouwer's life and work, held in Mainz in May 2018:

> On March 10, 1928, the mathematician L.E.J. Brouwer lectured in Vienna *[at the invitation of Menger]*. The title of his *[first]* lecture on the philosophical background of his views on the foundations of mathematics was '*Mathematik, Wissenschaft und Sprache*' [Mathematics, Science and Language]. Herbert Feigl wrote later that he managed after much resistance to coax Ludwig Wittgenstein to attend the lecture. Wittgenstein came and something remarkable happened. After the lecture, Feigl and Friedrich Waismann spent several hours with Wittgenstein in a café and Feigl wrote later: 'A great event took place. Suddenly and very volubly Wittgenstein began talking philosophy—at great length. Perhaps this was the turning point, for ever since that time, 1929, when he moved *[back]* to Cambridge University, Wittgenstein was a philosopher again, and began to exert a tremendous influence'.

The coffeehouse where Feigl, Waismann and Wittgenstein held their *Nachsitzung*—their 'meeting after the meeting', where they could discuss what they had heard in a congenial environment, with plentiful coffee to forestall drowsiness—was most likely the *Café Josephinium*, located in the building at *Währinger Str.* 33 (see Fig. 5.2), conveniently close to *Strudlhofgasse* 4 and *Boltzmanngasse* 5, where the philosophy and mathematics lectures were given, and the Vienna Circle held its bi-weekly sessions. Kurt Gödel was living with his brother at the time in that same building on the *Währinger Str.* The *Josephinium* was one of three coffeehouses where the members of the Circle met for discussions and for recapping what they had experienced in the more formal meetings. We will hear more about those three cafés in the following chapter, and also about the fact that the Gödel brothers lived in the building that housed the *Café Josephinium* during the critical period when Kurt was finding his way to the field in which he did his most important work.

Ludwig Wittgenstein had retreated from philosophy during most of the 1920s. After arranging the publication of his *Tractatus* in 1921/22, he renounced his (immense) inheritance, believing that it was unseemly for a philosopher to be rich, especially with money that he had not himself earned. He took training as an elementary-school teacher and spent several years in provincial villages, trying to bring culture to the local youth. During this period, Schlick and other members of the Circle invited him repeatedly to come to Vienna and attend their meetings, and Schlick even led an expedition to track him down in his current village—without success. After striking one of his pupils in a fit of frustration and annoyance, he retired from his teaching position and returned to Vienna, where he worked as an architect/designer, together with Paul Engelmann, designing a house for his sister, *Margaret Stonborough-Wittgenstein*. It is now known as '*Haus Wittgenstein*' and is a unique example of modern design, in particular the interior, on which Ludwig Wittgenstein had a major influence (see Fig. 4.7). During this period (1927–28), he had contact to some of the participants in the Vienna Circle, e.g., to Herbert Feigl and to Moritz Schlick. In 1929, he returned to Cambridge, where he was welcomed by Russell and Moore, and after completing his doctorate, he remained there for the rest of his life. Some of his students published his philosophical writings posthumously as '*Philosophical Investigations*'.[28]

Brouwer gave a second lecture four days later, a more 'mathematical' one, entitled '*Die Struktur des Continuums*' ('The Structure of the Continuum'). Both lectures probably contributed to Gödel's increased interest in mathematical logic, and soon after, he began working on his doctoral thesis in that area, as we shall see in Chap. 7. Dawson[31] mentions that Gödel was much

Fig. 4.7 *Left*: Interior of the '*Haus Wittgenstein*' (or '*Haus Stonborough*'), interior design by Ludwig Wittgenstein. The house now harbors the Bulgarian Embassy, Cultural Department. Photo: used with a CC A-SA 3.0 license.[29] *Right*: A painting of Margaret Stonborough-Wittgenstein in her wedding dress, by Gustav Klimt (painted 1905). *Photo* by Erwin Jurschitza, Wiki Commons, public domain[30]

later asked to contribute to a memoir of Brouwer's life and work following the latter's death in 1966, but he refused on the grounds that he had not known Brouwer well and had met him only once—during a visit of Brouwer to the IAS in Princeton in 1953. That does not, of course, exclude the possibility that he attended Brouwer's 1928 lectures in Vienna; attending a lecture by someone is not the same as 'meeting' that person. Very likely, Gödel in 1966 did not feel up to writing the memoir of Brouwer, and he used his (truthful) claim of having met him only once to 'beg off' from that request. But there can be no doubt that Brouwer's works influenced Gödel in his move to metamathematics and his choice of thesis topics. Even if he did not attend the Vienna lectures, he would have known about their content from his friends in the Circle (Karl Menger, for example, had worked with Brouwer for over two years in 1925–27, and had arranged his 1928 lectures in Vienna). Gödel maintained a similar distance to Wittgenstein, saying that he had never met him, although they certainly attended meetings together in Vienna after 1926, probably including Brouwer's (first) 1928 lecture.

Thus, Brouwer's lectures were apparently decisive for the careers of two younger philosophers: Ludwig Wittgenstein and Kurt Gödel.

Notes

1. The *Karlsuniversität* in Prague was founded in 1348 by the Emperor Karl IV of the Holy Roman Empire.
2. Stachel (1999).
3. Map: public domain, source: http://www.vidiani.com/large-detailed-map-of-center-vienna-city/.
4. The Emperor Franz, who acceded to the throne in 1792, was Franz II, Emperor of the Holy Roman Empire, and after 1804 he was Franz I, Emperor of Austria (later the *k.-und-k.* Empire). Compare Chap. 1.
5. Schimanovich-Galidescu (2002) (cf. Dawson (1997), CRC Edition (2005) p. 24).
6. Wang (1987), p. 21.
7. See https://mathematik.univie.ac.at/ueber-uns/geschichte/ (consulted May 2021).
8. Philipp Furtwängler had a brother, also named Wilhelm, who is not to be confused with their cousin, the orchestra conductor.
9. Wiki Commons, photographer unknown, public domain; reused from: https://commons.wikimedia.org/wiki/File:Philipp_Furtw%C3%A4ngler.jpg
10. Goldstein (2005), eBook edition, p. 58.

11. Photo by Theodor Bauer. Used under Creative Commons A-SA 4.0 International license. Original source: https://geschichte.univie.ac.at/de/bilder/hans-hahn-1879-1934-mathematik. Reused from Wiki Commons, https://commons.wikimedia.org/wiki/File:Hans_Hahn.jpg.

12. Stölzner & Uebel (2006).

13. WikiMedia, public domain. Owned by the Austrian Nationalbibliothek. {{PD-1996}} – public domain in its source country on January 1, 1996 and in the United States. Reused from: https://commons.wikimedia.org/wiki/File:Schlick_sitting.jpg.

14. Menger (1994), Chapter II, Sect. 6. See also Chap. 5 in the present book on Menger and his friendship with Gödel.

15. *Ibid.*, Final section on '*Memories of Kurt Gödel*'.

16. Sigmund (2015–17), p. 208.

17. Taußky-Todd (1988), p. 24.

18. *Ibid.*, p. 24.

19. See Neurath 1973.

20. Wang 1987, p. 77.

21. See Uebel 2013, p. 64.

22. Image © University of Chicago Photographic Archive, Digital Item No. apf1-01559r, Hanna Holborn Gray Special Collections Research Center, University of Chicago Library, cropped. From: http://photoarchive.lib.uch icago.edu/db.xqy?one=apf1-01559.xml.

23. Published in 1928 by Felix Meiner Verlag, Leipzig. English translation by Rolf A. George, 1967: *The Logical Structure of the World. Pseudoproblems in Philosophy*. University of California Press, Berkeley 1967. ISBN 0–812-69,523–2. This is one of Carnap's important works and is often cited as 'the *Aufbau*'. His '*Pseudoprobleme*' was also originally published in 1928.

24. Wang (1987, p. 80) points out that Gödel bought the PM books in July, 1928. Budiansky (2021, p. 110) suggests that he read the first volume at his parents' home in Brünn later that summer, and he quotes a letter to Feigl in late September in which Gödel expresses disappointment in it.

25. Russell & Whitehead 1910–13 (PM).

26. Hilbert & Ackermann 1928 (HA).

27. See https://www.geschichte.mathematik.uni-mainz.de/brouwer-symposium/. Consulted July 26th, 2021.

28. Wittgenstein, Ludwig (2001), [1953]: *Philosophical Investigations*. Blackwell Publishing, ISBN 0–631-23,127–7.

29. Image reused from: https://socks-studio.com/img/blog/Wittgenstein-haus-06.jpg, color reduced; licensed under a Creative Commons Attribution-ShareAlike 3.0 license.

30. Photo by Erwin Jurschitza, ejurschi@directmedia.de; released to the public domain. Reused from: https://commons.wikimedia.org/wiki/File:Gustav_Klimt_055.jpg.

31. Dawson (1997), CRC Edition (2005), p. 55; see also Endnote [124].

5

Private Life in Vienna

The young Kurt Gödel arrived in Vienna in October of 1924, as we know. He shared an apartment with his brother Rudolf, then beginning his fifth year of medical studies. Their apartment on the *Florianigasse* (Figs. 4.1 and 4.2) was conveniently near to the University and to its Medical Faculty, and it had sufficient space so that Kurt and Rudolf each had his private bedroom/workroom. They often went out together in their free time, but worked, and otherwise came and went, independently of each other. This arrangement was apparently quite satisfactory to both of them. Their father came to Vienna occasionally on business and they frequently met with him, and their mother also visited, more to participate in the cultural life of the city. They sometimes accompanied her to theater and operetta performances.

The two brothers remained in that same dwelling until April 1927; then they moved to an apartment on the *Frankgasse* in Vienna's 9th District (*Alsergrund*; see Fig. 5.1), still closer to the University and its hospital; but they stayed there only four months, until July 1927. At that point, Rudolf had completed seven years of medical studies and had presumably passed his *Staatsexamen* (State examinations), so that he could begin an internship and residency which would lead to his field of specialization (radiology; he later headed the X-Ray Department in a large Vienna hospital). Kurt could take advantage of the long semester break (from the end of July until mid-October) to visit their family back in Brünn, no doubt using the time to continue his studies in private, while Rudolf could also enjoy a certain amount of rest and relaxation before beginning 'life in earnest' at the hospital.

© The Author(s), under exclusive license to Springer Nature Switzerland AG 2022, corrected publication 2023
W. D. Brewer, *Kurt Gödel*, Springer Biographies,
https://doi.org/10.1007/978-3-031-11309-3_5

Fig. 5.1 *Left*: The building at *Frankgasse* 10, Wien-9 (*Alsergrund*), where the Gödel brothers lived briefly in 1927. *Right*: The plaque at *Frankgasse* 10. The text reads: 'Kurt Gödel (1906–1978), the most important logician of his time, lived here as a student of mathematics and philosophy from April 8th until July 20th, 1927'. The symbolic-logic formula is discussed in Chap. 7. Photos: AMZ, 2021

Thus, they left the *Frankgasse* in July and moved into their next apartment in Vienna (*Währinger Str.* 33—see Fig. 5.2) only in October[1] of 1927. It was also in the 9th District, somewhat further north, on a major street close to the University Hospital (*Allgemeines Krankenhaus Wien*) and also closer to the mathematics/physics building where the Vienna Circle held its meetings (*Boltzmanngasse* 5) and to the building in which the philosophy lectures were given (*Strudlhofgasse* 4)—see Figs. 4.2 (map) and 5.3. This latter building houses a lecture room now called the '*Kurt-Gödel-Hörsaal*'. Both now belong to the Physics Institute.

The imposing building at *Währinger Str.* 33 currently houses a hotel, the 'Hotel Atlanta', whose entrance is sheltered by a marquee that can be clearly seen in the photo in Fig. 5.2. The corner door now leads to the restaurant '*Forky*'. In Gödel's time there, the upper floors all housed private apartments, and the corner door was the entrance to the *Café Josephinium*, named for an institute belonging to the Medical Faculty of the University, which now holds its anatomical and other collections. That large institute building is located just 150 m further south on the same side of the *Währinger Straße*. It must have been particularly convenient for Kurt Gödel to live in the building that

Fig. 5.2 *Left*: *Währinger Str*. 33, where Kurt and Rudolf lived from the autumn of 1927 until July 1928. *Right*: The plaque at *Währinger Str.*, giving the dates of the Gödel brothers' residence. The last line is symbolic logic from Gödel's work; cf. Chap. 8. Photos: AMZ, 2021

Fig. 5.3 *Left*: The entrance to *Boltzmanngasse* 5, where meetings of the Vienna Circle were held. *Right*: *Strudlhofgasse* 4, which now harbors a lecture room named in Kurt Gödel's honor. Compare the map in Fig. 4.2. Photos: by Andreas Marschler, Vienna 2022. Used with permission and thanks to their author

harbored one of the principal meeting places for members of the Circle, and for other mathematicians and philosophers.

Besides the *Café Josephinium*, also the *Café Arkaden*, in the *Universitätsstraße* opposite the *Votivekirche*, the large neogothic church just north of the University main building, and the *Café Reichsrat*, located near the corner of *Stadiongasse* and *Rathausplatz*, close to the Austrian Parliament building, were also frequented by members of the Vienna Circle. The former was quite near the Gödel brothers' previous dwelling at *Frankgasse* 10. That short street begins just behind the *Votivkirche* and extends two blocks to the west, and it is also close to the location where Ludwig van Beethoven died in the year 1827.

Rudolf Carnap's diaries mention meetings with Gödel, Feigl, Waismann and other members of the Circle at the *Café Arkaden*.[2] Two important meetings of Gödel and Carnap (Feigl and Waismann joined them later) took place in August 1930 in the *Café Reichsrat*, at which Gödel disclosed his (new) results on incompleteness to Carnap[3] in preparation for the '*Second Conference on the Epistemology of the Exact Sciences*', held in Königsberg in early September of that year. We will come back to those meetings and their significance for the reception of Gödel's most important work in later chapters.

Personal recollections of Kurt Gödel from around this time have been published by Olga Taußky-Todd[4] and by Karl Menger,[5] young mathematicians at the University of Vienna in the second half of the 1920s who both knew Gödel relatively well and later recalled him in memoirs; we met up with both of them in the previous chapter in connection with the Vienna Circle.

Karl Menger (Fig. 5.4) was four years older than Gödel and had completed his doctoral thesis under Hans Hahn in 1924. Menger,[6] born in Vienna in 1902, attended the *Gymnasium* in *Döbling*, Vienna's 19th District (from 1913 to 1920), where he was a classmate of Wolfgang Pauli and Richard Kuhn, both later Nobel prize winners. He studied mathematics at the University of Vienna from 1920–1924. His doctoral thesis under Hahn's guidance was on '*The Dimensionality of Point Sets*'. He completed his doctorate in record time, in spite of more than a year spent in a sanatorium in southern Austria (in *Aflenz*, where the Gödels had often stayed during Kurt's youth, and where Kurt retreated several times during his 'most difficult year', 1936). Menger was there to treat his tuberculosis (successfully). In 1925, Menger went to Amsterdam on a fellowship, at the invitation of *L. E. J. Brouwer*, a Dutch mathematician who had been a protegé of David Hilbert's (but later became immersed in endless quarrels with him). We met up with Brouwer in the previous chapter, as the lecturer who inspired

Fig. 5.4 Karl Menger, in the late 1920s. Photo: Private collection[9]

Wittgenstein's return to philosophy in 1929, and probably influenced Gödel's choice of meta-mathematics as his field of specialization. Menger remained as Brouwer's assistant for two years, then returned to Vienna as professor of geometry at Hahn's invitation in the fall of 1927 (by which time he was also immersed in quarrels with Brouwer). It was then that he began attending the meetings of the Schlick Circle and met Kurt Gödel (who furthermore attended Menger's lecture course on dimension theory in 1927/28).[7] In his recollections of the Vienna Circle (and of his own Mathematics Colloquium, in which Gödel was an active participant from 1929 to 1937), he describes Gödel[8]:

> He was a spirited participant in discussions on a large variety of topics ... Orally, as well as in writing, he always expressed himself with the greatest precision and at the same time with the utmost brevity. In nonmathematical conversations he was very withdrawn.

Olga Taußky (later Olga Taußky-Todd; cf. Fig. 5.5) was born in *Olmütz* (now *Olomouc*) in Moravia, another linguistic island less than 50 km (about 30 mi.) to the northeast of Brünn, on August 30th, 1906—just 4 months after Gödel. Her family moved to Vienna when she was only 3 years old, then later to *Linz*, on the Danube around 150 km upriver from Vienna, in

Fig. 5.5 Olga Taußky (-Todd) in 1932. Photo: by Konrad Jacobs, used under CC A-SA 2.0 license[10]

western Austria. She studied and worked as an assistant to Hahn in Vienna from 1925–1934, and she wrote her doctoral thesis on number theory under Furtwängler. She had contact with Gödel during that time—Menger notes that the two of them often conversed after the meetings of the Circle.

Of Gödel's personal life, she writes,[11]

> There is no doubt that Gödel had a liking for members of the opposite sex, and he made no secret about this fact. Let me tell a little anecdote. I was working in the small seminar room outside the library in the mathematical seminar. The door opened, and a very small, very young girl entered. She was good-looking, with a slightly gloomy face (maybe *[due to]* timidity), and wore a beautiful, quite unusual summer dress. Not much later Kurt entered, and she got up and the two of them left together. It seemed a clear show-off on Kurt's part… You could talk to him about other things *[besides mathematics]* too, and his clear mind made this a rare pleasure…

Relationship with Adele

Just how and when Kurt Gödel and *Adele Thusnelda (Nimbursky) Porkert* met is unclear. It was probably sometime in the second half of 1928, when he was 22 years old (in April) and she turned 29 (in early November); however, Wang (1987) suggests that it was in 1927. After July 1928, both were living on the *Lange Gasse*, a small street, mostly residential, in the *Josefstadt* district of Vienna, west of the Ring and behind the *Rathaus* (see Fig. 5.6). Kurt and Rudolf had moved there from their *Währinger Strasse* apartment in July 1928, apparently looking for a larger place where their parents could visit them comfortably. The street begins in the south at the *Lerchenfelder Str.*, a major thoroughfare that starts behind the Parliament Building on the Ring and leads westwards, and it extends around 900 m (a bit over ½ mile) northwards, ending near the University campus.

Kurt and Rudolf lived at the northern end of the *Lange Gasse* in No. 72, and Adele with her parents in No. 65 (Fig. 5.7),[12] both solid *belle-époque* edifices, with shops on the ground floor and 5 floors of apartments above. Both are still in existence, as can be seen in Figs. 5.6 and 5.7. Number 65 now houses a shop for dental supplies on its ground floor. The *belle étage*, the first upstairs floor, has higher ceilings than the others, as was the custom in the early 20th century, when the buildings were probably constructed. Number 72, somewhat grander, currently has two small shops on its ground floor, flanking the large entrance, and carries an imposing memorial plaque from the Kurt Gödel Society and the City of Vienna, recalling that '*Kurt Gödel (1906–1978), the most eminent logician of his time, lived here as a student of mathematics and philosophy from July 4ᵗʰ, 1928 until November 5ᵗʰ, 1929*', followed by a formula on the relative consistency in set theory from his 1939 paper[13] (see Fig. 5.6, below right). Interestingly, No. 65 now also houses a dance school, which might be construed as recalling Adele's residence there, although unintended.

At the time when Kurt and Adele met, she had been working as a dancer,[14] cloakroom girl and ticket cashier in the nightclub '*Der Nachtfalter*' (the Night-Moth), near the center of Vienna.

It was located in the half-basement of the building at *Petersplatz* 1, in the central 1st District (*Erster Bezirk*, Wien-1), opposite the Church of St. Peter (Fig. 5.8). The building was constructed in 1873, and after a reconstruction in 1906, the nightclub was established there by Johann and Marie Prumüller, and it continued as the *Nachtfalter* until 1937. That venue is now occupied by a theater, earlier (1958–2009) called the *Theater am Petersplatz*, now known as "*Werk-X Petersplatz*". Information about its history can be found on its website.[15]

The nightclub was situated about 1.5 km (just under one mile) from *Lange Gasse* 65, so that it was a brisk 15-min. walk from the club to Adele's home. One is tempted to interpret its name as the image of a mothlike creature flitting through the night, looking for excitement—a cliché which was popular in the late 1920s.[16] However, given its founding in 1906, the club's name probably refers instead to a famous species of moth, the *Wiener Nachtfalter*

Fig. 5.6 *Above*: The house at *Lange Gasse* 72, Wien-8. This house was built by the architect Arpad Mogyorosy in 1910. *Below left*: The entrance to *Lange Gasse* 72, where Kurt and Rudolf Gödel lived from July 1928 until November 1929. *Below right*: The plaque at *Lange Gasse* 72 (see Chap. 11). Photos: (Upper image): by Andreas Marschler, Vienna, 2018. Used with permission and thanks to the author. (Lower images): AMZ, 2021

Fig. 5.7 Diagonally across the street from *Lange Gasse* 72 is the building at *Lange Gasse* 65, where Adele Porkert lived with her parents in 1928–29. Photo: AMZ, 2021

or *Wiener Nachtpfauenauge* (*Saturnia pyri*), which is native to the Vienna area and is the largest night-flying moth in Europe.

It is quite possible that Kurt spent one or more evenings at the *Nacht-falter* and noticed Adele there; perhaps he was in the company of some of his university friends or of his brother, since he was not above partaking of light entertainment and often did so in the company of his brother (and his mother, after she also moved to Vienna in 1929). But Kurt also might well have noticed Adele on the street in his own neighborhood, as their apartments were practically across from each other.

The entertaining and intelligent book '*La Déesse des petites victoires*' ('The Goddess of Small Victories') written by Yannick Grannec,[17] a scientific-historical docunovel, uses a similar device to that introduced by Louisa Gilder[18] in her '*Age of Entanglement…*': reconstructing real conversations between famous scientists, but with fictitious dialogue, to tell the history of a scientific development. Grannec however goes a step further and makes use of additional, fictitious characters to fill in the background story and even the historical action, which in her book is the relationship between Kurt Gödel and Adele Porkert over its nearly 50 years, as related by the venerable Adele in her last year at the care home in New Jersey where she

Fig. 5.8 *Petersplatz* 1 in Vienna's Central District (Wien-1). In the half-basement of this building, the nightclub *'Nachtfalter'* was situated from 1906 to 1937. Adele Porkert was working there when she met Kurt Gödel in the second half of 1928, and for some time thereafter. Photo: AMZ, 2021

moved after Kurt's death in early 1978. Grannec tells a story of Kurt and Adele's acquaintanceship which combines chance meetings on the street (in the very early mornings when Adele is returning from work at the *Nachtfalter*, while the sleepless Kurt is pacing the sidewalk, working out his formulas), complemented by his later visit to her club, where she flirts with him at the cloakroom.

Just how their meeting actually occurred is uncertain; one source[19] even suggests that she attended a session of the *Wiener Kreis* (Vienna Circle), where Kurt was a regular (if not very active) participant in the years 1926–1929. Given her background and interests in 1927/28, however, that version seems rather unlikely.

However they met, it is unquestioned that they became intimate friends by 1929, much to the disapproval of Kurt's parents. Adele (Fig. 5.9) was 6½ years his senior and, in their view, came from an inferior class. She had little

Fig. 5.9 Adele Porkert, photographed around 1920, when she was about 20 years old. It was 8 years later that she met Kurt Gödel. Photo: photographer unknown, public domain[20]

formal education; her father was a photographer who made a passable but not luxurious living from his art, and her family had arrived in Vienna from western Germany, the *Rheinland-Pfalz* region, when she was a child. In addition, she had married early (to the photographer Nimbursky); an unhappy marriage from which she was already separated when she met Kurt in 1928.

She divorced Nimbursky in 1933.[21] In the first few years of their relationship, Kurt kept a 'low profile' around Adele, since his family, predictably, did not approve of her; this was especially true of his father.

The Occult and *Esoterica*

In 1920s Vienna (as elsewhere at the time), there was a renewed interest in the occult and the paranormal, as well as a certain opening up to less-established religions. Those movements did not leave some members of the Mathematics and Philosophy departments completely indifferent, including Kurt Gödel. Karl Menger, in his '*Reminiscences of the Vienna Circle...*' (1994, p. 14), describes this scene:

In the first post-war years, numerous mediums had appeared in Vienna and they were viewed by the intelligentsia with the utmost skepticism… a group including Wagner-Jauregg, Schlick, Hahn, Thirring, and many others (most of them scientists) formed a committee for the serious investigations of mediums. Very soon, however, members of the group began to drop out: first, Wagner-Juaregg; soon after him, Schlick; so that by 1927 only two of the scientists were left, my friends and former teachers Hahn and Thirring. They were not fully convinced that any of the phenomena produced by the mediums were genuine; but they were even less sure that all of them were not…

He goes on to describe a séance where he 'filled in' for Hahn, who was ill. In the end, he was also uncertain as to whether it was all just a swindle, and vowed to never again attend a séance. Kurt Gödel was also interested in the supernatural, although he apparently did not take part in the 'paranormal' scene himself. Hao Wang (1987, p. 159) says about Gödel's attitude towards the supernational,

G[ödel] and I agree, I think, in believing that empirical inductions are important for doing philosophy … but also in evaluating what are (to be) taken as facts (and their degree of importance, certainty, etc.). But … our priorities differ, and we make different 'probability calculations'. The striking instances have to do with G[ödel]'s bidding defiance to the widely shared 'modern' skepticism toward the supernatural (which is to him a part of the 'prejudices of the time'). He says, I am not sure how seriously, that the 'spirits' were more active in antiquity than they are today. More significant and relevant is his prediction of the emergence of (religious) metaphysics as an exact theory before long …

We will hear more about this tendency of Gödel's in the chapters dealing with his philosophy.

The Fateful Year 1929

The year 1929 was a significant one for the Rudolf August Gödel family in several ways: early in that year, Rudolf August died suddenly in Brünn, on February 23rd, 1929. He was apparently the victim of an abscess of the prostate[22] following a failed operation.[23] He would have been 55 years old at the end of February, five days later. He had devoted himself to his business and apparently was not physically active, tending to chubbiness, as photographs from the time can testify.

His unexpected death left Marianne a widow and nearly alone in her large house on the *Spielberggasse*. Rudolf's Aunt Anna and Marianne's older sister

Pauline were still living on its upper floors; and of course, Marianne had other relatives and many friends in Brünn—but apparently those were not sufficient attachments to keep her there, since she moved to Vienna to live with her two sons in November 1929, taking Aunt Anna with her; and they all moved into a larger apartment at *Josefstädter Str.* 43–45, a few blocks west of the Parliament Building in the *Josefstadt*, in late 1929 or early 1930 (see Figs. 5.10 and 5.11). They lived there together until 1937. This was Kurt Gödel's longest stay in any dwelling in Vienna.

Kurt thus now had a 'Watchbird', in the person of his own mother, looking after him around the clock in his chosen city. This no doubt put a damper on his relationship with Adele, which had been established for over a year when his mother arrived in Vienna; and her presence may have been a principal reason why the relationship continued on an 'informal' basis for the next 9 years.

Kurt's mother Marianne and his lover/friend Adele apparently had few or no contacts during that time, and Kurt, with his tendency to avoid conflicts, would have seen to it that friction between them was kept to a minimum. His

Fig. 5.10 A modern view of the building at *Josefstädter Str.* 43–45, where Kurt Gödel lived with his mother and brother from 1930 to 1937. The façade has been modernized in the center on the ground floor. The entrance door, at lower left, as well as the upper floors, are still original. Photo: AMZ, 2021

IN DIESEM HAUSE WOHNTE
VON 1930 — 1937
DER GROSSE
MATHEMATIKER UND LOGIKER

KURT GÖDEL
1906 — 1978

HIER ENTDECKTE ER
SEINEN BERÜHMTEN
UNVOLLSTÄNDIGKEITSSATZ
DIE BEDEUTENDSTE
MATHEMATISCHE ENTDECKUNG
DES ZWANZIGSTEN JAHRHUNDERTS

2006

Fig. 5.11 The plaque at the entrance to the house at *Josefstädter Str.* 43–45: "In this house, the great mathematician and logician *Kurt Gödel* (1906–1978) lived from 1930–1937. Here, he discovered his famous incompleteness theorem, the most significant mathematical discovery of the twentieth century". (Plaque mounted in 2006). Photo: AMZ, 2021

brother Rudolf had also not met Adele before her marriage to Kurt. Adele showed great patience and remained loyal to her '*Kurtele*' during all those years, and her patience was rewarded finally on September 20th, 1938, when she and Kurt were married in Vienna.

By the time of their marriage, Kurt's mother Marianne had moved back to Brünn (in the autumn of 1937), perhaps missing her family and friends there, worried about the condition of her villa, which she had left in the hands of renters/caretakers, or tired of her life as a well-off widow in Vienna; and, in particular, in a move to save money, as life in Vienna was expensive and her economic assets were all in Czechoslovakia.

Rudolf, in his interview for a 1986 television documentary,[24] made the following remarks about this period:

> …Two events then rapidly opened our eyes *[to the political situation]*: the assassination of the Federal Chancellor Dollfuß *[1934]* and the murder of the

philosopher Prof. Schlick *[1936]*, in whose circle my brother had been active. This latter event was in the end the reason that my brother suffered a severe nervous breakdown and had to spend some time in a sanatorium, which was naturally a cause for much concern on the part of our mother. Soon after his recovery, my brother received an invitation to go as a guest professor to the USA. This, and the fact that life in Vienna had become increasingly expensive for us, since our fortune was in Czechoslovakia, was the reason that our mother moved back to her villa in Brünn; indeed in 1937, a year before Hitler occupied Austria. There had also been difficulties with the Czech caretaker of the villa, since the hatred of the Czechs for the Germans was naturally great at that time…

Their mother remained in Brünn throughout most of World War II, going back to Vienna and her older son Rudolf only in 1944 (at his—wise—request, for her own safety). We will return later to Kurt and Adele's marriage, and to Marianne's dwelling places, in particular in Chaps. 9 and 11.

Just what was the nature of the relationship between Kurt and Adele, who would seem at first glance to be so completely mismatched? She was certainly aware of his almost unearthly intelligence in certain areas, where she could not follow him—but at the same time, she was also aware of his susceptibility, his near helplessness in the face of aggressive or contradictory behavior on the part of other people, and his sensitivity. Adele, 6½ years older than Kurt, no doubt felt some motherly feelings toward him, or in any case protective feelings; these were evident in her terms of endearment for him: '*Kurtele*', a diminutive form usually applied to children, and '*mein strammer Bursche*' (roughly 'my strapping lad'), also usually applied to a youth in early manhood, and not too appropriate for Kurt, who was sleight of build. She did not have his kind of analytic intelligence, but she had 'a way with words and phrases', as was later attested to by some of his colleagues, and evidently a good portion of empathy—according to Plato and Hannah Arendt, the highest form of intelligence.

It was noted by various friends and colleagues that Kurt seemed more at ease with other people when Adele was also present, and there can be no doubt that her support and assistance with 'everyday life' (especially during their long journey across Asia and the Pacific to America in early 1940) were instrumental in keeping him alive and relatively healthy over many years. Without her, his life would certainly have taken a different route, most likely a briefer and more tragic one. In the event, they remained together for nearly 50 years, and while the end of his life may be considered tragic, and had much to do with Adele's temporary absence, as we shall see in later chapters, it came only after he had passed the age of 70 and was no longer able to work. His life was nevertheless reasonably long, and he was able to work and interact with others for many years, a fact in which Adele played a significant role.

Other important events in Kurt's life in 1929 were his release from Czechoslovakian citizenship, on Feb. 26th, 1929 (his application for release from citizenship had been submitted in 1928), and his acceptance in June as a citizen of Austria. He also noted in a letter to Wang[25] many years later that he, "*...had completed his formal studies at the University before the summer of 1929*", and he spent the late spring finishing his doctoral thesis, on which he had presumably been working since mid-1928. In Chap. 7, we take a closer look at that work.

Notes

1. Wang (1987), p. 79 suggests that they may have moved into the *Währinger Str.* apartment in September of 1927.
2. See the blog at: https://praymont.blogspot.com/2008/08/cafes-arkaden-and-josephinum.html (consulted on August 16, 2021).
3. Dawson (1984c); see p. 255.
4. Taußky-Todd, in *Gödel Remembered. Proceedings of the Gödel Symposium*, Salzburg, July 1983. See Weingartner & Schmetterer (1987). A pdf of Taußky-Todd's article is available; cf. Taußky-Todd (1988) in the Literature List.
5. Menger (1994), Chapter 'Memories of Kurt Gödel', Sect. 1.
6. See e.g. the MacTutor Biography of Karl Menger, online at https://mathshistory.st-andrews.ac.uk/Biographies/Menger/. See also the *Stanford Encyclopedia of Philosophy* article on Logical Empiricism, online at https://plato.stanford.edu/entries/logical-empiricism/.
7. Menger, *op. cit.*, 'Memories of Kurt Gödel', Sect. 1.
8. *ibid.*, Sect. 1, p. 201.
9. Photo: photographer unknown, public domain; from the private collection of Karl Menger, property of Rosemary Menger Gilmore, Chicago and her siblings. Reused here from [Leonard (1998)].
10. Photo: from the *Oberwolfach Photo Collection*, by Konrad Jacobs, cropped. Licensed under Creative Commons. Attribution Share-Alike 2.0 license—reused from: https://opc.mfo.de/detail?photo_id=4151
11. Taußky-Todd (1988).
12. According to Dawson (1997). Budiansky (2021) gives the house number as 67, but this seems unlikely because that building is nearly at the corner of *Alster Str.*, the next cross-street, and at some distance from No. 72. Nos. 65 and 67 are in any case very similar buildings.

13. Gödel's (1939a) on consistency in set theory; see Dawson 1983 and Chap. 11.
14. Whether Adele was in fact a dancer has been the subject of some controversy. Schimanovich (2005, pp. 401 ff.), in a letter to Martin Davis (dated Nov. 2001), claims that she was not, 'dancer' being practically synonymous with 'prostitute' in early twentieth century Vienna. Kurt Gödel himself affirmed that she *had* been a dancer (in a conversation recorded by his psychiatrist Dr. Philip Erlich on May 21st, 1970, and quoted by Budiansky (2021), Footnote 48 on p. 106)—but *not* a prostitute. She was, in any case, working as a masseuse by 1929, and was registered as such in the city commercial directory in 1929–31.
15. Website *Werk-X-Petersplatz*: https://werk-x.at/werk-x-petersplatz/spielstaette/. (Consulted in January, 2021).
16. Moth image 1920s: "*Ich bin von Kopf bis Fuß…* ", song by Friedrich Holländer (1930), for the film '*The Blue Angel* '. English title: '*Falling in love again*'; translated text:"*Men flit around me, like moths around a flame/And if they should burn up, I'm not the one to blame*".
17. Grannec (2012).
18. Gilder (2007).
19. See the broadcast for Gödel's 100th birthday, April 2006, by the ÖRF (Austrian Public Radio/TV channel), Science News Service; summary online at https://sciencev1.orf.at/science/news/144363.html. Quote:"*Der Wiener Kreis beeinflusste nicht nur die Forscherlaufbahn des Mathematikers. Er lernte bei den Treffen auch seine spätere Frau Adele Porkert kennen. 1938 heirateten die beiden.*" ("The Vienna Circle influenced not only the research career of the mathematician [Gödel]. He got to know his future wife Adele Porkert at its meetings. In 1938, they were married ".) See also https://dewiki.de/Lexikon/Kurt_G%C3%B6del#Studium_in_Wien: "*Auch für sein Privatleben waren die Treffen des Zirkels von Bedeutung, da er hier 1927 zum ersten Mal seine spätere Frau Adele Nimbursky traf.*" ("For his private life, also, the meetings of the [Vienna] Circle were significant, since there in 1927, he met his future wife Adele Nimbursky for the first time".
20. Photo: photographer unknown, public domain; reused from: https://www.privatdozent.co/p/kurt-godels-brilliant-madness (July 2021).
21. But compare Budiansky (2021): he mentions that divorce was in fact not recognized in Austria between 1934 and 1938, because control over marriages had been given to the Catholic Church (during the *Austrofascism* period in 1933/34).
22. Wang (1987), p. 69; Budiansky (2021), p. 118.
23. Budiansky (2021), p. 118.
24. Schimanovich and Weibel (1997). See page 48, Rudolf's interview, on politics in Vienna in the later 1930s. Translation by the present author.
25. Wang (1987), p.82.

6

An Introduction to Mathematical Logic

In order to understand and appreciate Kurt Gödel's important contributions to mathematical logic and set theory, readers should have at least a modest acquaintance with the history and development of those fields before and leading up to his time in the 1920's and 30's, when he carried out his most significant work. Those readers who are already knowledgeable about these topics, or who are in a hurry to get to Gödel's own work, can skip over this chapter without losing the continuity of our story.

The topic of 'mathematical logic' is generally agreed to have originated with the ancient Greek philosopher-mathematicians. Names which are often mentioned in this context are those of *Euclid* (of Alexandria; Greek *Εὐκλείδης*), who was active as a mathematician during the late 4th and early third centuries BCE (he lived presumably between about 325 and 270 BCE). His most famous work, which survived to modern times and served as a geometry textbook over many centuries, is his *Elements*, in which he sets out the principles of geometry and number theory in terms of *theorems*, all derivable from a few *axioms* or given premises, which are presumed to be intuitively evident or verifiable by observation. Other Greek mathematicians often associated with the origins of logic are *Thales* (late 7th to mid-sixth century BCE), *Pythagoras* (sixth century BCE), *Archimedes* (third century BCE); *Hipparchus* of Nicaea (second century BCE), the 'father of trigonometry'; and, of course, the philosophers *Plato* (late fifth century to mid-fourth century BCE) and his pupil *Aristotle* (fourth century BCE), whose writings, rediscovered in the Middle Ages, were instrumental in preserving ideas from the ancient Greeks to modern times.

© The Author(s), under exclusive license to Springer Nature
Switzerland AG 2022, corrected publication 2023
W. D. Brewer, *Kurt Gödel*, Springer Biographies,
https://doi.org/10.1007/978-3-031-11309-3_6

Geometry

Many authors[1] hold that geometrical ideas are innate or intuitive to the human mind, and it is certainly true that we form geometrical models within our brains which guide us through our 3-dimensional world (think of finding your way through a familiar room on a dark night with no artificial light). The origin of this intuition is of course empirical, our experience with the world, and making it into pure mathematics requires a certain leap, summarized by Hilbert[2] in his statement, "*The task of geometry is the logical analysis of our spatial intuition*". Geometry is evidently also applicable, in fact essential to the physical sciences, and just where the distinction between geometry as pure mathematics and geometry as a physical science should be located is not immediately obvious.

The word '*geometry*' is derived from the Greek for 'measuring the earth', and the subject is important for understanding objects in space and their relations to each other. Many readers will be familiar with geometrical logic from school classes, even if they are not themselves mathematicians or scientists.

In his *Elements*, Euclid gives the first complete exposition of the axiomatic method, characteristic of the logical-mathematical aspect of geometry. He begins with plane geometry, i.e., the geometry of lines and figures on a flat two-dimensional surface, a plane, such as the smooth top of a table. Based on five axioms and a number of 'procedures', which describe how lines and figures can be constructed in the plane, one is presumed to be able to construct figures and derive theorems about them using just two instruments: an (unmarked) straightedge for drawing lines, and a compass with two arms whose angle can be fixed, each ending in a sharp tip which can be used to define a point in the plane or to draw arcs of constant radius. Every school child is familiar with the use of a compass to draw circles (whose radii are equal to the spacing of the arms) with their centers fixed by the point on one arm, and the other point—perhaps equipped with a pencil lead or pen tip—used to inscribe the circle.

Euclid's five axioms of *plane geometry*, loosely translated, are the following:

(i) Objects that are equal to the same thing are equal to each other (*transitivity*).
(ii) Addition of equals to equals results in equal sums (*additivity*).
(iii) Subtraction of equals from equals results in equal differences (*subtractivity*).
(iv) Objects that coincide (congruent objects) are equal to each other (*reflexivity*).

(v) Only one straight line can be constructed through a point outside another straight line in such a way that the two lines, indefinitely extended, never intersect (*parallel axiom*). (6.1)

The first four axioms are uncontroversial. The fifth, the 'parallel axiom', was long disputed—the question being whether it was independent of the other four, or could perhaps be deduced from them. In the meantime, it is known that the existence of *at least one* parallel line (in a plane) is implied by the first four axioms, but that there is *only one* is a separate assumption, independent of the other axioms. Furthermore, this axiom is not applicable to curved spaces. In positively-curved spaces (e.g. spherical geometry), there are *no* parallel lines in the above sense, and in negatively-curved spaces (hyperbolic geometries), there is an unlimited (*in-finite*, without end) number of them.

In addition, Euclid gave some procedures or 'common-sense rules' for constructing lines and figures, etc. Some examples of these rules are: constructing a line segment between two specified points; extending a line segment *infinitely* (again, 'without end') from either end; drawing a circle or circular arcs with a compass (known to every school child); the statements that all right angles are equal, that two right angles produce a parallel (inverted) straight line, and four yield a congruent straight line (i.e. rotation through a full circle, or 360° in modern notation).

Euclid's proofs of proposed theorems proceed geometrically, i.e., 'by construction', as we shall see in a few examples. The tools used for the constructions, as mentioned, are the straightedge which allows us to draw (inscribe, or 'construct') straight line segments, and a compass for inscribing circles or arcs around a given center point (or for transferring fixed distances from one line segment to another).

In his *Elements*, Euclid included not only plane geometry, but also its extension to a three-dimensional (flat) space, known as *solid geometry*, as well as some basics of *spherical geometry* (essential for 'measuring the earth', since the ancient Greeks already knew that the earth is roughly spherical in shape), and some number theory (necessary if one wants to express geometric theorems numerically, i.e. in terms of lengths and angles given in numerical units).

Example Proofs

As a simple example of a theorem, we consider the construction of the *perpendicular bisector* of a given line segment AB (here, we denote the *endpoints* of the segment by Roman capital letters, A, B, C etc. This is not necessary, but it simplifies reference to diagrams). Straight line segments, or their lengths, are denoted by small Roman letters, a, b, c…, while angles are given as small Greek letters α, β, γ… The perpendicular bisector is a straight line CD which meets the given line AB at its midpoint M and makes right angles with it. First, the procedure for constructing the perpendicular bisector is given (cf. Fig. 6.1):

i. Set the spacing of the compass points to a value greater than half but less than the full length of the line AB (by estimation). Place one compass point at the point A and use the other to inscribe a circular arc above the line and another below it (or to inscribe a semi-circle through the line AB).

ii. Repeat this procedure with the compass point fixed at the other end of the line, at point B. The arcs (or semicircles) will intersect at two points C and D above and below the line AB. Construct the line segment CD. This is the perpendicular bisector of AB, intersecting it at the point M.

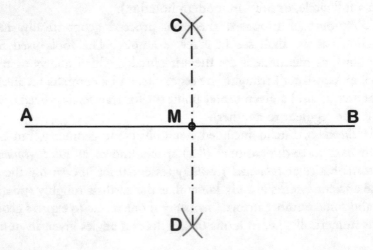

Fig. 6.1 Construction of a perpendicular bisector to the given line AB. The bisector is shown as the dash-dotted line CD. *Source* private, WDB

Proof:

i. The salient fact is that the four diagonals AC, BC, AD, and BD are all equal 'by construction', since they are the radii of the arcs, which were fixed by the spacing of the compass points, kept constant during the construction. In Fig. 6.2, they are drawn in as dashed lines. Note that they define four large and four small triangles, $\triangle_{ACB}$, $\triangle_{ADB}$, $\triangle_{ACD}$, $\triangle_{BCD}$; and $\triangle_{ACM}$, $\triangle_{BCM}$, $\triangle_{BDM}$, $\triangle_{ADM}$.

ii. Compare the two large triangles $\triangle_{ACD}$ and $\triangle_{BCD}$, to the left and right of the line CD. They are *isosceles triangles* (they have two equal sides), and share a common baseline, CD. In fact, they are *congruent*, since their corresponding sides are all equal (their outer sides by construction, their common bases by reflexivity; this property is sometimes abbreviated as '*sss*', representing the equal corresponding sides). A similar argument can be applied to $\triangle_{ACB}$ and $\triangle_{ADB}$, above and below line AB.

iii. Now use a property of congruent triangles: corresponding parts of congruent triangles are congruent (abbreviated *CPCTC*). This means that each of the corresponding internal angles of the two triangles are equal, and their *altitudes* (AM and MB) are also equal. It follows immediately that M is indeed the midpoint of AB: it divides AB into two equal line segments, AM and MB. (Likewise, CM = MD; see above)

iv. Next, consider the four smaller triangles $\triangle_{ACM}$, $\triangle_{BCM}$, $\triangle_{BDM}$, and $\triangle_{ADM}$ (Fig. 6.2) which are grouped around their common vertex M.

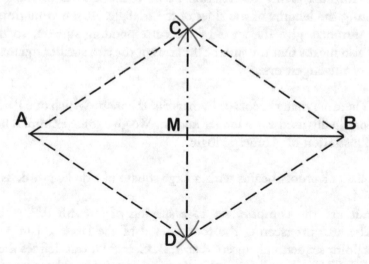

Fig. 6.2 The perpendicular bisector construction with equal diagonal line segments connecting the four endpoints (shown as dashed lines). *Source* private, WDB

Their outer sides are all equal (by construction), and they share pair-wise common sides (CM, MB, MD, and AM). They can thus be shown pairwise to be congruent (*sss*), and are therefore *all* congruent (by transitivity). This means that their corresponding angles around the vertex M are all equal. Those four angles add up to a full circle, and are therefore all right angles (by the rule for right angles, see above). Thus, the line CD makes right angles with AB, and is its *perpendicular* bisector. Q. E. D.

Pythagoras' Theorem

Another useful property (of right triangles in a plane) is given by *Pythagoras' law* (or *theorem*), dating from several centuries before Euclid collected geometric theorems and proofs in his *Elements*. Pythagoras' law relates the length of the hypotenuse (the long side of a right triangle, opposite its right angle) to the lengths of the other two sides; more precisely, the squares of these quantities. One formulation is "The square on the hypotenuse is equal to the sum of the squares on the other two sides", where 'square' refers both to a closed planar figure with four equal sides and four internal right angles, and to the product of a number with itself. This double use of the word derives from its relation to Pythagoras' law and to the formula for the area enclosed by a square. This is illustrated by Fig. 6.3.

As seen in the figure, Pythagoras' law can be expressed compactly by an algebraic formula giving the relation between the squares of the numbers representing the lengths of the sides of the triangle. At the same time, these squared numbers give the areas of the corresponding squares, so that the formula also means that the sum of the areas of the two smaller squares equals the area of the largest one.

Proof: There are many proofs of Pythagoras' theorem. Which one Pythagoras himself originally used is no longer known. We give some examples here as a further illustration of geometric logic.

One class of proofs begins with a large square of side length d, as shown in Fig. 6.4.

We can use the compass, set to a spacing of b (with $d/2 \leq b < d$), to inscribe arcs intersecting the four sides d of the large square, with the compass point sequentially inserted at A, B, C, and D; this defines the points K, L, M, and N. Constructing the diagonals connecting these latter points produces four triangles within the square. They are right triangles, since they share their internal angles γ with the large square, and they are congruent,

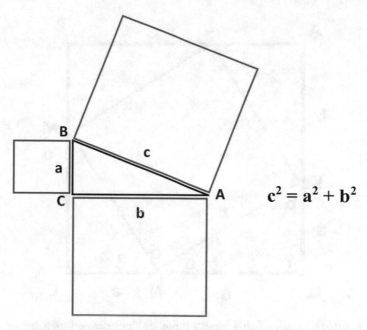

Fig. 6.3 A right triangle ABC with its sides a, b, c; showing "the squares on the sides" which have the same side lengths as each side of the triangle. The resulting algebraic formula is also shown. *Source* private, WDB

since they all have altitude a, base b, and the same included angle γ. Thus, their hypotenuses c are also all equal, defining the smaller square KLMN. Its internal angles δ are all right angles, as can be seen from Fig. 6.4: the sum of angles α, δ, and β gives a straight line (180°), but α + β + γ (internal angles of the triangles) also gives 180° (triangle theorem), so δ = γ = right angle. The area of the inner square, c^2, plus the areas of the four triangles, must equal the area within the large square, d^2.

With this information, we can prove Pythagoras' theorem by rearranging the triangles within the large square, changing nothing else. One possible rearrangement is shown in Fig. 6.5. It can be obtained from Fig. 6.4 by simply sliding the two lower triangles diagonally upwards until their hypotenuses coincide with those of the two upper triangles, producing two rectangles, one at upper left with altitude b and base a, one at upper right with altitude a and base b. We note that the area of the triangles has not changed during this rearrangement, so the 'left over' area within the large square must be the same as before. It can be proved in detail, but is apparent by inspection of Fig. 6.5, that this 'left over' area consists of two squares, one of side a and the other of side b. Using the expression for the area of a square (which we have not proven, but have already used in the discussion of Fig. 6.3), we have

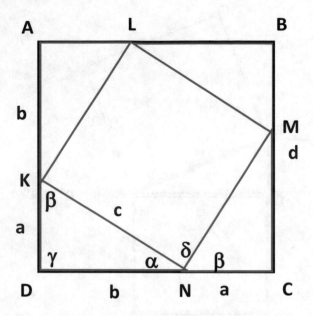

Fig. 6.4 A square $\square_{ABCD}$ of side length d has its sides divided into segments a and b (a + b = d), ending at points K, L, M, N. Constructing the diagonals KL, LM, ... produces four right triangles within the square, e.g., $\triangle_{KND}$, with internal angles α, β, γ and hypotenuse c. This in turn defines a smaller square $\square_{KLMN}$ within the larger square, with side length c and internal angles δ = right angles. See text for details. *Source* private, WDB

for the 'left over' area in Fig. 6.5: $a^2 + b^2$, which is equal to the area of the inner square in Fig. 6.4, i.e. c^2. We have thus found that $c^2 = a^2 + b^2$, and that is the Pythagorean formula. Since we chose the sides of the triangles in Figs. 6.4 and 6.5 arbitrarily, the result holds for all right triangles.

The result can also be obtained *algebraically* by noting that the area of the large square is d^2, where d = a + b, i.e., $d^2 = a^2 + 2ab + b^2$. 2ab is the combined area of the two rectangles in Fig. 6.5, so the remaining area within the large square is $a^2 + b^2$, and is again equal to the remaining area in Fig. 6.4, c^2, yielding $a^2 + b^2 = c^2$, Pythagoras' formula.

Finally, in Fig. 6.6, we show another route to proving Pythagoras' theorem: we first draw a large right triangle ABC, then construct a line segment from its right-angle vertex C and perpendicular to its hypotenuse, which it intersects at the point D. That line divides the original triangle into two smaller ones, $\triangle_{BCD}$ and $\triangle_{ACD}$. They are both also right triangles, by construction. Each shares an inner angle with $\triangle_{ABC}$, and the hypotenuse of each is a side of $\triangle_{ABC}$. This is sufficient to prove that all three triangles are 'similar', i.e., their

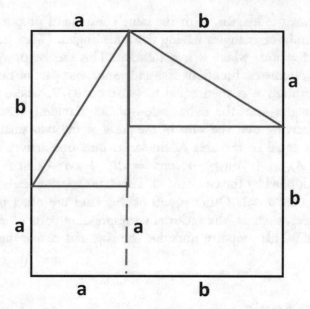

Fig. 6.5 A rearrangement of the inner triangles in Fig. 6.4. The two lower triangles save been slid diagonally upwards, allowing their hypotenuses to coincide with those of the two upper triangles and yielding two rectangles with the same area as the four triangles in Fig. 6.4. We could just as well have rotated the two lower triangles by right angles upwards around their mutual vertices K and M with the upper triangles to obtain this same rearrangement. Source private, WDB

three corresponding inner angles are all equal. Now we can use theorems about similar triangles to derive Pythagoras' theorem.

One such proof is attributed to the young Albert Einstein.[3] He knew the theorem that the areas of similar right triangles are proportional to the squares

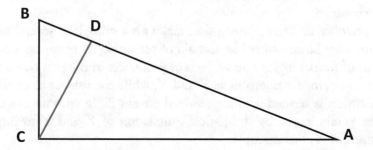

Fig. 6.6 A large triangle ABC is divided into two smaller triangles by constructing a line from the opposite vertex C to its hypotenuse, which intersects it at D, making two right angles with it. Thus, the two smaller triangles are also right triangles, and each shares an inner angle with $\triangle_{ABC}$. Then their corresponding angles are all equal, and all three triangles are *similar*. *Source* private, WDB

of their hypotenuse lengths, with the same constant of proportionality for the whole family of triangles having the same angles. [That is, the area A is given by $A = \kappa c^2$, where κ is a constant. This can be proven easily by using the trigonometric functions *sine* and *cosine*, not part of Euclid's repertoire. The constant κ is then equal to $(\sin\alpha\cos\alpha)/2$, where α is one of the interior angles—not the right angle—of the triangles]. Given that relation, and observing that the sum of the areas of the two smaller triangles in Fig. 6.6 is equal to the area of the larger one, one arrives immediately at $A_{ABC} = A_{ADC} + A_{BDC} \rightarrow \kappa c^2 = \kappa b^2 + \kappa a^2$—that is Pythagoras' theorem (multiplied by the constant κ). This same relation leads to the identity $\sin^2\alpha + \cos^2\alpha = 1$. Other proofs of this kind use other properties of similar triangles, such as "the ratios of corresponding parts of similar triangles are equal". This property underlies the sine and cosine functions used above.

Aristotle's Logic

We now turn to the ancient Greek philosopher *Aristotle*, who lived about 50 years before Euclid. He is generally considered to be the founder of formal logic, having worked out a system of logic based on the *syllogism*. This word is derived from a Greek term referring to *deduction*, and Aristotle's logic is deductive. It was summarized by some of his followers in a book called the *Organon* (the '*instrument*').[4] A deduction, according to Aristotle, consists of some suppositions, termed *premises*, which lead to a logically-compelling *conclusion*. He also considered the possibility of reaching conclusions by *induction*, a process of extrapolation from known or observed phenomena/facts to a generally-applicable conclusion. We shall consider both in later chapters.

Two premises and their conclusion make up a syllogism; several or many syllogisms may be combined as a chain of reasoning to make up a complex argument of formal logic. The subjects or predicates of the premises are often denoted in a compact notation as Y and Z, while the subject or predicate of the conclusion is termed X. Frequently, the term Z, in common to the two premises, is eliminated by the logical connection of Y and Z to imply the conclusion for X. For example:

Major premise: 'All men are mortal'.
Minor premise: 'Epimenides is a man'. (6.2)
Conclusion: 'Epimenides is mortal'.

Here, 'man/men' are the 'middle term', the subject/predicate Z; 'mortal' is the 'major term' Y; and 'Epimenides' is the 'minor term' X. Various combinations of the terms in the premises and the conclusion are possible and were catalogued by Aristotle, who called them 'figures'.

If the premises (containing Y and Z) are *true*, and the conclusion is logically compelling from them, then the conclusion (about X) is also *true* and the syllogism is *logically valid*. (The terms '*true*' or '*false*', '*correct*' or '*incorrect*', and '*provable*' or '*unprovable*' are often used interchangeably, but in modern formal logic, they are distinct and should not be conflated. We will hear more about this in later chapters). Aristotelian logic makes use of the 'Law of Excluded Middle' (LEM), sometimes characterized by the Latin phrase '*tertium non datur*' ('there is no third'), which asserts that a proposition or sentence must be either true or false;[5] there is no third possibility. Aristotle goes on to classify the 'figured' arguments in terms of possibility or necessity of their premises, leading to what is called *modal logic*. It is in many ways a precursor to modern formal logic, but it has various shortcomings when analyzed in terms of the latter.

Aristotle also presented a detailed classification of methods of deduction in the different figures, and was then able to give some *metatheoretical* (or metalogical) results; i.e. results which generalize over all the methods of deduction and proof and are therefore 'beyond' ('*meta—*') the logical methods themselves. This presages the modern field of *metamathematics*, which makes general statements about mathematical methods, beyond those methods themselves. The methods deal with mathematical objects, such as numbers, variables, operations, functions, functionals; while metamathematics deals with the fundamentals behind those methods and questions of their general validity; i.e., what is possible or permissible in terms of mathematical methods.

Paradoxes

It was noted in ancient times, in particular by *Epimenides of Knossos* (sixth century BCE), that certain sentences contain or imply logical contradictions, and thus result in *paradoxes* (known in modern philosophy as *antinomies*). A typical rendering of Epimenides' paradox consists of two statements:

All Cretans are liars,
    I am a Cretan.

If we assume that both statements are true (and that being a liar implies that one cannot tell the truth), then they must be false, since they are spoken by a liar. On the other hand, if one or both are false, then they may be true; in either case a logical contradiction. It has been pointed out [6] that the contradiction is not inevitable, since the first statement could be a lie, in the sense that there exist one or more Cretans who are not liars, even if the majority (including the speaker) may be. Then the first statement may be false without contradiction. We see also that the paradox arises due to the self-reference (reflexivity) in the second statement; it calls the first statement into question, since it was then presumably made by a liar. If the second statement referred to some third party, e.g. "*Ergoteles* is a Cretan", there would be no problem; each might be true or false without causing a logical contradiction. They would simply be logically independent (leading, as a syllogism, to the conclusion: "Ergoteles is a liar"). An even simpler form is the single sentence,

This sentence is false,

which even more clearly illustrates the role of self-reference. Many paradoxes have been discovered in the intervening years, and some of them played a significant role in the development of formal logic and metamathematics, as we shall see later.

Modern Developments

As mentioned above, portions of the teachings of Aristotle and of Euclid were saved for posterity and formed the basis for the development of modern logic in the late Middle Ages in Europe. Major advances were made in the seventeenth century, in particular by *René Descartes* (1596–1650) and *Gottfried Wilhelm Leibniz* (1646–1716). Descartes, among other things, introduced scaled coordinate axes into geometry; these are mutually perpendicular (orthogonal) straight-line axes which are divided into equal (unit) segments that are numbered (e.g., by the natural numbers 0, 1, 2, …, with the origin, where the axes intersect, denoted as '0'). Any point in the n-dimensional space defined by the axes can then be uniquely specified by an n-tuple of numbers—its 'coordinates' (n is e.g., $n = 2$ for a 2-dimensional (planar) space, $n = 3$ for a flat 3-dimensional space etc.). Geometrical figures can then be defined in terms of their coordinates and the relations between them, i.e., by algebraic expressions. This leads to *analytic geometry*, in which methods of algebra and calculus can be applied to the study of geometry, greatly extending its possibilities.

Leibniz, whose thinking had a certain fascination for Kurt Gödel, developed the differential and integral calculus, parallel to but independently of Isaac Newton, and contributed to rationalist philosophy. He was a very early proponent of dual (binary) arithmetic, and he noted its advantages for mechanical computation. In this vein, he designed various mechanical calculators and cipher (cryptography) machines, and he was thus a precursor of modern computer science. Leibniz approximated the area of a circle by calculating the areas of polygons of more and more sides, and adding or subtracting the areas of smaller and smaller triangles to approximate the circle ever more closely. This led him to a slowly-converging series which gives the value of the transcendental number π, the ratio of the circumference to the diameter of a circle.

But in the present context, he also developed formal logic, although it was not expounded in his published works during his lifetime; and he rejected absolute space and time—all topics which came to fruition only 200 years after his death. In 1900, Bertrand Russell wrote '*A Critical exposition of the Philosophy of Leibniz*',[7] based on Leibniz's published and unpublished works, and details in it many of Leibniz's ideas which presaged much later developments in several fields. We shall meet up with Leibniz again in later chapters when we consider the metamathematics and the philosophy of Kurt Gödel.

In the first half of the nineteenth century, geometry, which had previously been restricted mainly to flat spaces, was generalized to *curved spaces* by *Carl Friedrich Gauss* (1777–1855), *Bernhard Riemann* (1826–1866), and *Nikolai Lobachevski* (1792–1856), among others. This led to the development of differential geometry, which later found application in general relativity, a topic also taken up by Gödel in the 1940s.

The early nineteenth century also saw the development of *Boolean algebra* (George Boole, 1815–1864), based on a two-valued number system (the modern binary system used in electronic computing), which had already been explored by Leibniz, and the first practical mechanical computing machines (Babbage, Lovelace). Boole's *Mathematical Analysis of Logic* was published in 1847, and is considered to be the first work of modern mathematical logic.[8] Boole showed that Aristotle's syllogistic logic can be translated into an algebraic calculus. His system is now called called *sentential* (or *Boolean*) *logic*. It was used and extended by several other authors in the mid-nineteenth century, including Augustus De Morgan, Charles S. Peirce (and Gottlob Frege, working independently of Boole's followers). They introduced multi-valued predicates and symbols for universal and existential quantification (the modern symbols are ∀ and ∃, denoting 'for all…' and 'there exists…'. These modern symbols were established later by David Hilbert and Giuseppe Peano, respectively).

Formal Logic

In the latter part of the nineteenth century, there was a concerted effort by mathematicians and philosophers to arrive at a fundamental, logical basis for the many newer developments as well as for the older branches of mathematics. Important contributors to that effort were the German *Gottlob Frege* (1848–1925) and the Italian *Giuseppe Peano* (1858–1932). About this development, Gödel [9] later wrote:

> Mathematical logic, which is nothing other than a precise and complete formulation of formal logic, has two quite different aspects. On the one hand, it is a section of mathematics treating of classes, relations, combinations of symbols, etc. *[i.e., metamathematics]*, instead of numbers, functions, geometric figures, etc. On the other hand, it is a science prior to all others, which contains the ideas and principles underlying all sciences. It was in this second sense that mathematical logic was first conceived by Leibniz in his '*Characteristica universalis*', of which it would have formed a central part. But it was almost two centuries after his death before his idea of a logical calculus really sufficient for the kind of reasoning occurring in the exact sciences was put into effect (in some form at least, if not the one Leibniz had in mind) by Frege and Peano.[1] Frege was chiefly interested in the analysis of thought and used his calculus in the first instance for deriving arithmetic from pure logic. Peano, on the other hand, was more interested in its applications within mathematics and created an elegant and flexible symbolism, which permits expressing even the most complicated mathematical theorems in a perfectly precise and often very concise manner by single formulas.
>
> It was in this line of thought of Frege and Peano that Russell's work set in. Frege, in consequence of his painstaking analysis of the proofs, had not gotten beyond the most elementary properties of the series of integers, while Peano had accomplished a big collection of mathematical theorems expressed in the new symbolism, but without proofs.
>
> [Footnote: (1) Frege doubtless takes precedence, since his publication on the subject, already including all the essentials, appeared ten years before that of Peano.].

Friedrich Ludwig Gottlob Frege was born on November 8, 1848 in Wismar, in the northern German region of *Mecklenburg*. After 15 years of schooling at the *Gymnasium* there, in 1869 he matriculated at the University of Jena in southern Thuringia.[10] There, he became a student of *Ernst Karl Abbe*, later known for his theoretical model of optical microscopes. Frege moved to the University of Göttingen after two years, obtaining his doctorate in mathematics there in 1873 with a thesis on '*a Geometrical Representation of Imaginary Forms in the Plane*', referring to imaginary components in the complex plane (and not to virtual images, as one might at first think). Only a year later, he submitted his *Habilitation* thesis, entitled '*Calculational Methods based on an Extension of the Concept of Magnitude*'. The 'extension' that he had in mind was from real to complex numbers. In his *Habilitationsschrift* [11], he points out a conclusion to which he was led by considering a complex representation of originally geometrical quantities such as the length of a line, a surface area, etc.: "*The elements of all geometrical constructions are intuitions, and geometry refers to intuition as the source of its axioms. Since the object of arithmetic does not have an intuitive character, its fundamental propositions cannot stem from intuition...*". This presages his later work in mathematical logic.

With the aid of Abbe, Frege (Fig. 6.7) soon became a *Privatdozent* at the University of Jena, where he gave lecture courses on analytic geometry, Abelian and elliptical functions, functions of complex variables, algebraic analysis and other topics, learning as he went along. (As *Privatdozent*, he would have received no regular salary from the University, but he had the privilege of offering courses for which he could collect fees from the students who enrolled in them. It was thus in his own interest to make his courses interesting and effective for the students).

Frege's first major work, the book '*Begriffschrift...*' ('*Concept script: A formula language of pure thought, following that of arithmetic*'), was published in 1879, when he was 31, and marks the beginning of his work as an independent scholar. Its publication led to his promotion to *außerordentlicher Professor* (roughly equivalent to Associate Professor, a tenured position at the University of Jena). The book contained many of his new ideas on mathematical logic, but was not well received, probably because it was not understood (or read) by many contemporaries. It was this work to which Gödel was referring when he stated that Frege was 10 years ahead of Peano in publishing on the mathematical logic of arithmetic. Frege's second book, '*Die Grundlagen der Arithmetik*' ('Fundamentals of Arithmetic', 1884), was a redefinition of number theory and a philosophical precursor to his *magnum opus*, '*Grundgesetze der Arithmetik*' ('Fundamental Laws of Arithmetic'), a two-volume work which was published in 1893 and 1903. Frege was promoted to *ordentlicher Honorarprofessor* (regular honorary Professor, or Full Professor) in 1896, and

Fig. 6.7 Gottlob Frege, ca. 1879. Photo: Courtesy of Uni Münster, with permission[12]

was by then an established authority on mathematical logic and philosophy of mathematics.

Frege's great contribution was to develop a symbolic language for dealing with mathematical logic. He was interested in language in general, and he also developed a philosophy of language, expounded in his article '*Über Sinn und Bedeutung*' ('On Sense and Meaning', 1892),[13] in which he pointed out that words in a sentence obtain their meanings not only from their *denotations*, but that a *sense* is required in addition in many cases. This was an important motivation for Ludwig Wittgenstein's philosophy, among others. A visit to Frege early in his career, when he was trying to decide between continuing his work in engineering or turning to philosophy, convinced Wittgenstein that the latter should be his goal, and he moved to Cambridge, where he came under the influence of Bertrand Russell and G. E. Moore.[14]

In the summary of Michael Bleaney's essay on Wittgenstein and Frege,[15] Bleaney states:

Of all philosophers, it is Frege whom Wittgenstein held in greatest esteem. The aim of philosophy, Wittgenstein wrote in the *Tractatus*, 'is the logical clarification of thoughts', a characterization that might well be taken to be true of Frege's philosophy. The clarity that Wittgenstein saw as an important philosophical virtue is arguably nowhere better illustrated than in Frege's writings, even if one disagrees with the substantive philosophical claims that Frege makes. Peter Geach reports a remark that Wittgenstein made to him when they were discussing Frege's essay '*On Concept and Object*'.[16] Wittgenstein may have envied Frege's style, but he nevertheless felt it had a strong effect on his own writing. Frege is explicitly cited as an influence on the *Tractatus*, but although he is rarely mentioned by name in his *[Wittgenstein's]* later writings, his views continued to be a major source of inspiration to the very end of Wittgenstein's life.

A late, perhaps tragic development in Frege's work was the discovery in 1901 by the British mathematician and philosopher *Bertrand Russell* that a paradox that he had found[17] (now known as 'Russell's paradox'; also found independently by some other logicians, e.g., Zermelo, who however failed to consider its importance) would invalidate part of Frege's conclusions in his first volume of the *Grundgesetze*. Russell communicated this to Frege in 1902, who was at the time preparing the publication of the second volume of his major work. Russell's paradox is of a similar kind to those described by the ancient Greek philosophers, and its paradoxical nature is contingent on its *self-referentiality*.

Its most straightforward formulation is in terms of set theory: consider sets which are members of themselves. This is entirely possible for some types of sets; however, for e.g., 'the set of all natural numbers greater than 50', it is of course not possible, since the elements of that set are numbers, not sets. But for 'the set of all sets containing more than 50 elements', it is quite possible and in fact imperative that it be a member of itself, since there are certainly many more than 50 sets containing more than 50 elements. Now, says Russell, consider the set of all sets which are *not* members of themselves. If that set *is* a member of itself, then it is—by construction—*not* a member of itself; we have a logical contradiction, or paradox. Conversely, if it is *not* a member of itself, it *must* be a member of itself, by definition. Either way, we arrive at a paradoxical conclusion. The 'law of excluded middle' is applicable here, since a set either *is*, or *is not*, a member of itself—there is no third possibility.

Russell's paradox showed that the axioms proposed by Frege in his formal logic were not consistent. Frege had suggested a set of axioms for arithmetic, and in particular, his Axiom V (which he called a 'Basic Law') involved a property of functions (their *course-of-values*, a history of the functional values for different arguments or 'objects'), which Frege called their *extension*, and which he also applied to concepts [see the *Stanford Encyclopedia of Philosophy* articles on 'Gottlob Frege', by Edward N. Zalta, (2022, backnote [10]), and on 'Russell's Paradox', by Andrew David Irvine & Harry Deutsch (2021)].

Russell saw that his paradox would undermine this axiom, since some extensions can be elements of themselves (like the sets mentioned above). Frege quickly recognized that his whole logical structure would be invalidated if the paradox applied. He made some attempts to solve the problem by adding an appendix to his second book, stating that the effects of the paradox were not clear. But he later gave up on much of his own understanding of mathematical logic.

Russell later commented on Frege's reaction:[18]

> As I think about acts of integrity and grace, I realise that there is nothing in my knowledge to compare with Frege's dedication to truth. His entire life's work was on the verge of completion, much of his work had been ignored to the benefit of men infinitely less capable, his second volume was about to be published, and upon finding that his fundamental assumption was in error, he responded with intellectual pleasure clearly submerging any feelings of personal disappointment. It was almost superhuman and a telling indication of that of which men are capable if their dedication is to creative work and knowledge instead of cruder efforts to dominate and be known.

For his own part, Russell also wrote an appendix to his *Principles of Mathematics* (soon to be published; the precursor to Russell and Whitehead's *Principia Mathematica*). The appendix contained his 'Doctrine of Types', intended to avoid self-referential statements and therefore his paradox.

[A note about Russell's paradox: He later gave an everyday analogy to make it more understandable, using as an example a village where there is only one barber. The men in the village then either shave themselves (and are not shaved by the barber), or they *are* shaved by the barber. The question is, who shaves the barber? If he is a member of the first group, he is also 'shaved by the barber', contradicting the original assumption. If he is not a member of the first group, he cannot shave himself, again a contradiction. This analogy is imperfect, since in real life, there is a 'third possibility': a man in the village can choose to grow a beard, and not shave or be shaved at all. Or he may shave himself sometimes and be shaved by the barber on other occasions. The 'law of excluded middle' then does not apply. If this is the case for the barber, the paradox is avoided.].

We will return in later chapters to Frege's logic, which was more influential than that of Peirce and the Boolean school, although Pierce's writings were well known in continental Europe. Parallel to Frege's works, those of *Giuseppe Peano* were also an important source for Gödel.

Giuseppe Peano[19] was born in a village in the Piedmont region of northern Italy in August, 1858, just about 10 years after Frege's birth and a few months after that of Max Planck. After finishing the *Liceo* (high school) in Turin in 1876, he entered the University of Turin as a student of mathematics and obtained his doctoral degree with honors in 1880. He then joined the faculty of the University of Turin as an assistant to his geometry professor, *Enrico D'Ovidio*. After his former professor *Angelo Genocchi* died in 1889, Peano succeeded him in his professorship the following year.

In 1878, the young Peano discovered an error in a standard definition in the textbook by *Serret*, which he was using for a course that he was teaching. This aroused his interest in precise mathematical logic. In 1884, he assisted in the publication of a new textbook based on the lectures of his former calculus teacher, Angelo Genocchi. Genocchi later stated publicly that most of the content in the book was due to Peano. That same year, Peano received his *libera docenza*, qualifying him as a university teacher. He began independent teaching at the university and later also offered lectures at the Turin Military Academy. During this period, he published several proofs involving differential equations, and in 1888 a book entitled *Geometrical Calculus* appeared, his first work containing a section on mathematical logic. It was based on work by *Ernst Schröder* and by the Boolean school (*George Boole, Charles Peirce*). It was also the first work to give a clear definition of a vector space.

His definitive work in mathematical logic was begun with the publication in 1889 of a pamphlet called '*Arithmetices principia, nova methodo exposita* '(Principles of Arithmetic, the new method expounded') which, according to his biographer H. C. Kennedy [20] was "*... at once a landmark in the history of mathematical logic and of the foundations of mathematics*". The pamphlet was written in Latin for unknown reasons, except that "*... it appears to be an act of sheer romanticism, perhaps the unique romantic act in his scientific career*" (*ibid.*).

Peano (Fig. 6.8) in his pamphlet introduced a system of symbols for expressing logical statements and gave a set of axioms as the foundation of a formal-logical system underlying arithmetic, known as 'Peano's axioms for the natural numbers'. They can be seen for example online at the St. Andrews mathematical history site[22]—and we quote them verbatim here:

1. 1 is a natural number.
2. Every natural number n has a natural number n' as a successor.
3. 1 is not the successor of a natural number.
4. Natural numbers with the same successor are the same. (No two different numbers have the same successor).
5. If the set X contains 1 and for every natural number n also its successor n', then all the natural numbers belong to X.   (6.3)

Fig. 6.8 Giuseppe Peano, ca. 1920. Photo: Wiki commons[21]

The 'successor function' s gives $s(n) = n + 1$. Peano later altered his axioms so that $\mathbb{N}$ included 0. Then Axiom 3 becomes '0 is not the successor of a natural number', and 1 is the successor of 0.

These axioms, and Peano's system of symbolic logic, together with the earlier system of Frege, were an important motivation for Kurt Gödel's doctoral thesis, which we will consider in detail in the following chapter. For a symbolic version of Peano's axioms, see e.g., Franzén (2005), p. 167.

In fact, in the later nineteenth century, Frege's work was not at all well known, and it was Peano, for example, who introduced not only symbolic logic, but also Frege's work on it, to Bertrand Russell[23] in 1900 following the two major conferences in Paris, the 1st International Congress of Philosophy (Aug. 1st–5th, 1900) and the 2nd International Congress of Mathematicians (ICM, Aug. 6th–12th, 1900). As we saw above, Gödel held Frege's work to be much more detailed and careful, but (necessarily) less complete, while Peano gave a rather complete list of theorems, but no rigorous proofs. Russell much later wrote in his autobiography.[24]

The Congress was the turning point of my intellectual life, because there I met Peano. I already knew him by name and had seen some of his work, but had not taken the trouble to master his notation. In discussions at the Congress *[on Philosophy]* I observed that he was always more precise than anyone else, and that he invariably got the better of any argument on which he embarked. As the days went by, I decided that this must be owing to his mathematical logic. ... It became clear to me that his notation afforded an instrument of logical analysis such as I had been seeking for years ...

During their discussions, Peano also made Russell aware of Frege's works, and motivated him to read Frege's *Begriffschrift*, which Russell had, in turn, been given earlier by his former philosophy teacher James Ward, but did not read until 1901, after being prompted by Peano.[25]

The 1900 Paris Congress of Mathematicians is also famous for David Hilbert's lecture,[26] in which he set out 23 important open problems in mathematics.[27] We will return to some of his problems, later repeated in his 1928 book with W. Ackermann,[28] which influenced Gödel in his significant work on mathematical logic.

In the years following 1890, Peano began working on what he considered to be his *magnum opus*, entitled *Formulario Mathematico*, a compendium of mathematical theorems and formulae. He attempted to use it as a textbook, causing a certain alienation from his students and colleagues and resulting in his dismissal as a teacher at the Military Academy. After 1903, he had a second project: developing an international language which he called '*Latino sine flexione*', a simplified version of Latin. He wrote his own papers and held lectures in that language, which he was continually perfecting. Peano died in April 1932 in Turin.

Set Theory, Infinities: Georg Cantor

A further important figure in mathematics in the later nineteenth century was the German mathematician *Georg Ferdinand Ludwig Philipp* **Cantor** (1845–1918). Cantor[29] was born into a German merchant's family in St. Petersburg, Russia. He went to primary school there, but the family moved to Germany when he was 11. He attended the *Gymnasium* in Wiesbaden, then a *Realschule* in Darmstadt, graduating in 1860. He then spent a year at the *Höhere Gewerbeschule* (Advanced Vocational School) in Darmstadt, because his father wanted him to become an engineer. In 1862, he transferred to the *Polytechnic* (later ETH) in Zurich (where 35 years later, *Albert Einstein* would matriculate). After his father's death in 1863, he moved to the University of Berlin, where he was a classmate of *Hermann Schwarz*. Among his professors were *Weierstrass*, *Kummer* and *Kronecker*. In 1866, he spent a semester at the University of Göttingen, then returned to Berlin where he completed his dissertation on number theory, entitled *De aequationibus secundi gradus indeterminatis* (Indeterminate Equations of 2nd degree), in 1867.

Cantor (Fig. 6.9) later became famous for the introduction of set theory and for his discussion of infinities. Hilbert once referred to the 'mathematical paradise of Cantor'.[31] After obtaining his doctorate in 1867, Cantor remained for a time in Berlin, teaching at a secondary school and then at a seminar for mathematics teachers. He worked on the side on his *Habilitationschrift*, also on number theory, and in 1869, he went to the University in *Halle an der Saale* in Saxony-Anhalt, some 150 km southwest of Berlin, where he submitted his thesis and obtained the *Habilitation*, allowing him to become a *Privatdozent* at that university. He published several papers on trigonometric series in the period from 1870–1872, having solved a long-standing problem on the representation of functions through such series. He was promoted to *Außerordentlicher* (Associate) Professor in 1872 and published a paper that year on irrational numbers as convergent series of rational numbers.

His interest shifted again, back to number theory, and the following year, he proved that the *rational* numbers (expressible as ratios of natural numbers) are countable, that is they can be put into a one-to-one correspondence with the natural numbers $\mathbb{N}$ (0,1,2,3...). He proved the same property for the

Fig. 6.9 Georg Cantor, ca. 1885. Photo: Wiki Commons, public domain[30]

algebraic numbers A, which are roots of polynomial equations with integer coefficients. Later in 1873, he showed that the *real* numbers R are *not* countable in this sense, and so there must be many of them which are not algebraic numbers (and are thus 'transcendental' in the sense of Liouville (1851)).

During this period, Cantor often corresponded with *Richard Dedekind* (1831–1916), whom he had met in Switzerland. He proved in 1874 that there is a one-to-one correspondence between points on a line from 0–1 and points in a *p*-dimensional space (e.g., a square of unit side length), a result at which he himself was surprised. This aroused his interest in set theory, and he considered sets which are *denumerable*, i.e., those whose elements can be put in a one-to-one correspondence with the natural numbers. In the years 1879–84, he published a series of papers on basic set theory. [32] This led him to considerations of infinities.

Cantor developed the concept of *transfinite* numbers; considering the natural numbers $\mathbb{N}$ to be finite, one can call their successor ω: 0,1,2,3, ... ω, $\omega + 1$, $\omega + 2$ ... 2ω, 3ω ... The set of numbers ω is called the transfinite numbers.[33] They represent an infinity of a higher order than that of the natural numbers $\mathbb{N}$. This led him to the concept of *cardinals*: a one-to-one correspondence with $\mathbb{N}$ represents a *counting*, the *ordinality* of a set of objects. The order of their infinity is their *cardinality*, later symbolized by the Hebrew letter *aleph*, $\aleph$. The natural numbers $\mathbb{N}$ have a cardinality $\aleph_0$ (*card* $\mathbb{N} = \aleph_0$), while *card* $\omega = \aleph_1$. Cantor conjectured that the *real* numbers have the cardinality $\aleph_1$: *card* $\mathbb{R} = \aleph_1$. Later, he proposed (but could not prove) that there is nothing between these two cardinalities, i.e., the real numbers 'fill in the spaces' between the natural numbers, resulting in a once-higher order of their infinity. This is one form of the *Continuum Hypothesis*, which was considered extensively by Gödel (and later Paul Cohen) in the 1940s 1960s. We shall take up this topic again in Chap. 11.

Cantor, and Dedekind, continued publishing on set theory throughout the decade of the 1890s. Their results were however controversial among contemporary mathematicians. In 1895–97, Cantor published two summary papers[34] on his results for set theory (at the time still called the 'theory of manifolds', following Riemann). This was later termed 'naïve set theory' (and still later was referred to by Gödel as 'the dichotomy conception'). However, Cantor failed to publish the third paper in the series, in which he considered paradoxes that arise in set theory (the original manuscript is lost, but he discussed it in letters to Dedekind). After 1899, he had recurring health problems and gradually lost touch with mathematics, retaining his interests in literature and philosophy. He died in a sanatorium in January 1918, before the end of the 1st World War.

The Twentieth Century

The foundations of formal mathematical logic were laid in the last quarter of the nineteenth century by Frege, Peano and others, as we have seen above. Their project was by no means complete at the turn of the century. By then, a new, younger generation took over the task of formulating and establishing a formal-logical basis for mathematics. Important participants in that effort were *Bertrand Russell* (1872–1970), *Alfred North Whitehead* (1861–1947), *David Hilbert* (1862–1943), *Ernst Zermelo* (1871–1953), and *Luitzen Egbertus Jan Brouwer* (1881–1965), among others. A philosophical participant, if not himself a mathematical logician, was Ludwig *Wittgenstein* (1889–1951). All of them will play a role in our further story of Kurt Gödel; he interacted in some way with each of them, was influenced by them and in turn affected their lives and careers with his work.

By the early 1930s, when Gödel published his most important mathematical-logical works, there were four major movements within fundamental mathematical logic, termed '*logicism*' (Frege/Russell/ Whitehead), '*formalism*' (Hilbert, Bernays, von Neumann), '*idealism*' (Brouwer, Heyting), and Wittgenstein's program as understood by the Vienna Circle. Each of these was represented at the '2nd Conference on the Epistemology of the Exact Sciences' ('*Zweite Tagung für Ekenntnislehre der exakten Wissenschaften*') held in *Königsberg*, then in northern Germany (now the Russian city of *Kaliningrad*) in September 1930. Gödel attended that conference and presented there the results of his doctoral thesis (1929) as well as his then new results on the incompleteness of (consistent) formal-logical systems (now called his First Incompleteness Theorem). In the next two chapters, we will follow the developments of each of these movements and try to show how they affected and related to Gödel's own work. They will be recurring themes in later chapters, when we consider his further work on set theory and the continuum hypothesis, and his philosophical writings, as well as the reception of his work in other fields.

Notes

1. See e.g. Husserl (quoted in Majer 1995): '*Geometry, Intuition and Experience: From Kant to Husserl* ', Ulrich Majer, in *Erkenntnis*, Vol. 42, No. 2 (1995), pp. 261–285.
2. Hilbert, quoted by Majer 1995, *ibid.*; cf. his Ref. 13.

3. See Strogatz 2015: '*Einstein's First Proof* ', by Steven Strogatz, in *The New Yorker*, Nov. 19th, 2015. Online at https://www.newyorker.com/tech/annals-of-technology/einsteins-first-proof-pythagorean-theorem#:~text=Inasense%2CEinsteincontinued,outlikearipeavocado. Consulted on April 16th, 2021.

4. See e.g. the *Stanford Encyclopedia of Philosophy*, entry on Aristotelian Logic, online at https://plato.stanford.edu/entries/aristotle-logic/. Consulted April 19th, 2021.

5. See Church (1928), '*On the Law of Excluded Middle*', Alonzo Church, in *Bulletin of the American Mathematical Society*, Vol. 34 (1928), pp. 75–78.

6. Epimenides' paradox: Weisstein, Eric W., "*Epimenides Paradox*", from *Math-World*—A Wolfram Web Resource: https://mathworld.wolfram.com/EpimenidesParadox.html. Consulted April 23rd, 2021.

7. Russel (1900),'*A Critical exposition of the Philosophy of Leibniz*', Bertrand Russell. Cambridge University Press.

8. See the *Stanford Encyclopedia of Philosophy*, entry on 'the Emergence of First-Order Logic', online at https://plato.stanford.edu/entries/logic-firstorder-emergence/#GiusPean. Consulted July 16th, 2021.

9. Gödel, *On Bertrand Russell*: an essay which appeared originally in Schilpp (1944).

10. See e.g. the *Stanford Encyclopedia of Philosophy*, entry on Gottlob Frege, online at https://plato.stanford.edu/entries/frege/. Consulted July 15th, 2021.

11. Frege, *Habilitationschrift* 1874, translation in McGuinness (ed.), 1984, pp. 56–92.

12. Photo from Uni Münster, public domain. Reused from: https://www.ulb.uni-muenster.de/sammlungen/nachlaesse/sammlung-frege.html.

13. Frege (1892a):'*Über Sinn und Bedeutung* ', Gottlob Frege, in *Zeitschrift für Philosophie und philosophische Kritik*, Vol. 100, pp. 25–50. An English translation ('On Sense and Reference', by M. Black) is in the book by Geach and Black (eds. and transl.), 1980, pp. 56–78. See [Geach & Black, eds. (1980)].

14. Eilenberger (2018), Geier (2017).

15. Bleaney (2016): Chapter 4, *Wittgenstein and Frege* by Michael Bleaney, in '*A Companion to Wittgenstein*', eds. Hans-Johann Glock and John Hyman. John Wiley and Sons, Ltd. 2017, ISBN 978–1-118–64,116-3.

16. Frege (1892b): '*Über Begriff und Gegenstand*', Gottlob Frege, in *Vierteljahresschrift für wissenschaftliche Philosophie*, Vol. 16, pp. 192–205. Translated as 'Concept and Object' by P. Geach in Geach & Black (eds.) 1980, pp. 42–55. See Geach & Black (eds.) 1980.

17. Russell (1903): *The Principles of Mathematics*, Bertrand Russell. 2d. ed. Reprint, New York: W. W. Norton and Co., 1996. (First published in 1903.).

18. Quoted in van Heijenoort (1967), p. 127: *Frege, Gottlob (1902)*. "Letter to Russell," in '*From Frege to Gödel*', Jean van Heijenoort (ed.), Cambridge, MA, Harvard University Press, 1967, pp. 126–128.

19. See the MacTutor Biography, online at https://mathshistory.st-andrews.ac. uk/Biographies/Peano/. See also the entry on Peano by H.C. Kennedy in the *Dictionary of Scientific Biography* (New York 1970–1990); online at https://www.encyclopedia.com/history/historians-and-chronicles/historians-miscellaneous-biographies/giuseppe-peano#2830903321.

20. Kennedy, H.C. (1980). See also Kennedy, H.C. (1974).

21. Photo: from Wiki Commons, public domain. {{PD-US-expired}} for works published before January 1, 1927. Reused from: https://commons.wikimedia. org/wiki/File:Giuseppe_Peano.jpg.

22. Online at https://mathshistory.st-andrews.ac.uk/Extras/Peano_axioms/ (consulted July 15, 2021).

23. Kennedy, H.C. (1973).

24. As quoted in [18]. See Russell's *Autobiography*, Russell (1967–69). Also quoted in Kennedy (1973).

25. See Kennedy, H.C. (1973).

26. Cf. https://mathshistory.st-andrews.ac.uk/ICM/ICM_Paris_1900/, (consulted July 21, 2021).

27. Hilbert's lecture is described in the reference of backnote [26]: "*In many ways it was David Hilbert's lecture on the future problems of mathematics which became the most famous lecture delivered at any International Congress of Mathematicians… Hilbert was invited to give a plenary lecture at the 1900 Congress. He could not decide on a topic and discussed what he should talk on with Hermann Minkowski and Adolf Hurwitz. Eventually he decided to discuss outstanding mathematical problems but took a long time to write his lecture. The programme for the 1900 Congress was published before he had things ready and so his talk was not advertised as a plenary lecture. It was delivered at the Congress in a joint session of the Teaching Section and of the History Section. By the time the Proceedings of the conference appeared the organisers decided that, because of its importance, they would put Hilbert's paper in the section with the plenary lectures… because of lack of time, Hilbert only spoke about 10 of his 23 problems in the lecture although all 23 appear in the published paper*". He had originally formulated 24 problems, but left the last one off in his paper.

28. Hilbert & Ackermann (1928).

29. See MacTutor Biography of Georg Cantor, online at https://mathshistory.st-andrews.ac.uk/Biographies/Cantor/. (consulted on July 21, 2021), and also backnote [31].

30. Photo: Wiki Commons, public domain; reused from: https://commons. wikimedia.org/wiki/File:Das_Fotoalbum_f%C3%BCr_Weierstra%C3%9F_033_(Georg_Cantor).jpg

31. Cf. Hilbert (1926) ('*Cantor's Paradise*').

32. See the *Stanford Encyclopedia of Philosophy*, article on 'the Early Development of Set Theory'; online at https://plato.stanford.edu/entries/settheory-early/. (Consulted on July 22, 2021).

33. Cantor (1883).

34. Cantor (1895/97).

7

Gödel's Doctoral Thesis, 1928–30: The Completeness of First-Order Logic

The exact moment when Kurt Gödel decided on a thesis topic and began writing his dissertation is not known—nor is it known whether his mentor Hans Hahn or some other associate influenced his choice of thesis topic. It *is* clear, from his answers to the Grandjean questionnaire, his correspondence with Hao Wang, notes in his legacy, and the introduction to his dissertation, that his choice of thesis topic was influenced by Brouwer's lectures in Vienna in March 1928 and by his readings of Russell and Whitehead (1910–13) (PM) and of Hilbert and Ackermann (1928) (HA) in 1928 and early 1929. Most likely his contacts with Rudolf Carnap also played a role: not only Carnap's seminar on the fundamentals of theoretical logic, which he offered in 1928/29 and which was attended by Gödel, but also their many discussions on mathematical logic, begun in 1927 and documented in Carnap's diaries.

Gödel probably started thinking about his thesis topic by mid-1928, and he wrote the actual dissertation in the first half of 1929 (when he also took the examinations to complete his formal studies at the University). His thesis was approved by Hahn and by Furtwängler in July 1929, who returned it with their and other colleagues' comments, so he must have submitted it to them by the end of June at the latest.[1] Gödel then apparently spent September polishing the thesis before submitting the final version (Gödel 1929) to the faculty, and wrote up a compact version for publication, which was sent to a journal in October but appeared only in late September of the following year (Gödel 1930). In the meantime, the thesis was approved by the University,

© The Author(s), under exclusive license to Springer Nature Switzerland AG 2022, corrected publication 2023
W. D. Brewer, *Kurt Gödel*, Springer Biographies,
https://doi.org/10.1007/978-3-031-11309-3_7

so that he was formally granted the doctorate in early February 1930. He also presented the results of his thesis at a meeting of the *Mathematical Colloquium* in March 1930, and again, along with an initial version of his first incompleteness theorem, at the Königsberg Conference in early September, and in addition at a meeting of the Vienna *Mathematical Society* in November of that year.

Just who *was* this obviously talented and intelligent young man? What his inner feelings and goals may have been remains a mystery. He kept meticulous records of superficial facts like the books that he borrowed or bought, but he did not write a diary or even many notes about his plans, hopes or fears, nor did he mention them in his correspondence. (He did, somewhat later in the 1930s through the 1950s, write copious notes about his philosophy, both general and personal (ethics)—see Item 14 in Appendix A—but they were not a personal diary in the usual sense. He also wrote personal notes, rather self-reflective (his '*Protokolle*'), possibly due to his psychiatric problems, in 1936–37). He was very private and revealed himself to no-one, not even to the members of his immediate family. Stephen Budiansky, in his recent biography,[2] makes a serious attempt to investigate Gödel's relations with others and his inner life, using many sources; in the main from his time at Princeton, but also from Gödel's own notes and correspondence in Vienna. The memoirs of Menger[3] and of Carnap[4] offer some insights into Gödel's life in the years 1926/27–1935/36, while he, and they, were still in Europe.

Gödel was helpful and polite to his fellow students, and as we know, e.g., from Olga Taußky and Karl Menger, he was always ready to answer questions and offer help to those who had less knowledge and understanding; he explained things clearly and in detail when asked. Until early 1929, his goals had been set for the most part by the educational system through which he was passing—and he was, if anything, an overachiever, who could easily meet the standards of his schools and the university. After he had completed the 'formal requirements' in the spring of 1929, as he put it in a letter to Hao Wang, he was however on his own—he would have to choose a topic and write his dissertation with minimal external guidance and present his results in a way that would be convincing to others.

The publication of his thesis and his public presentations thereafter were the beginning of a more stressful time for Kurt Gödel, and the more he became a 'public figure', the greater the stress, since interacting with other people, particularly in controversial ways, had never been his strong point. That stress took its toll over the next few years; but in the meantime, he was able to achieve more within a brief period than many people in a whole lifetime. His remarkable achievements began with his doctoral dissertation.

As with many other exceptional scientists, one of Gödel's special talents was asking the *right questions*—in particular, those that other people hadn't thought of, or were hesitant to ask. Many well-known scholars have emphasized this point; *Albert Einstein*, for example, said, "The important thing is not to stop questioning".[5] *Richard Feynman* wrote that "Curiosity demands that we ask questions".[6] In an essay on Gödel's life in his book *In the Light of Logic*,[7] Solomon Feferman, the chief editor of Gödel's *Collected Works*, quotes Ernst Strauss, Einstein's assistant from 1944–1948, who compared the two friends Einstein and Gödel, whom he had been able to observe from close up. After noting how different their personalities and their ways of dealing with everyday life were, Strauss pointed out, "*...But they shared a fundamental quality, both went directly and wholeheartedly to the questions at the very center of things*".[8] This special ability of Gödel's, of asking the right questions, was first apparent in his choice of thesis topic (at which he evidently arrived on his own; he thanks his mentor, Hans Hahn, for 'valuable suggestions for formulating this work', but not for suggesting the thesis topic[9]; see below).

The topic that he chose resulted from the program of placing number theory (specifically, arithmetic) on an axiomatic-logical basis, begun by Frege and Peano (see Chap. 6) and continued by Russell and Whitehead, and by Hilbert and his school. Peano had defined a set of axioms for arithmetic (Chap. 6; a more modern version is *Robinson arithmetic* (1950)[7], a somewhat weaker form, which however lacks the axiom of induction). There is also a system which is weaker than Peano arithmetic but stronger than Robinson arithmetic, called Primitive Recursive arithmetic. It was proposed by Skolem in his (1923) and has been used for demonstrating consistency proofs. (These logical systems are denoted as PA, Q, and PRA).

In order to develop their program and understand its underlying structure (metalogic, meta-mathematics), Russell and Whitehead first investigated the fundamentals of mathematical logic, and they made use of a 'toy model' called *propositional logic*. It deals, as the name suggests, with propositions, which have truth values of 'T' or 'F' ($\top$ or $\bot$), and with the relations between them, as well as the construction of proofs (in the form of sequences of propositions and relations). Combinations of propositions can be formed by connecting individual propositions through logical connectives (these include *implication, conjunction, disjunction* and *negation*; cf. Table 7.1). Individual propositions are called 'atomic' propositions, and they are often denoted by Roman letters. For example, consider two atomic propositions P and Q. They can form a syllogism, e.g.

$$P \rightarrow Q \quad \text{(1st premise : 'If } P \text{ then } Q')$$

P (2nd premise : $'P$ is true$')$

Q (conclusion : $'Q$ is true$')$

The conclusion is reached from the premises by applying a *rule of inference* (here: *modus ponens*). The propositions can of course be interpreted in terms of natural-language expressions (e.g., P = 'The sun is shining'; Q = 'It is daytime'). These interpretations however do not belong to the formal system represented by the above syllogism, which is valid independently of the meanings or interpretations of the symbols P and Q; this is what is meant by *formal* logic.

Propositional logic dates from ancient times and is considered to be an advance over syllogistic logic (Aristotle). It was lost in the Middle Ages and rediscovered in the twelfth century by Peter Abelard. Later developments were carried out by Leibniz (and again rediscovered by Boole) and by Frege (cf. Chap. 6). Proofs of the completeness of propositional logic were given by Bernays (1917–1926) and by Post (1921).

Frege developed a further advance, *predicate logic* or first-order predicate calculus. Its history is recounted by Ewald.[10] He traces its beginnings to Boole and his follower Pierce, then to Frege, Schröder, and Peano, and, in the twentieth century, to Russell and Whitehead, Löwenheim, Hilbert and Bernays, Skolem, and finally to Gödel. We will meet up with these names again later in this chapter.

Gödel, in his thesis, concerned himself with the more differentiated *predicate logic* (in Gödel's time, it was called 'the first-order predicate calculus'). This was both a natural (and successful) choice, but also a limiting one: as Hintikka (1999, p. 18) puts it,

> Gödel's completeness proof for the ordinary first-order logic thus, in effect, served to reassure logicians and philosophers that they could happily go on practicing their proof-theoretical problems... *[But: his proof]* is therefore unrepresentative of the conceptual situation of logic in general. His Einsteinian question, the question concerning semantical completeness, was asked about a wrong logic.

Antecedents

Gödel himself mentions Russell and Whitehead (*Principia Mathematica*, 1910–13: PM) and Hilbert and Ackermann (*Grundzüge der theoretischen Logik*, 1928: HA) as sources for his thesis topic. The problem of the completeness of the first-order predicate calculus was raised in the latter work, which Gödel apparently read soon after its publication. The specific question asked by HA[11] was,"...*whether a certain axiom system for the first-order predicate calculus is complete, in the sense that from it all logical formulas that are correct for each domain of individuals can be derived.*" This question was not new, and, as noted by van Atten & Kennedy (2009), it is remarkable that it was first formulated clearly in print by HA only in 1928. The question had been considered already in 1915 by Löwenheim, and later by Skolem (1920) (quoted by Gödel in the introduction to his thesis); and Gödel pointed out still later[12] that his conclusion follows "almost trivially from Skolem" (1923),[13] who however failed to reach it himself. Completeness was simply 'not a hot topic' in the 1920s, although it later became one, not least through Gödel's publications. Other antecedents were the work of Bernays (1917–1926) and of Post (1921), both of whom, as mentioned above, proved the completeness of the propositional calculus as set out in PM.

What Gödel did in his thesis was to consider the *consistency* and the *completeness* of first-order logic, two properties considered important for establishing the validity of a logical system. Of particular note is that he proved them using a well-specified logical scheme, i.e., formally, as opposed to the informal methods (e.g., appealing to *tertium non datur*, and the presumed solvability of all mathematical problems: Hilbert). Along the way he introduced the *compactness theorem*, considered by some to be the most important result of his thesis.[14] It was, however, not mentioned specifically in the thesis itself, but was stated in the resulting publication (Gödel 1930). We will take up the meaning of these terms: 'first-order logic', 'consistency', 'completeness', 'soundness', and 'compactness', in what follows.

If Gödel did indeed begin thinking about his thesis topic in mid-1928, this was initially confined to 'asking the right questions'. His readings of PM and HA (evidently in late 1928 and early 1929) may have crystallized his thinking. The actual writing of the dissertation must have taken place in the period from January–May/June of 1929.

Aside from his father's sudden death in February 1929, which left him and his brother with many worries concerning their mother, he also had to complete his formal studies, i.e., take his examinations at the University (which he had completed by early summer, as he wrote later to Hao Wang). As we saw above, he submitted his thesis to Hahn before July 1929, and wrote the final version in September, after receiving the comments of Hahn and Furtwängler; then he also began writing the published version (Gödel 1930). It appeared only after some delay, and it contained the results of the thesis (but *not* its introduction, in which he had questioned certain assumptions of Brouwer and of Hilbert). This publication was received by the journal

Monatshefte für Mathematik und Physik on October 22nd, 1929, but it appeared in print only in late September of 1930.

First-Order Logic

First-order logic is applied to *individuals*, members of a *domain of individuals* (or the *domain of discourse*), abbreviated D. What or who the individuals are depends upon what system the logic is being applied to; they could be names, e.g., the names of persons; or numbers; or sentences. In the case of simple arithmetic, they are the natural numbers, members of the set $\mathbb{N}$, and the domain is defined as D: $\mathbb{N}$. The individuals can occur as *variables*, typically denoted by $x, y, z\ldots$ whose values are (to be) determined by the sentences in which they occur. If their values are fixed in advance (i.e., they are *constants*), they are often denoted as $a, b, c\ldots$ The relations listed in Table 7.1 are also termed 'constants'.

A formal logical system in first-order logic is expressed by a *language* **L** and may contain a set of *axioms* (as we have seen in the examples given in Chap. 6) and *rules of inference*, specifying how new sentences (*theorems* when proved) are to be derived from the axioms. There are also relations between individuals, specified by operators or functions, and quantifiers, which limit the values that variables can take. The *language* is the collection of the relations, functions, and constants which define the logical connections between the individuals that are the objects of the formal system of first-order logic. These are usually denoted by symbols, giving a compact notation (we have seen previously how both Russell and sometime later also Gödel were impressed by the brevity and effectiveness of the notations introduced by Frege and Peano).

In mathematical logic, the following symbols are useful (Table 7.1). They are not unique, and various authors in various areas of specialization often use other symbols to denote similar meanings. A helpful overview can be found online.[15] A more complete list is given by Enderton[16] (1972), pp. xi–xiii. Using these (or other) symbols, we can express axioms and theorems, relations and functions and their relationships briefly and precisely.

Some properties of formal logical systems which are of interest are their *consistency, completeness, soundness*, and *compactness*. We will investigate just what those properties mean with some examples.

Note that *second-order logic* deals with sentences, theorems etc. about more complex entities—classes, sets, functions, properties of individuals—rather than the individuals themselves; it is thus more powerful in reaching general

Table 7.1 Logic symbols

Symbol	Name	Read as	Explanation
$\supset$ or $\rightarrow$	implication	'if P, then Q'	$P \supset Q \rightarrow$ 'false if P is true and Q false; otherwise true'
$\equiv$ or $\Leftrightarrow$	equivalence	'if and only if': iff	$P \equiv Q \rightarrow$ 'P and Q are both true' '(... both false)'
$\neg$	negation	'not'	$\neg P \rightarrow$ 'not P', 'converse of P'
$\wedge$	conjunction	'and', 'plus'	$x \wedge y \rightarrow$ 'both x and y'; 'x + y'
$\vee$	disjunction	'both or either'	$x \vee y \rightarrow$ 'either x or y or both'
$\underline{\vee}$	exclusive disjunction	'either or (not both)'	$x \underline{\vee} y \rightarrow$ 'x but not y'; 'y but not x'
$\top$	tautology	truth	$\top(P) \rightarrow$ 'P is unconditionally true'
$\bot$	contradiction	falsity	$\bot(P) \rightarrow$ 'P is unconditionally false'
$\forall$	universal quantifier	'for all'	$\forall x: P(x) \rightarrow$ 'P(x) is true for all x'
$\exists$	existential quantifier	'there exists'	$\exists x: P(x) \rightarrow$ 'P(x) is true for at least one x'
.	definition	'is defined as'	$P := Q$, 'P and Q are logically equivalent'
( )	grouping	brackets	'the expression within ( ) is to be evaluated with precedence'
$\vdash$	turnstile	'proves'	$P \vdash Q \rightarrow$ 'P proves (syntactically) Q'
$\nvdash$	negated turnstile	'does not prove'	$P \nvdash Q \rightarrow$ 'P does not prove Q'
$\vDash$	double turnstile	'models'	$P \vDash Q \rightarrow$ 'P models (semantically) Q'
$\in$	membership	'belongs to'	$x \in S \rightarrow$ 'x is a member of the set S'

conclusions, but more complex in the metamathematical sense, and moreover semantically incomplete, as was shown by Gödel's famous incompleteness theorems.

In his delightful little book '*On Gödel*', Jaakko Hintikka[17] gives an example of a formal system in which the individuals are *persons*, their names abbreviated by their first letters. To make the example concrete, he uses the names of members of the Bloomsbury Group, a kind of 'Vienna Circle' which met in the London district of Bloomsbury. Its members were mostly graduates of Cambridge University, and they discussed in particular aesthetics and art and their roles in society.

The group persisted over two generations, from ca. 1905 through the 1930s, and included among others Bertrand Russell, Virginia and Leonard Woolf, Lytton Strachey, Dora Carrington, and John Maynard Keynes (Hintikka's b = *Bertie*, m = *Maynard*, v = *Virginia*, c = *Carrington*, l = *Lytton*, and e = *Leonard*). He defines two relations, L(x,y) ('x loves y') and C(x,y) ('x is cleverer than y').

One of the 6 statements (the 'axioms' of the system) representing the rules for gossip among and about the members of the group takes the form $\forall$ x: L(b,x) $\equiv$ ¬ L(x,x), i.e. Bertie loves only (all) those who do not love themselves. This statement appears at first glance to be harmless, and simply to describe a particular (perhaps peculiar) preference which Bertie exhibits in his love life. However, it leads to a contradiction when it is taken to be *self-referential*: Ask the question, "Does Bertie love himself?" Then he must either belong to the set of those who love themselves (but then cannot love himself according to the above sentence); or he belongs to the set of those who do *not* love themselves (making the first part of the above sentence invalid). The sentence is self-contradictory when 'b' is substituted for 'x': L(b,b) $\equiv$ ¬ L(b,b). But since the original sentence holds for *all* x, it must hold for any particular x, i.e., also for x = b. We see that this formal system is *not consistent*. It contains a potential internal contradiction—a paradox which is reminiscent of the paradox in set theory which Bertrand Russell himself recognized in 1901. (It is well worth reading Hintikka's book to see the rest of the 6 'axioms' that he proposes to represent the rules of gossip within and about the Bloomsbury Group, among many other things).

Consistency thus means that a formal system contains no (implicit) contradictions (which may not be at all evident on first view); and specifically: no contradictions can be revealed by any finite chain of logical conclusions within the system. This was the definition used by Gödel in his thesis.

What about *completeness*? This is a question which neither Frege nor Peano bothered to ask. In the nineteenth century, the emphasis was on finding a set of axioms that would enable the precise description of an area of mathematics (e.g., number theory, initially elementary arithmetic; or geometry). Whether *all possible* theorems or sentences that could be derived from those axioms using the rules of inference could be proven, or disproven (i.e., their converses could be proven) was not an issue. But that is just what is meant by *completeness*.

Strictly speaking, we must distinguish between the *syntax* ('proof theory') of the formal system, i.e., the formal-logical rules which govern its language, relations, functions, and how proofs are carried out; and its *semantics* ('model theory'), i.e., what the system *means* or represents. Hilbert took up the question of completeness in considering the axiomatization of geometry, at first not making the distinction between syntactic and semantic completeness (cf. Sieg 1999). In the course of time over the period 1900–1920, Hilbert and his school (especially Bernays) developed several concepts of completeness, including *descriptive* completeness and *semantic* completeness. Hilbert also mentioned what he called *maximal* completeness (i.e., the logical system has a *maximal model*). This means that no new element (e.g., from a particular axiom) can be added to the system without producing a contradiction with the other axioms. This property is closely related to *categoricity*—a system that has only one model up to isomorphisms is called *categorical* (see backnote [29]). The theorem proved by Löwenheim and by Skolem limits the occurrence of this type of completeness. A completeness criterion was also developed by Emil Post in his thesis (published 1921; see below).

When a set of individuals is specified, and an interpretation function is defined which can apply the relations and functions of a language to that set of individuals, the result is a '*model*', a realization of the language (or of some set of its relations, functions and constants). The model contains the set of individuals, the results of the language acting upon them, and the interpretation function used. It represents a 'universe' or 'world' generated by the language on the set of individuals.

The Thesis

What did Gödel himself say about completeness in the introduction to his thesis [Gödel (1929)]? Note that he gave a relatively informal introduction, mentioning some aspects of the history of mathematical logic and his own path to the problem that he dealt with in the thesis. It was not included in the later publication [Gödel (1930)] of his thesis results. We quote from [Gödel (1929)], Introduction (English translation by the present author):

> The principal object of the following investigation is the proof of the completeness of the axiomatic system of what is called the restricted functional calculus as given in Russell *[Fig. 7.1, right; and Whitehead]*, *Principia Mathematica* (PM), P-I, No. 1 and No. 10, and in a similar form in Hilbert *[Fig. 7.1, left]* and Ackermann's (HA) *Principles of Mathematical Logic* III, §5. Here, 'completeness' is taken to mean that every generally valid formula that can

be expressed in the restricted functional calculus (every generally valid *quantity statement*, in the terminology of Löwenheim)[18] can be deduced by means of a finite series of formal conclusions from the axioms. This assertion can readily be shown to be equivalent to the following: Every consistent axiomatic system[19] which consists only of quantity statements has a realization (a model) ('consistent' means here that no contradictions can be derived by means of any finite series of formal conclusions). The latter formulation would seem to offer considerable interest for its own part, since the solution of this question represents in a certain sense a theoretical complement to the usual method of proving consistency (naturally only for the special kind of axiomatic system considered here); for it would indeed offer a guarantee that this method would in any case achieve its goal, i.e. that either a contradiction could be found, or the freedom from contradictions must be provable by means of a model [1]. The fact that one cannot necessarily conclude that a model is constructible simply from the consistency *[freedom from contradictions]* of a system of axioms has been stated with emphasis, in particular by L.E. Brouwer. One might however think that the existence of concepts derived from a system of axioms could in fact be defined by their consistency, and thus reject a proof from the outset. However, such a definition (for which one would only need to insist upon the self-evident requirement that the concept of existence thus defined would obey the same operational rules as the elementary one) evidently presupposes the axiom of the solvability of every mathematical problem. Or, more precisely spoken, it presupposes that one cannot prove the non-solvability of any problem. For if the non-solvability of some problem (for example from the theory of real numbers) were to be proven, then from the above definition, it would follow that two non-isomorphic realizations *[models]* of the system of axioms for the real numbers must exist; while on the other hand, one can prove the isomorphism of any two models. Now, however, a proof of the non-solvability of some problem cannot be excluded *a priori*, if one considers that it would be a non-solvability only using certain *precisely specified formal* rules of implication. For all the concepts that are considered here (provable, consistent, and so on) have an exact meaning only when we have precisely defined the means of inference that are employed.

[Footnote: 1) 'To be sure, the existence of this alternative is not proven in the intuitionistic sense (i.e., through a decision process); see below'. (Gödel's footnote; cf. footnote 1) in the German original).]

We will return to Gödel's remarks on Brouwer's (und indirectly, Hilbert's) positions later. The two theorems on '*completeness*' and '*soundness*' (sometimes called '*correctness*') which Gödel proved are converse, and together they give the connection between the syntactic (proof-theoretical) properties of the language and its semantic (model-theoretical) properties.

Fig. 7.1 *Left*: David Hilbert, before 1912. *Right*: Bertrand Russell, about 1910. Photos: WikiCommons, public domain[20]

Note that the formula shown at the bottom of the memorial plaque at the entrance to *Frankgasse* 10:

$$\vdash A \Leftrightarrow \vDash A \tag{7.1}$$

(cf. Fig. 5.1, right side) is a symbolic expression of these two concepts. Reading from left to right, Formula (7.1) says that 'a system **S** or language **L** (not specified) proves *only* sentences which are satisfied (valid)'—this is *soundness*—and reading from right to left, it says that '*all valid* sentences are proved by **L**'—*completeness*. In words, it might be expressed as "… proves **A** iff … models **A**".

What precisely is meant by a *model*? As we saw above, it is a set of individuals for the language to act upon, and the structure and properties of that set when acted upon by the language. It is sometimes called a 'universe' or a 'world'—that world which results when the individuals are acted upon by the language. It thus expresses the *meaning* (semantics) contained in the language, realized in terms of a particular set of individual objects. One way of stating Gödel's completeness theorem is the following: "Suppose *S* is a set of formulas (derived from some language **L**). If *S* is consistent, then it has a model".

Gödel's Proof of Completeness

Note that a simplified and more general proof was later set out by Henkin (1949), and it is summarized for example in Chang and Keisler (1973), p. 61 ff., and in Enderton (1972), pp. 129–136. Henkin's proof can be roughly sketched as follows[21]:

There are two equivalent formulations of the completeness theorem:

(i) If $S \vDash \varphi$ then $S \vdash \varphi$ (this is Formula (7.1), read from right to left); or

(ii) Any consistent set of formulas is satisfiable (it has a model).   (7.2)

Henkin's proof is given for the second statement, for a countable language (one whose elements can be put into a one-to-one correspondence with the natural numbers). This is similar to what was done by Gödel in his thesis and then published as Gödel (1930). The extension to non-countable languages was first shown by Mal'cev in 1936; see below. The elements of the language are, in modern terms, 'well-formed formulas' (wff), i.e., formulas which conform to the syntax of the language.

Proof: Consider a consistent set S of sentences of a language **L**. The language **L** can be expanded (extended) by adding for example a set of new constants, to give a new language $\underline{\text{L}}$, and by adding corresponding sentences to S, yielding $\underline{S}$. Now $\underline{S}$ has a model U that is a model for $\underline{\text{L}}$. It can then be shown that M, the reduction of U to the original language $\overline{\text{L}}$, is a model for S. This demonstrates the formulation (ii) of the completeness theorem given above.

We summarize Gödel's original proof from [Gödel (1930)]. His longer, historical introduction in the thesis was replaced in this publication by a short statement of intention (quoted here from [Gödel (1930)]) (English translation by the present author with permission from the publisher):

*"**The Completeness of the Axioms of logical Functional Calculus**[1)]*

by Kurt Gödel, *Vienna*

> Whitehead and Russell structured logic and mathematics, as is well known, in such a way that they placed certain evident theorems as 'axioms' at the beginning and, making use of some precisely-formulated rules of inference in a purely formal manner (i.e. without paying further attention to the meanings of the symbols), they deduced the theorems of logic and mathematics *[from them]*. With a method of this type, the question arises immediately as to whether the chosen system of axioms and principles of inference is complete,

i.e., whether it in fact suffices for deducing every logical-mathematical theorem, or whether perhaps true (and possibly, using other principles, provable) theorems are thinkable which cannot be derived within the system considered. For the domain of logical propositional formulas, this question has been answered positively, i.e., it has been shown [2] that in fact every valid sentential formula follows from the axioms given in the *Principia Mathematica*. Here, the same is to be demonstrated for another domain of formulas, namely for the restricted functional calculus [3]; i.e., we will show that:

Theorem I: Every generally-valid formula of the restricted functional calculus is provable...

[Footnotes]: 1) Thanks are due to Prof. H. Hahn for some valuable suggestions relevant to formulating this work.

2) Cf. P. Bernays, *Axiomatic investigation of the propositional calculus of Principia Mathematica. Mathematische Zeitschrift* **25**, 1926.

3) The terminology and symbols used in the following work are those of Hilbert-Ackermann, *Principles of Mathematical Logic*, Berlin 1928. They include for the restricted functional calculus the logical expressions which are obtained from the propositional variables: X~ Y~ Z "...

Gödel continues in his article (1930) by enumerating the theorems that he proposes to prove[22]:

Theorem I: Every generally-valid formula of the restricted functional calculus is provable.   (7.3)

Here, we presume that the following system of axioms forms the basis for this theorem: ...

Formal Axioms:

1. $X \vee X \rightarrow X$
2. $X \rightarrow X \vee Y$
3. $X \vee Y \rightarrow Y \vee X$
4. $(X \rightarrow Y) \rightarrow [Z \vee X \rightarrow Z \vee Y]$
5. $(x)\, F(x) \rightarrow F(y)$ (7.4)
6. $[X \vee F(x)] \rightarrow X \vee (x)\, F(x)$.

[Here, $X, Y, \ldots$ are terms containing variables, $x, y, \ldots$ are individual variables, and $F(x)$ is a function of the variable x; and the relations $\vee$ and $\rightarrow$ are as given in Table 7.1. These axioms are very similar to those given in HA—see e.g., Zach (1999).]

Rules of inference:

1. *The conclusion scheme* [modus ponens]: *From A and A → B, we can conclude the validity of B.*
2. *The substitution rule holds for quantity and functional variables.*
3. *From A(x), we can conclude (x) A(x)* [i.e., ∀x: A(x)].
4. *Individual variables (free or bound) can be replaced by different, arbitrary individual variables, so long as no overlap with the range of action of other, similarly-denoted variables occurs.*

[Gödel then goes on to list a number of abbreviations and symbols that he will use in his proofs, as well as a series of (7) auxiliary theorems that he will need, which he does not prove, as they are "*in part known, in part easily completed...*"] ...

Theorem II: Every formula of the restricted functional calculus can either be contradicted [1] or else it is valid (to be sure only within a countable domain of individuals).

   1). 'A is contradicted' means that '¬ A is provable'.

[His two theorems (I) and (II) are equivalent to the two formulations used by Henkin in his simplified proof (Formula (7.2), (i) and (ii), above). We paraphrase the outline of Gödel's proof as given in van Atten and Kennedy (2009), based on the thesis [Gödel (1929)]. Gödel in fact proved the second version (his *Theorem II*), as did Henkin].

Gödel expresses formulas in the 'prenex normal' form—that is, they are written as a string of entries, each preceded by a prefix containing quantifiers and variables [e.g. (∀x), (∃y)]. The number of such prefixes in the string is called the 'degree' of the formula. Gödel proves that if each formula of degree n is either satisfiable (has a model) or refutable (contradictory), then so is each formula of degree $n + 1$.

He then shows that this assertion is valid for formulas of degree 1 [the lowest degree, and thus the first formula in the sequence], having the general form:

$$\forall x_1 \ldots \forall x_r \exists y_1 \ldots \exists y_s A(x_1, \ldots, x_r, y_1, \ldots, y_s) \qquad (7.5)$$

[van Atten and Kennedy show the proof for the simpler special case of a sentence of the form (P) $A = \forall x_1 \exists y_1 A(x_1, y_1)$, where $A(x_1, y_1)$ has no quantifiers]. Gödel defines a series of formulas

$$(P_n) \ A_n = \exists x_0 \exists x_1 \exists x_2 \ ... \ \exists x_n \ (A(x_0, x_1) \wedge (A(x_1, x_2) \ ... \ \wedge A(x_n-1, x_n)),$$

and then uses induction on n to show that $(P) \ A \rightarrow (P_n) \ A_n$ is provable. In the case that, for some n, the propositional formula A_n is *not* satisfiable, i.e. $\neg \ A_n$ is provable, then the original sentence is refutable. The case that *no* A_n is refutable is more complex. Here, Gödel forms a tree of all possible assignments of truth values satisfying the various A_n. This is a finitely-branching tree (there are only two truth values and a finite number of terms in each A_n), but the tree is itself infinite, since there are infinitely many n-levels. Gödel concludes that there must be an infinite branch[23] containing truth levels $s_0 < s_1 < ...$ in the tree. The assignments of this branch form a single overall assignment satisfying each A_n. This allows him to define a model of the original sentence $(P) \ A$, and thus to show that $(P) \ A$ is satisfiable.

Constructing the tree is more complicated in the general case (7.5), and the order in which the sequences are dealt with becomes important. Van Atten and Kennedy (2009) mention that the above lemma showing that (P) $A \rightarrow (P_n) \ A_n$ is provable is the crucial step missing from the earlier work of Löwenheim and of Skolem, and that it provides the link from semantics to provability (syntax). Making clear the distinction between proof theory and semantics (model theory, in modern terms) was also an important aspect of Gödel's thesis work (1929, 1930).

In the publication (1930) containing the results of his thesis, Gödel omits the relatively long and informal introduction, as mentioned above. Whether this was his own idea, or was suggested by Hans Hahn or someone else; and whether it was done simply to save space (the published article had 11 printed pages, while the thesis had about 35 pages), or because Gödel did not want to expose some of his comments in his thesis introduction to a wider audience (e.g. about the disagreements between Hilbert and Brouwer, for example on the question of whether *consistency* implies *existence*), we do not know. We will take up this conflict and Gödel's reading of it in the following. Another important difference between the thesis and his 1930 article is that in the latter, the *compactness theorem* was given explicitly (without calling it by its modern name, however). It may be that Hahn recognized its importance and suggested to Gödel that he should mention it in the published paper, rather than leaving it merely implicit as in his thesis. We will consider the subject of this theorem in the following section.

The Compactness Theorem

This theorem was proved by Gödel in his thesis [but demonstrated in detail only in the later publication of his thesis work, Gödel (1930)], and it is an important result. For a simple description, see the paper by Call (2013). It was later found to be very useful for various applications. It is perhaps even more significant than his 'principal result', the completeness of first-order logic; that result was not unexpected, and it confirmed what Hilbert and his school had assumed (in particular in the book HA (1928), quoted by Gödel).

The compactness theorem is in the first instance a theorem of model theory, in which it plays a significant role; but in the meantime, it has found applications in very diverse fields, including topology, set theory, graph theory as well as first-order and propositional logic. Gödel originally proved the theorem only for countable languages; it was later proved for the uncountable case by Anatoly Mal'cev (1936). A brief statement of the theorem as proved by Gödel, but in modern language, would be: "A set of first-order sentences has a model if and only if each of its finite subsets has a model". In his own language, from Gödel (1930), the compactness theorem (his *Theorem* X) states that, "*For a countably infinite system of formulas to be satisfiable* [have a model], *it is necessary and sufficient that each of its finite subsystems be satisfiable*".

The proof of the compactness theorem follows from the completeness theorem. If the whole system S is satisfiable, it is clear that each of its subsets must also be satisfiable. In the reverse direction, assume that *every subset* is satisfiable, but that the *whole system S* is not, in contradiction to the compactness theorem. But then, by the completeness theorem, S is not provable and is thus inconsistent; it would contain some sentence φ such that $S \vdash \varphi$, but at the same time, another sequence leading to $S \vdash \neg \varphi$. We can define a set of sentences S' which is the subset of S containing the sequence that deduces $\vdash \varphi$, and another, S'', which is the subset that deduces $\vdash \neg \varphi$. Then the combined set $S' \cup S''$ (the union of S' and S'', also a subset of S) is *inconsistent* and contains a contradiction, and therefore, it is not satisfiable (has no model). There would thus be a finite subset of S which *also* has no model, contradicting the above assumption, and thus proving the compactness theorem.

In addition to explicitly proving the compactness theorem in his 1930 paper, Gödel, as another 'bonus' of his thesis work, shows the independence of the axiom system of first-order logic. This makes up the final section of his (1930).

The Intellectual Setting of Gödel's Thesis

Some perspectives on Gödel's intellectual achievement in his thesis are given by Wang (1987), (1997), Dawson (1997), and Hintikka (1999). Several evaluations of the thesis have appeared more recently: for example the one quoted above, by Mark van Atten and Juliette C. Kennedy (2009); and another by J.C. Kennedy (2011) entitled 'Gödel's Thesis: An Appreciation'—the latter is a sensitive and thorough analysis of Gödel's achievement in his thesis, which he wrote at the early age of 23—in particular its philosophical background. We will consider her analysis in more detail below. In addition, memoirs by Gödel himself (as told to Hao Wang, and quoted in the latter's book 'A Logical Journey', [Wang (1997)]); by Paul Bernays, about his earlier work on the completeness of the propositional calculus, quoted e.g. in Zach (1999); and also a memoir by Leon Henkin (1996) about his discovery of a new proof of the completeness of first-order logic—have been published in the late twentieth century. All of them shed light on the complex evolution of the topic of *completeness* over 4 decades. That evolution is summarized compactly in the article on 'The Emergence of First Order Logic' by William Ewald in the Stanford Encyclopedia of Philosophy ([Ewald (2018)]; see backnote [10]).

We first have a look at Gödel's memories, as told much later to Hao Wang. In his last year as an active professor at the IAS in Princeton (1975/76— he retired in the summer of 1976), Gödel arranged for Hao Wang to be a visiting scholar at the Institute. Interestingly, they had only one longer in-person meeting, in Gödel's office in December 1975. But they often spoke on the telephone, a habit of Gödel's in his later years which has been mentioned by a number of people. He avoided personal, eye-to-eye contacts whenever possible, but was often accessible by telephone. Wang collected the results of various conversations with Gödel under the title 'Some Facts about Kurt Gödel in his own Words', an appellation approved by Gödel himself. They were interpreted by Wang (1981), and they are reproduced verbatim on pp. 81–90 of Wang (1997). Concerning the time when he was working on his dissertation, Gödel recalls,

> ...When I entered the field of logic, there were 50 percent philosophy and 50 percent mathematics. There are now 99 percent mathematics and only 1 percent philosophy. I doubt whether there is really any clear philosophy in the models for modal logic.
> Shortly after I had read Hilbert-Ackermann, I found the proof [of the completeness of predicate logic]. At that time, I was not familiar with Skolem's 1922 paper [reprinted in Skolem 1970, pp. 137–152; the relevant part is remark

3, pp. 139–142]. I did not know König's lemma either—by the same man who had the result on the power of the continuum.[24]

(Remarks added by Wang).

This quote shows Gödel's early interest in philosophy and the philosophical implications of mathematics. It also demonstrates that he was somewhat naïve about his thesis topic when he began working on it. He was however aware of the early work of Paul Bernays, an assistant to Hilbert in Göttingen, on the completeness of propositional logic (he quotes Bernays' 1926 paper on the subject, derived from his *Habilitationsschrift* of 1918)—but not of the work of Emil L. Post (1921) on the same topic. And although he quotes the Löwenheim-Skolem theorem [Löwenheim (1915), Skolem (1920)], he was also unaware of Skolem's later (1923)[25] paper on the axiomatic foundation of set theory, relevant to Gödel's own work in his thesis. Gödel later recognized that Skolem had come very close to the proof that he found in 1929; he wrote in 1963 to van Heijenoort,[26]

As far as Skolem's paper is concerned, I think I first read it about the time when I published my completeness paper. That I did not quote it must be due to the fact that either the quotations were taken over from my dissertation or that I did not see the paper before the publication of my *[own]* paper. At any rate I am practically sure that I did not know it when I wrote my dissertation. Otherwise, I would have quoted it, since it is much closer to my work than the paper of 1920, which I did quote.

We examine Gödel's thesis introduction once more, on the subject of *Beweismittel* (means of proof) (English translation by the present author):

...In conclusion, one more remark on the means of proof which are used in the following work. No sorts of limitations are placed on them. In particular, substantial use is made of the law of excluded third *[excluded middle, LEM;* or tertium non datur]* for infinite universes (the uncountable is in contrast not used in the main proof). It might appear as if this renders the whole completeness proof useless; for what was to be proved can be taken to be a sort of decidability (every expression of the restricted functional calculus can either be recognized as generally valid, through a finite number of logical conclusions, or else its general validity can be refuted by means of a counter-example). On the other hand, the law of excluded third would appear to imply precisely the decidability of every problem. But we can counter this point with the following arguments:

1. The law of excluded third is interpreted in that way only by the intuitionists.

2. Even if one accepts that interpretation, it supports the solvability only by *all imaginable* general methods, and not at all by *particular* means[1], while in the following work, it is directly proved that every generally valid expression can be deduced by means of *certain precisely specified* rules of inference. Seen from the intuitionistic point of view, the whole problem would be quite different, since the meaning of the statement, "A system of relations fulfills a logical expression" (i.e., the sentence which results from substitution is true) would be fundamentally different…".

1). Whether indeed such a generalized concept of solvability, and thereby the interpretation of the law of excluded third as considered above, makes any sense at all remains questionable [Gödel's footnote].

Here, Gödel has anticipated an apparent contradiction in the program of his thesis, namely the possible objection that the use of *tertium non datur* presupposes the completeness of any system, since it maintains that any expression is either provable or non-provable (i.e., its negation is provable). He suggests that this position might be taken by the intuitionists (Brouwer's school), and then refutes it by noting that he is dealing with a *formal system*, which has precisely specified axioms and rules of inference, and not the informal (intuitive) case.

The Brouwer-Hilbert Controversy

In the introduction to his thesis, Gödel briefly mentions the question of whether *consistency* implies *existence*, as maintained by Hilbert—see the first excerpt quoted above. *L.E.J. Brouwer* denied the validity of that proposition; he was a supporter (the principal and most prominent supporter, in fact) of a school of thought on mathematical logic known as *intuitionism*, which was to a considerable extent at odds with Hilbert's school, called *formalism*.

In fact, the controversy dates back much further, to a time well before Brouwer became active. It originated with a paper of Hilbert's on algebraic geometry, containing the *Hilbert Basis Theorem* and published in 1890 [Hilbert (1890)]. His proof aroused opposition from the *constructionists*, in particular Leopold Kronecker, who held that mathematical ideas are not innate in the universe, as proposed by Plato and Kant, but rather must be 'constructed' and proved. Kronecker abandoned his opposition to Hilbert shortly before his death, but other constructionists, in particular Henri Poincaré, and later Luitzen Egbertus Jan Brouwer, Arend Heyting and Hermann Weyl, continued the dispute. Both Brouwer and Weyl were originally protegés of Hilbert: Hilbert had helped Brouwer (Fig. 7.2, right) to obtain a professorship in Amsterdam, and Weyl was a student of Hilbert's in Göttingen. Both of them later became *intuitionists*, a school which evolved from constructionism. They held that mathematical ideas are neither innate in nature nor purely logical entities that can be proved by abstract, formal means (as maintained by the *logicists*—Frege, Russell, Wittgenstein and others, and the *formalists*, members of Hilbert's school (David Hilbert, Paul Bernays, Wilhelm Ackermann, Johann/John von Neumann, and Gerhard Gentzen, among others). Instead, they arise, or must be confirmed, by human intuition—an insight into the correctness or falsity of mathematical propositions and procedures.[27] And they avoid the *law of excluded third* (or *middle*, LEM), as well as proof by double negation.

The controversy between the once congenial colleagues Hilbert and Brouwer, originally restricted to the rather abstract level of the proper interpretation and subsumption of mathematical logic, became more personal following World War I. German and Austrian scientists and mathematicians were to a considerable extent shunned for their support of their national governments and military during the war, which was deemed morally reprehensible after the defeat of the Central Powers. They were initially not

Fig. 7.2 *Left*: Paul Bernays, 1914. *Right*: L.E.J. Brouwer, at a conference in Zurich, 1932; cropped. Photos: from Wiki Commons[28]

welcome at international conferences and meetings. This icy treatment gradually thawed, but resentments persisted on both sides.

The combination of purely scientific, intellectual disagreements with personal resentments, wounded pride and latent nationalism came to a head for Brouwer and Hilbert at the 1928 International Congress of Mathematicians (ICM) in Bologna, Italy. It would have been the 8th in the series of conferences that began in Zurich in 1897 and were held at four-year intervals, sponsored by the International Mathematical Union. They were interrupted by WW I, but they resumed in 1920 (Strasbourg) and 1924 (Toronto). German, Austrian, Hungarian and Bulgarian mathematicians were however excluded from those latter two meetings, and they were later deleted from the series as not being truly 'international'. Thus the 1928 meeting, where again mathematicians from all nations were to be welcomed, became the 6th in the series. Serious disputes between groups supporting internationality and those still wanting to punish former Central Power nations arose during the preparation of the 1928 conference. Brouwer opposed German participation in the conference, much to Hilbert's annoyance.

After the conference, Hilbert, who was in poor health, wanted to break off all connections with Brouwer. Hilbert was editor-in-chief of the prestigious journal *Mathematische Annalen*, with headquarters in Göttingen. Other members of its Editorial Board included Otto Blumenthal, an earlier Hilbert student who served as Managing Editor, as well as Albert Einstein and Constantin Carathéodory (who had replaced Felix Klein when the latter retired in 1924). Brouwer was an Associate Editor and was very active in supporting the journal and editing papers that were submitted to it. Hilbert decided to dismiss Brouwer as Associate Editor, and a fierce campaign of correspondence between Hilbert (Göttingen), Blumenthal (Aachen), Einstein (Berlin), Carathéodory (Munich) and Brouwer (Amsterdam) ensued. Even Hilbert's physicist colleague in Göttingen, Max Born, was drawn into the fray.

Einstein's efforts to defuse the conflict and settle the matter peacefully are documented in the most recent volume of the *Collected Papers of Albert Einstein* (CPAE), Vol. 16 (2021). For example, on October 19th, 1928, Einstein wrote to Carathéodory: "*I am sending you herewith the letter that I wrote to Hilbert. It would be best to pay no attention to this whole Brouwer matter. I would never have thought that Hilbert was capable of such outbursts of emotion—*".

He was unsuccessful: Blumenthal had to choose between his old mentor, Hilbert, and the brilliant (and diligent, but eccentric) mathematician Brouwer; and he finally sided with Hilbert. Brouwer was dismissed, and both the journal and Brouwer were seriously damaged as a result. This was the situation in 1929, when Gödel was writing his thesis, and it required tact on his part to justify his methods of proof, mainly aligned with the formalist school of Hilbert, as opposed to the ideas of Brouwer, whom he admired and did not want to offend, and whose intuitionist school he respected, even if he did not agree with it. This is probably a principal reason why he did not include the introduction to his thesis (1929) in the published version (1930). This view is supported by J.C. Kennedy (2011), who writes, "*The introductory remarks were never included in the publication based on the thesis, and indeed, Gödel would not publish such unbuttoned philosophical material until 1944, with his* 'On Russell's Mathematical Logic'."

Kennedy, in her (2011), discusses Gödel's notions of consistency, considering the prevailing views among logicians at the time, and in particular the distinction between *consistency* and *categoricity* (Gödel does not actually use that term, but adopts results on the isomorphism of models from the categoricity discussion. This was an early phase of a long development which began to reach maturity in the 1960s with the advent of a formal model theory. See e.g., [Grzegorczyk (1962)]).

The term *categorical* in connection with models of a formal logical system was apparently first used by *Oswald Veblen* (see [Veblen (1904)],[29] where he introduced a set of axioms for geometry). He defines 'categorical' to mean that the system has only one model (up to isomorphisms). In his own words, "*...In more exact language, any two classes K and K' of objects that satisfy the twelve axioms are capable of a one-to-one correspondence such that if any three elements A, B, C of K are in the order ABC, the corresponding elements of K' are also in the order ABC. Consequently, any proposition which can be made in terms of points and order either is in contradiction with our axioms or is equally true of all classes that verify our axioms.*" Here, K and K' are models, in modern terms, and his statement about the elements A, B, C denotes an isomorphism.

Tales of Bernays, Post and Henkin

As we have seen, *Paul Bernays* (Fig. 7.2, left), at the time (1917–1922) assistant to Hilbert in Göttingen (later untenured associate professor), and *Emil Leon Post* (Fig. 7.3, left), at the time (1920–21) a postdoctoral fellow in Princeton—both set out proofs of the completeness of propositional logic, a forerunner of the predicate logic treated in Gödel's thesis. Gödel was aware of Bernays' work (first published in 1926) but not of Post's when he wrote his thesis. Leon Albert Henkin (Fig. 7.3, right), then also a graduate student in

Princeton, presented a simplified and more general proof of the completeness of first-order predicate logic in 1947–49, which has essentially supplanted Gödel's original proof in modern textbooks (without diminishing Gödel's achievement in any way). We quote from some of their memoirs, which give insights into the situation in formal logic around the time of Gödel's thesis, and how that situation had changed by the later 1940's.

Paul Isaac Bernays was born in 1888 in London to a German-Jewish family. He spent his early years in Berlin, where he completed the *Gymnasium* and studied mathematics, physics and philosophy at the University of Berlin (the *Friedrich-Wilhelms-Universität*) and at the *Georg-August-Universität* in Göttingen. He wrote his doctoral dissertation in Berlin in 1912 under Edmund Landau, on the number theory of binary quadratic forms. He was awarded the *Habilitation* (unusually, since generally a minimum of two years must pass between the doctorate and the *Habilitation*) in that same year by the University of Zurich for a thesis on complex analysis (examined by Ernst Zermelo).

From 1912–1917, he was *Privatdozent* in Zurich. In 1917, he became an assistant to David Hilbert in Göttingen, where he submitted a second *Habilitation* thesis in 1918 (most likely, the 'rapid' one from Zurich was

Fig. 7.3 *Left*: Emil Leon Post, June 1924. Photo: from the *American Philosophical Society* archives, used with permission.[30] *Right*: Leon Albert Henkin, Berkeley, 1990. Photo: from the *Oberwohlfach. Photo Collection*, by George M. Bergmann. Used with permission[30]

not recognized in Göttingen). It was on the axiomatics of the propositional calculus from Russell and Whitehead's *Principia Mathematica*, and it
contained, among other things, his proof of the completeness of the propositional calculus. However, he made no attempt to publish his thesis for over
8 years, a fact which he explained later in an interview,[31] given on July 25th,
1977 (English translation by the present author):

> [It *(his* Habilitation *thesis)]* indeed had a mathematical character, but the
> prevailing opinion at the time was that such investigations into fundamen-
> tals, which tied in with mathematical logic, they were not taken very seriously
> as mathematics, you know: 'Yes, that is very pretty, it's almost like a play-
> thing, isn't it…'—and I went along with that spirit *[…]*, and didn't take [my
> work] very seriously either, and so *[…]* I wasn't especially eager to publish it
> promptly, and it was quite a while later, and still not even complete, just some
> parts were published *[…]*. That's how it is, some things that I had in that work
> *[his 2nd* Habilitation *thesis]* aren't properly taken into account, for example in
> articles on the development of mathematical logic, you know….

In other words, Bernays did not get credit for much of the work that was
contained in his *Habilitation* thesis for Göttingen, although in fact he had
done very novel and important work. In some circles, Post is given priority,
e.g., for the completeness proof for propositional logic; and indeed, he did
publish his work (his doctoral thesis of 1920) promptly, as Post (1921), and
thus before Bernays, who published only in 1926 [Bernays (1917–1926)].
This is detailed in an article by Richard Zach [Zach (1999)] with the appropriate title '*Completeness before Post: Bernays, Hilbert, and the Development of
Propositional Logic*'. And, in turn, Emil Post 'got there first' in some themes
later taken up by Gödel and by Alan Turing, as recognized in an article by
John Stillwell [Stillwell (2004)], entitled '*Emil Post and His Anticipation of
Gödel and Turing*'.

This is of course a recurring theme in the history of science: many developments are achieved at nearly the same time—or sometimes with considerable
delay—by different scientists working independently in different places—and
who is later credited with 'priority' is often a matter of chance: the speed with
which journals put papers into print, the slowness of publication of conference proceedings, the obscurity of some journals, and often enough simply
the inertia of authors who see no need to expedite the publication of their
work. One could make a long list of such 'injustices', which sometimes even
lead to the loss of a Nobel prize, and sometimes produce severe regrets and
rancor in their 'victims'. In this sense, both Bernays and Post seem to have
had healthy attitudes about the vagaries of 'who got there first'.

Emil Leon Post was born in 1897 in *Augustów*, Poland, of Polish-Jewish parents. He emigrated with his family to the USA (New York) at age 7, in May 1904. He received a B.S. in mathematics from City College in 1917 (his original ambition to become an astronomer was frustrated by an accident which resulted in the loss of his left arm, so that he turned to the (presumably more 'theoretical') field of mathematics). He obtained the doctorate in mathematics from Columbia University in 1920, and it was his thesis work on the completeness of propositional logic (as set out in Russell and Whitehead's *Principia Mathematica*) which he published the following year (Post 1921), while a postdoc at Princeton. He also independently developed truth tables (often attributed to C.S. Peirce and/or Ludwig Wittgenstein). While at Princeton, Post also considered the possibility of the incompleteness of the logic of *Principia Mathematica*, as proved later, and famously, by Gödel (1931). Post however was doubtful that his ideas would be convincing without a more complete analysis, so he made no attempt to publish them at the time. Years after Gödel's 1931–32 publications of his incompleteness theorems, Post (who later carried on a correspondence with Gödel over several years) met Gödel for the first time in 1938 (in October, in New York; see Chaps. 9 and 12), and soon after wrote on a postcard to Gödel (quoted in Stillwell 2004),[32]

> I am afraid that I took advantage of you on this, I hope but our first meeting. But for fifteen years I had carried around the thought of astounding the mathematical world with my unorthodox ideas, and meeting the man chiefly responsible for the vanishing of that dream rather carried me away. Since you seemed interested in my way of arriving at these new developments, perhaps Church can show you a long letter I wrote to him about them. As for any claims I might make, perhaps the best I can say is that I would have proved Gödel's theorem in 1921, had I been Gödel.

Post later worked on recursion theory and computability, and anticipated some of the work of Alan Turing in the mid-1930s. We will return to this topic in connection with Gödel's incompleteness theorems and their connection with computability in Chap. 12.

We now turn to a later generation—Bernays and Post were 18 and 9 years older than Gödel, respectively, a half-generation ahead of him, and their anticipation of some parts of his work is thus not too surprising. *Leon Albert Henkin*, in contrast, was born in 1921, 15 years after Gödel, so that he was a half-generation younger. He was born in Brooklyn, NY, of Russian-Jewish parents who had immigrated a generation before, and he was thus a 'native

American', in contrast to nearly all of the other participants in our story thus far, many of whom however ended up in the USA as immigrants.

In many respects, though, Henkin's life story was bizarrely parallel to Post's, with a 20-year delay. Both studied mathematics at Columbia University, Post as a graduate student, Henkin as an undergraduate. Henkin developed an interest in mathematical logic while at Columbia, and took some courses from Ernest Nagel, in the Philosophy Department (logic was not taught in Mathematics). Nagel's first course introduced him to Russell's works, and in particular to the axiom of choice and to the theory of types as described by Russell in his *Principles of Mathematics* and in *Principia Mathematica*. He later also took another course from Nagel which treated Hilbert and Ackermann's *Principles of Mathematical Logic*; and he read, at Nagel's suggestion, W.v.O. Quine's paper giving a modern proof of the completeness of propositional logic (Quine 1938). In his memoir (1996), Henkin also tells of hearing a lecture by the Polish logician *Alfred Tarski*, who will play an important role in later chapters. Tarski was on a lecture tour when Poland was invaded by Nazi Germany on September 1, 1939, beginning World War II. He remained in the US, later joining the faculty in Berkeley, where Henkin became his colleague in 1953 (at Tarski's invitation).

At Columbia, another of his professors suggested to Henkin that they read Gödel's recently-published monograph on the consistency of the axiom of choice and the generalized continuum hypothesis [Gödel (1940)]. Of that experience, Henkin (1996) writes:

> This event was probably my most important learning experience as an undergraduate. I gained much more of the content of Gödel's monograph than I had in reading Quine's paper the year before. I admired the metamathematical treatment whereby the comprehension schema of set formation is obtained from finitely many axioms, and the sophisticated handling of innermodel constructions by means of the notion of the 'absoluteness' of various set-theoretical notions. I was intrigued by the creation of a universal choice function in the realm of constructible sets, while none had been at hand in the realm of sets described by the original axioms, drawing my attention to a class of functions which were to be the starting point of my dissertation investigations five years later.

After graduating from Columbia in 1941, Henkin went on to Princeton for graduate work (as had Post as a postdoc in 1920–21). He was attracted to Princeton because the logician *Alonzo Church* was on its faculty, and Church later became his mentor. In his first year as a graduate student, he attended Church's lectures on logic and noted how skewed towards logicism (syntax) they were, treating semantics (the term 'model theory' had not yet been introduced) as a less-important footnote.

In December 1941, the US naval base at Pearl Harbor was attacked by Japanese forces, precipitating the entry of the USA into WW II, and bringing Henkin's graduate-student career to an abrupt halt. He was allowed to prepare for the qualifying exams (which he passed the following May) and to take a Masters degree as a 'consolation prize', and then spent 4 years working as a mathematician on military projects relevant to the war. When he returned to Princeton in early 1946, he took up the thread of his PhD thesis, and he describes in his (1996) how he was led to finding a simpler and more general proof of the completeness of first-order predicate logic (published in 1949 and subsequently). The details are fascinating but too intricate to reproduce here. Whether he ever met Gödel personally is not clear—the two of them were both in Princeton during Henkin's graduate student days (1941–42, 1946–48) and his subsequent two years as a postdoc there. But Gödel was at the IAS, where he led a very private life. Henkin may well have heard Gödel's lecture on his rotating universe (May 1949).

In the next chapter, we take up the complex topic of Gödel's *Incompleteness Theorems*, the subject of his *Habilitation* thesis, first published in 1931. They are at the origin of his fame as a logician and metamathematician. It was on the basis of that work (in addition to his talk on the relation of classical to intuitionistic arithmetic, heard by Oswald Veblen in June, 1932; see Chap. 8) that he was invited to the Institute for Advanced Study (IAS) in Princeton, at first as a visitor and temporary member, and later as a permanent member and professor, where he spent essentially the second half of his life. In the following chapter, we rejoin him in Vienna in October of 1929.

Notes

1. Budiansky (2021, p. 117) suggests that he submitted it "around February"; but given how many things he had to deal with just then, it seems more likely that he submitted it to Hahn – who had not seen it previously – in May or June. Gödel probably also saw the text of Hilbert's lecture at the Bologna Conference only after it appeared in the proceedings in 1929; he did not attend the conference himself, and that lecture probably played no role in his thesis as submitted.
2. '*Journey to the Edge of Reason: The Life of Kurt Gödel*'. Stephen Budiansky, W.W. Norton/Oxford Univ. Press/Ullstein, 2021. ISBN 978-0-19-886633-6. See Item 45 in Appendix B; [Budiansky (2021)].
3. See Menger (1994).

4. Carnap (1963).
5. Quoted in the memoirs of William Miller; see *Life* magazine, May 2nd, 1955.
6. From *The Feynman Lectures on Physics*, Vols. 1-2.
7. Feferman (1998).
8. Strauss (1982), p. 422.
9. See Wang (1987) and (1997); Dawson (1997, 2005 edition); Hintikka (1999; 2000 edition).
10. *Stanford Encyclopedia of Philosophy*, 'The Emergence of first-Order Logic', by William Ewald (2018), online.
11. Quoted from van Atten & Kennedy (2009), who in turn quote from the translation of van Heijenoort (1967), p. 48.
12. Quoted by Wang (1987), p. 271, from a letter written to him by Gödel in December 1967.
13. There is some confusion in the literature about the date of this paper. It was originally given at a conference in 1922, and was published in the conference proceedings in 1923. Here, it is referred to as Skolem (1923), cf. References.
14. Dawson 1997 (CRC edition, 2005), pp. 60 ff.
15. See https://en.wikipedia.org/wiki/List_of_logic_symbols
16. Enderton (1972).
17. Hintikka (1999); Item 16 in Appendix B.
18. The word '*Zählaussage*' is a *terminus technicus* apparently introduced by Löwenheim. It has no obvious English translation, and in the translation of Gödel's thesis in the *Collected Works* (Vol. I, 1986), it is simply left as the German term, yielding something of a *non-sequitur*. We translate it here as '*quantity statement*', a rough approximation but adequate to communicate the idea (in Ewald (2018), it is called a '*counting expression*'). In modern terms, it includes formulas, sentences, theorems which conform to the syntax of the language at hand; it may be considered to be a precursor to the modern term 'well-formed formula' (wff).
19. Gödel promises here, in a footnote, to give a more precise definition in a later section.
20. Photo of Hilbert: photographer unknown, taken before 1912, public domain. Reused from: https://commons.wikimedia.org/wiki/File:Hilbert.jpg. Photo of Russell: photographer unknown, public domain. Reused from: https://commons.wikimedia.org/wiki/File:Bertrand_Russell_transparent_bg.png.
21. Cf. Henkin (1949). Leon Henkin later wrote an account of how he arrived at his proof, originally part of his PhD thesis (Princeton 1947), and published in 1949: see Henkin (1996).
22. The following is quoted directly from the original article [Gödel (1930)]. Translation by the present author.
23. This is in fact '*König 's lemma*', published in 1927 by Dénes König. However, as he wrote much later to Hao Wang, Gödel was unaware of it when he wrote his thesis, and he rediscovered it on his own. Later, he thought it was due to

Dénes König's father, Dyula König, also a well-known mathematician. König calls the infinite branch a 'ray' (or 'simple path').

24. This was Dyula König, the father of Dénes König; the latter was in fact the author of the 1927 paper containing 'König's lemma'.

25. See backnote [13].

26. Quoted in Wang (1987), p. 271. The quotation given here is also included in Gödel's *Collected Works*, Vol. I, p. 51.

27. See the *Stanford Encyclopedia of Philosophy*, article on *Intuitionistic Logic*, by Joan Moschovakis, 2018.

28. Photo of Bernays: from WikiCommons, by Franz Schmelhaus, Zurich; public domain. Reused from: https://commons.wikimedia.org/wiki/File: (UAZ)_AB.1.0057_Bernays.tif. Photo of Brouwer: from Wiki Commons, public domain. Cropped and enhanced; reused from: https://commons.wik imedia.org/wiki/File:Stormer_Hille_Walsh_Giambelli_Fjeldstad_Gonseth_P olya_Hille_Mordell_Riesz_Fejer_Wilkosz_Stormer_Lovenskiold_Bohr_Brou wer_Zurich1932.tif.

29. Veblen attributes the terms '*categorical*' (and '*disjunctive*', for a system which can have more than one model, i.e. to which one can add new elements or axioms without creating a contradiction with the other axioms) to John Dewey, and dates the concept to Hilbert's lecture '*Über den Zahlbegriff*' (Hilbert 1900).

30. Photo of Post: from the *American Philosophical Society* archives, public domain. Used with permission. Reused from: https://diglib.amphilsoc.org/isl andora/object/graphics:3086. Photo of Henkin: from the *Oberwohlfach Photo Collection*, by George M. Bergmann. Reused from: https://commons.wikime dia.org/wiki/File:Henkin_leon_berkeley_1990.jpg. Used under a GNU Free Documentation License, Vers. 1.2, and permission from the *Mathematisches Forschungsinstitut Oberwolfach* (MFO); cf. https://opc.mfo.de/detail?photo_ id=5256.

31. Paul Bernays, Interviews with J.-P. Sydler and E. Clavadetscher, *Bernays Nachlaß*, WHS, ETH Zürich, T 1285 (1977). Quoted in Zach (1999).

32. See Stillwell (2004), p. 5. Stillwell has a copy of the postcard, from Gödel's legacy, provided by Martin Davis and John Dawson (see p. 14 of Stillwell's article).

8

The Mathematician in Vienna. *The Incompleteness Theorems*

By mid-October of 1929, Gödel had done all that he could to finish his doctorate and publish the results, and he could only wait for the degree to be formally conferred—which indeed it was, in early February of 1930. In the meantime, he had to consider how his career would proceed. It is clear that he intended to stay on in Vienna and become a docent, and eventually a professor, at his *alma mater*. Before 1929, he had probably not worried much about his income, having grown up in a well-to-do family where all his needs were taken care of. But after his father's death in early 1929, he perhaps began thinking more about the future. His mother was well provided for by his father's will, and his brother Rudolf already had a profession which promised at least a comfortable income. For the moment, there was still enough in Rudolf and Kurt's inheritance to cover their everyday needs, which were generous but not extravagant. But for the intermediate future, Kurt would have to continue his qualification process—aiming for the *Habilitation*, a kind of 'super-doctorate' degree which would permit him to apply for a teaching and research position at a university in Austria.

[A similar system was in place in many European countries in the twentieth century—in Germany, the *Habilitation* was also generally a precondition for obtaining a tenured university position, and it is only now, in the twenty-first century, being phased out in favor of qualification via a non-tenured teaching and research position, what is called 'assistant professor' in the

© The Author(s), under exclusive license to Springer Nature Switzerland AG 2022, corrected publication 2023
W. D. Brewer, *Kurt Gödel*, Springer Biographies,
https://doi.org/10.1007/978-3-031-11309-3_8

Anglo-Saxon countries, and '*Junior-Professor*' in modern-day Germany. Likewise, the *Doctor of Science* in England and the *docteur d'État* in France have been supplanted in the past decades by more modern procedures.]

In order to obtain his *Habilitation*, Kurt would have to carry out independent research beyond what he had done for his doctoral dissertation, and write it up as a thesis (*Habilitationsschrift*). The thesis would be judged by one or more experts in the field, who would submit written evaluations to the commission responsible for conferring the *Habilitation*. The formal procedure might also include a public defense of the thesis, as well as a second public lecture on an unrelated topic (to prove that the candidate had at least potential teaching ability, and could give a correct, coherent and understandable lecture on a topic outside his/her field of specialization). The first step was to find a suitable topic for the *Habilitation* research, one that would attract attention in the field and underscore the candidate's (Gödel's) originality and talent for research.

In October of 1929, the stock market crashed in New York, eventually plunging much of the world into an economic decline of enormous proportions—the great depression. The effects reached Austria only about two years later, and Kurt was most likely unaware of this development at the time (his brother Rudolf however probably noted it). Kurt seems to have been relatively immune to 'external' political, social and economic concerns at this point in his life [see Dawson (1997), 2005 edition, pp. 59 ff. on this subject].

Soon thereafter, in November, their mother Marianne moved from Brünn to Vienna, where she would live with her two sons in the coming 8 years. They moved out of their apartment at *Lange Gasse* 72, no doubt to have more space for their mother, into a large apartment on the *Josefstädter Str.* in the 8th District (cf. Figs. 5.8 and 5.9). The three of them remained there, living together in that apartment, until the autumn of 1937. Rudolf August Gödel's testament had been generous, and they had no immediate financial worries, as we have seen; most likely Kurt spent little time at that point worrying about his financial situation or future employment. After he had finished writing his doctoral thesis and the resulting publication, he must have devoted some time to family problems (his mother's move, finding a new apartment), and he also seemed to be carried along on a wave of enthusiasm for his research in mathematical logic. Indeed, in his recollections as told to Hao Wang, he makes no mention of the time between autumn 1929 and the summer of 1930, except to note that his doctorate was formally granted on February 6th, 1930.

We return to Kurt Gödel's memories as recorded by Hao Wang [Wang (1997), pp. 82–83]:

In the summer of 1930, I began to study the consistency problem of classical analysis. It is mysterious why Hilbert wanted to prove directly the consistency of analysis by finitary methods. I saw two distinguishable problems: to prove the consistency of number theory by finitary number theory, and to prove the consistency of analysis by number theory. By dividing the difficulties, each part can be overcome more easily. Since the domain of finitary number theory was not well defined, I began by tackling the second half: to prove the consistency of analysis relative to full number theory. It is easier to prove the relative consistency of analysis. Then one only has to prove by finitary methods the consistency of number theory. But for the former one has to assume number theory to be true (not just the consistency of a formal system for it).

However, the gap in his account can be filled in to some extent by consulting Rudolf Carnap's recollections, from his autobiography [Carnap (1963)]. During the period 1927–31, Carnap was perhaps the person closest to Kurt Gödel, at least with respect to Gödel's interests in fundamentals of mathematics and philosophy of science. Carnap relates[1] that:

> The first contact between the Vienna Circle and the Warsaw group was made when, at the invitation of the Mathematics Department, Alfred Tarski came to Vienna in February 1930, and gave several lectures, chiefly on metamathematics. We also discussed privately many problems in which we were both interested. Of special interest to me was his emphasis that certain concepts used in logical investigations, e.g., the consistency of axioms, the provability of a theorem in a deductive system, and the like, are to be expressed not in the language of the axioms (later called the object language), but in the metamathematical language (later called the metalanguage).

> Tarski gave a lecture in our Circle on the metamathematics of the propositional calculus. In the subsequent discussion the question was raised whether metamathematics was of value also for philosophy. I had gained the impression in my talks with Tarski that the formal theory of language was of great importance for the clarification of our philosophical problems. But Schlick and others were rather skeptical at this point. [...] My talks with Tarski were fruitful for my further studies of the problem of speaking about language, a problem which I had often discussed, especially with Gödel.

Alfred Tarski was a younger member (born January 14th, 1901) of a group of philosophers and logicians in Poland (University of Warsaw, *Lvóv-Warsaw group*). They were not very well known outside Poland, since they published almost exclusively in Polish. Tarski's visits to other European countries and his invitation of colleagues from other countries to Warsaw changed this situation to a considerable extent. Tarski himself was caught abroad, on a lecture tour in the USA, by the beginning of World War II on September 1st, 1939, when Poland was invaded by Nazi German forces from the west and by the Soviet Army (17 days later) from the east.[2] He remained in the U.S., and we met up with him in Chap. 7 in connection with a lecture that he delivered at Columbia University in 1940. He eventually obtained a professorship at UC Berkeley, where he spent the rest of his career. He died in 1983 in Berkeley. The names of Alfred Tarski and Kurt Gödel are often mentioned together, since they both made major contributions to mathematical logic and metamathematics. Tarski was later (1950s) one of the founders of *model theory*.

Kurt Gödel certainly attended the lectures given by Alfred Tarski in Vienna in early 1930; Tarski had been invited by the Mathematics Department (Menger), and he spoke initially at Menger's *Mathematical Colloquium*, in which Gödel was a regular participant. Tarski's later lectures were offered also to the Vienna Circle. It seems likely that hearing Tarski's lectures and speaking with him privately [a meeting arranged by Menger at Gödel's request; see Dawson (1997)[3]] had a similar catalytic effect on Gödel's thinking and possibly his choice of topics for his next work, which became his *Habilitationsschrift*, as had Brouwer's lectures two years before on the work for his doctoral thesis (compare Fig. 8.1).

Tarski's visit also had an energizing effect on Carnap, who later in 1930 visited Warsaw and met Tarski's older colleagues, *Stanislaw Lesniewski* and *Tadeusz Kotarbinski*. Carnap was particularly interested in the question of whether a statement made in a particular language could be verified (proven, validated, satisfied) within that *same* language. This question had been much discussed in the Vienna Circle, stimulated by Wittgenstein's view (expressed in his *Tractatus*) that it could *not*.

What Gödel demonstrated with his incompleteness results in 1930/31 was that it in fact *could*, within the 'language' of a formal logical system. He used the language of natural numbers to obtain conclusions about the logic of natural numbers—this procedure was at the core of his incompleteness proofs and was an important development, a tool for which Gödel would in later years become famous.

The procedure is a *mapping* of natural numbers onto statements about natural numbers, a kind of code in which every element of symbolic logic used in the axiomatic definition of arithmetic—whose objects are natural numbers—is itself represented by a number, and sequences of those symbols, terms, sentences and theorems about the natural numbers, are in turn represented as products of functions of the numbers for their individual symbols. The resulting (often rather large) 'Gödel numbers' are identifiable through

Fig. 8.1 Alfred Tarski and Kurt Gödel in Vienna; photographed in 1935, during Tarski's (second, longer) visit to the University of Vienna. Photo: by Maria Kokoszyńska-Lutmanowa, 1935. Reused from *calisphere* (Bancroft Library, UC Berkeley)[4]

their arithmetic properties and can be used to define relations and to prove or disprove the sentences and theorems they represent. Thus, the numbers play two roles: as 'themselves', the objects of number theory, the elements used in counting and simple arithmetic; and, on another level, as code symbols representing logical statements which can be used to manipulate and prove those statements. Some examples will be given below of how this 'coding' of logical statements works.

Hintikka (1999) compares Gödel's procedure to the different levels of language used by an *actor*: He has his own private life, where he uses language in the 'natural' way to communicate his personal thoughts, feelings and needs as a private human being; but when he is acting in a play, his language conveys the (fictitious) thoughts and feelings of the character he is playing, and those in turn reflect the thoughts of the playwright who wrote the lines for that character to speak. It thus represents a quite different level, although it makes use of the same words and phrases as his 'private' language.

In 'Gödel numbering' (also known as 'Gödelization', in more modern terms as 'arithmetization'),[5] the numbers have their 'everyday lives' as the individual elements of simple arithmetic, but also their 'metanumber lives' as code names for terms and sentences in a formal-logical description of arithmetic, in turn based on the natural numbers. This trick is the genial core of Gödel's incompleteness proofs. The basic idea—of using a mapping to generate a code representing words and sentences on quite a different level of language—was not new or original with Gödel; but the way he used it to obtain results in mathematical logic—to speak about properties of a language from *within* that language itself—*was* new, and it is at the heart of his 'claim to fame' as a metamathematician and logician.

Many years later, Gödel wrote a letter in reply to an inquiry from a graduate student, Yossef Balas, sent to him in May of 1970. Characteristically, he composed but never mailed his reply to Balas's inquiry; it was found among his papers after his death, and quoted by Wang[6] and by Dawson.[7] The original, unsent reply can be found in the archives of Gödel's legacy [GN 010,015.37, folder 01/20; see endnote 137 in Dawson (1997)]. An important aspect of Gödel's treatment was his distinction between *provability* and *truth*. He held that formal-logical arguments cannot determine *truth*, and he avoided speaking about the latter in his published work. In his letter to Balas, he mentions that at the time when he obtained his completeness and incompleteness results, "... *a concept of objective mathematical truth... was viewed with great suspicion and widely rejected as meaningless*". Carnap confirms this in his (1963), pp. 59/60:

Even before the publication of Tarski's article[8] I had realized, chiefly in conversations with Tarski and Gödel, that there must be a mode, different from the syntactical one, in which to speak about language. Since it is obviously admissible to speak about facts and, on the other hand, Wittgenstein notwithstanding, about expressions of a language, it cannot be inadmissible to do both in the same metalanguage. In this way it becomes possible to speak about the relations between language and facts. *[...]*

When I met Tarski again in Vienna in the spring of 1935, I urged him to deliver a paper on semantics and on his definition of truth at the International Congress for Scientific Philosophy to be held in Paris in September. I told him that all those interested in scientific philosophy and the analysis of language would welcome this new instrument with enthusiasm, and would be eager to apply it in their own philosophical work. But Tarski was very skeptical. He thought that most philosophers, even those working in modern logic, would be not only indifferent, but hostile to the explication of the concept of truth. I promised to emphasize the importance of semantics in my paper and in the discussion at the Congress, and he agreed to present the suggested paper.

At the Congress it became clear from the reactions to the papers delivered by Tarski ... and myself ... that Tarski's skeptical predictions had been right. To my surprise, there was vehement opposition even on the side of our philosophical friends.

Gödel was well aware of the need to differentiate between the syntactical and the semantic properties of a logical system, already from his thesis work. This was underscored by the later work of Tarski, as mentioned by Carnap in the above quote. In particular, the notion of *completeness* (and of *incompleteness*—which is *not* simply the converse of completeness) is different in the syntactical (proof theoretical) sense and in the semantic (model theoretical) sense. Gödel in fact used different terminology for those two concepts: he called completeness in the semantic sense '*Vollständigkeit*', which is directly translated by the English word 'completeness'; but completeness in the syntactic sense he termed '*entscheidungsdefinit*', which Dawson[9] translates as 'decisive'.[10]

By the beginning of summer 1930, Gödel had evidently settled on his topic and begun work. As he states in the above quote [Wang (1997)], that topic was "*the consistency problem of classical analysis*", dating from Hilbert's list of open problems (from his 1900 lecture), and also mentioned in Hilbert and Ackermann (1928). This was a logical (in both senses) continuation of his thesis work on the completeness of the predicate calculus. What he means by '*classical analysis*' is in fact second-order arithmetic, making use of variables which represent *sets of natural numbers*.

Gödel's unsent letter to Balas[11] tells the story of his plan (in modern language, since he wrote it in 1970 or later, 40 years after he set out to solve Hilbert's second problem using the logic of Russell and Whitehead's *Principia Mathematica*). Gödel writes,

The occasion for comparing truth and demonstrability was an attempt to give a relative model-theoretic consistency proof of analysis in arithmetic. That leads almost by necessity to such a comparison. For an arithmetical model of analysis is nothing else but an arithmetical $\in$-relation satisfying the comprehension axiom:

$$(\exists n)\,(x)\,[x \in n \equiv \varphi(x)]. \tag{8.1}$$

Now, if in the latter, 'φ (x)' is replaced by 'φ (x) is provable', such an $\in$-relation can easily be defined. Hence, if truth were equivalent to provability, we would have reached our goal. However (and this is the decisive point), it follows from the correct solution of the semantic paradoxes that 'truth' of the propositions of a language cannot be expressed in the same language, while provability (being an arithmetical relation) can. Hence true $\neq$ provable.

[Note that Gödel uses an abbreviated notation for the universal quantifier: (x) $\equiv$ ($\forall x$).]

Here, Gödel emphasizes the distinction between *truth* and *provability* (or '*demonstrability*'), important for his proofs of the incompleteness theorems. He wants to use first-order logic, treating sets of natural numbers as countable definable subsets which can be referenced by an index n (as in Formula (8.1)).

He in fact set out to prove a property which belonged to Hilbert's program of putting number theory (initially arithmetic) on a formal-logical basis, using the methods of *Principia Mathematica* (PM). In the end, he arrived at a different conclusion, which effectively derailed Hilbert's program as well as PM. Indeed, as we have seen, he had already considered the *possibility* that a formal logical system might give rise to sentences which were neither provable nor disprovable within that system, although they might be 'true' and even provable within another, more powerful system—that is, such statements would not be *decidable* within the original system. (See the first paragraph of his 1930 article on the completeness of the axiomatic system of the predicate calculus, quoted in Chap. 7).

Gödel's most important paper, his (1931), in which he demonstrated his proof of the First Incompleteness Theorem and gave a brief summary of the Second, was entitled '*Über formal unentscheidbare Sätze der Principia Mathematica und verwandter Systeme* I' ('*On Formally Undecidable Theorems in Principia Mathematica and Related Systems* I'), and it appeared in the *Monatshefte für Mathematik und Physik*, in its first issue of 1931. The Roman number 'I' in its title suggested that it would be the first in a series of papers, but Gödel later decided that a further paper (giving for example more details about the Second Incompleteness Theorem) would not be necessary.

About this paper [Gödel (1931)], Nagel & Newman (1958), p. 52, write, "*Gödel's paper is difficult. Forty-six preliminary definitions, together with several important preliminary propositions, must be mastered before the main results are reached ...*". We will summarize the principal conclusions and methods of that paper in the following, but those readers who want to gain a deeper insight into Gödel's proofs, as well as into the history and reception of the paper, should consult the more detailed books which deal specifically with this seminal article: *Nagel & Newman* (1958), re-issued (2001), Item

1 in Appendix B; the excellent, detailed and readable book by *Peter Smith* (2007), re-issued (2020), Item 32 in Appendix B; see also '*Gödel without (Too Many) Tears*', Peter Smith (2021), Item 49 in Appendix B; '*Gödel's Incompleteness Theorems*' by R.M. Smullyan (1992), Item 11 in Appendix B); and Torkel Franzén's '*Gödel's Theorem. An Incomplete Guide to its Use and Abuse*' (2005), Item 26 in Appendix B. Especially this last book offers a great deal of information on the reception and interpretation of the incompleteness theorems by mathematicians, philosophers, computer scientists and many other members of the well-informed (and less-well-informed) public.

Hilbert's Second Problem

As we have already seen, David Hilbert put forth a list of unsolved problems in mathematics at the Paris Mathematicians' Congress (ICM) in 1900, which became a program for twentieth century mathematics. Hilbert was of the opinion that there are no *unsolvable* problems in mathematics—which he called *non ignorabimus*.[12] (However, he did admit that some problems as formulated may have no solution if they are too imprecisely defined). Originally, he had proposed 24 problems, 10 of which he described in his lecture given at the Congress. In the published proceedings, he included 23 problems (having decided to leave out the 24th, a problem of finding criteria for simplicity and generality in proof theory).

Kurt Gödel tackled the 1st problem—the *continuum hypothesis*—in the later 1930s and early 1940s, and we will deal with it in later chapters. A further decisive step regarding this topic was made by *Paul Cohen* in 1963. Its status as a 'Hilbert Problem' is however still considered to be 'undecided'.[13]

Hilbert's 2nd problem—finding a 'finitary' proof that the axioms of arithmetic are consistent—was the subject of Kurt Gödel's *Habilitationsschrift*, submitted in 1932, where he showed that the answer was *negative*, i.e., no such proof is possible within the axiomatic system—but this holds only for certain specific formal systems. (*Finitary* here means that the proof consists of a finite number of lines and deals with a system having a finite number of axioms—i.e., that it in principle could be comprehended *in toto* by a human or carried out by a machine. This condition is also called *finitistic* by some authors; that appellation, strictly speaking, refers however to the philosophical school rather than to individual objects.)

Gerhard Gentzen, a student of Paul Bernays and assistant to Hilbert, found a proof of the consistency of the Peano axioms for arithmetic in 1936, but later showed that his proof could not be carried out *within* the system of the Peano axioms, confirming Gödel's incompleteness theorem.

Nagel and Newman, in their book '*Gödel's Proof*',[15] discuss the possibility of solving Hilbert's second problem:

> The sentential calculus *[a.k.a. propositional calculus]* is an example of a mathematical system for which the objectives of Hilbert's theory of proof are fully realized. To be sure, this calculus codifies only a fragment of formal logic, and its vocabulary and formal apparatus do not suffice to develop even elementary arithmetic. Hilbert's program, however, is not so limited. It can be carried out successfully for more inclusive systems, which can be shown by meta-mathematical reasoning to be both consistent and complete. By way of example, an absolute proof of consistency is available for a formal system in which axioms for addition but not multiplication are given. But is Hilbert's finitistic method powerful enough to prove the consistency of a system such as *Principia*, whose vocabulary and logical apparatus are adequate to express the whole of number theory and not merely a fragment?

The answer, found by Gödel and published in his (1931), is 'no'. "*Gödel showed that* Principia, *or any other* [consistent] *system within which arithmetic can be developed, is essentially incomplete.*"[15] He showed that, in fact, even extending the logic of *Principia* by adding new axioms would not solve the problem, since then there would be new unprovable statements derivable in the extended system.

The Königsberg Conference

By the late summer of 1930, Gödel (Fig. 8.2) had found his answer to Hilbert's second problem, although he had not yet fully formulated its proof. In September of that year, a conference was planned for the easternmost city in Germany, *Königsberg* in East Prussia, to be sponsored by the *Gesellschaft für empirische Philosophie* (the 'Society for Empirical Philosophy'), which in turn had been organized by the Berlin Circle (chaired by Hans Reichenbach) and by the *Verein Ernst Mach* in Vienna, with strong participation by Rudolf Carnap. Reichenbach and Carnap had acquired the journal '*Annalen der Philosophie und der philosophischen Kritik* ' ('*Annals of Philosophy and Philosophical Criticism*') earlier in 1930 and renamed it *Erkenntnis*. That journal also co-sponsored the Königsberg conference and published its proceedings.

The meeting itself was called 'the *Second Conference on the Epistemology of the Exact Sciences*', representing what were later termed *logical positivism*

Fig. 8.2   Kurt Gödel, Vienna 1935. Photo: picture-alliance.de, used with permission[14]

(Vienna Circle) and *logical empiricism* (Berlin Circle), and its local organizer was Kurt Reidemeister, of the University of Königsberg. It was a 'satellite conference' to the 91st annual meeting of the '*Gesellschaft deutscher Naturforscher und Ärzte*' (GDNÄ) and the '*Deutsche Physiker- und Mathematikertagung*', major conferences which were also held in Königsberg that year. The GDNÄ, together with the DMV (*Deutsche Mathematiker Vereinigung*; German Mathematicians' Union) and the DPG (*Deutsche Physikalische Gesellschaft*; German Physical Society) held joint annual conferences until (and including) 1931.

Sessions at the smaller conference were planned on the 'Foundation of Mathematics' (where the four major directions *logicism* (Russell et al., presented by Carnap), *intuitionism* (Brouwer et al., presented by Heyting), *formalism* (Hilbert et al., presented by von Neumann) and the *linguistic school* (Wittgenstein, presented by Waismann) were the subjects of the main session); and on 'Philosophical Questions arising from Quantum Mechanics', with presentations by Reichenbach and Heisenberg, among others. Gödel, along with several others from Vienna, planned to attend, and he would present his Completeness Theorem, the result of his doctoral thesis. But he also mentioned his new results on incompleteness, however only "*as an offhand remark during a general discussion on the last day*".

In preparation for the conference, Gödel met at least twice with Carnap, at one of their favorite haunts, the *Café Reichsrat*, near the end of August 1930 [see Dawson (1997, 2005 edition), p. 69]. Feigl and Waismann were also present during part of the meetings. They met mainly to discuss their participation in the Königsberg conference and the arrangements for their travel there, but Gödel, according to the recollections of Carnap, also mentioned (*for the first time*) his unfinished results on incompleteness. Carnap was apparently unimpressed at the time and only later realized the importance of what Gödel had accomplished.[16]

At the conference itself, Gödel gave a formal paper on his completeness results, still not published [his (1930) appeared a few weeks after the conference]; but during an informal discussion on the last day of the conference (September 7th, 1930), he briefly presented his new work.

Verena Huber-Dyson[17] quotes from Gödel's *Collected Works*, Vol. I, on his short announcement of the new results on incompleteness. She gives a witty and dramatic, if somewhat ironic reconstruction of the scene on the last day of the Königsberg conference, expressing her opinion that Gödel was in fact considerably more sophisticated than generally believed, and suggesting that he was 'playing his audience' to make a dramatic announcement of his results. After his first assertion that no consistent formal-logical system can express all the content of mathematical thought, there is a shocked silence of incomprehension. The silence is broken by von Neumann, who chimes in with his assenting remarks—again followed by baffled silence. Gödel caps off his brief presentation by stating that there are true sentences which cannot be proven by consistent systems of formal logic, and asserting that adding new axioms to such systems will still not make them complete.

This rendering may somewhat over-dramatize the occasion. Indeed, the only person present who seems to have immediately recognized its significance was *Johann* (John) *von Neumann*, who represented the Hilbert group at the conference. Von Neumann was noted for his perspicacity (he was later among the Hungarian *emigrés* in the USA, also including *Eugene Wigner*, *Leó Szilárd*, *Cornelius Lanzcos*, and *Edward Teller*, who were called 'the Martians' by their American colleagues because of their other-worldly intelligence). He met with Gödel privately after the discussion session, and he was very interested in Gödel's new results. He asked Gödel about demonstrating an undecidable proposition for natural numbers, and Gödel thought that possible, but only with more complicated operations than addition or multiplication. He soon discovered, however, that he could find such a proposition for polynomials with quantifiers, and in their later correspondence, he reported this to von Neumann.

Von Neumann went back to Berlin after the conferences, and apparently did not inform Hilbert of Gödel's results immediately; instead, he started work on the same topic himself. In November, he wrote to Gödel about his own (preliminary) results, but Gödel had in the meantime finished formulating his proofs and had submitted both a summary (to the Vienna *Academy of Sciences*), and the full paper (Gödel 1931) to the *Monatshefte* (a few days before von Neumann's letter arrived; it was received by the journal on November 17th, 1930).

He sent a preprint to von Neumann, who took the fact that Gödel had 'gotten there first' in good grace and left the field to Gödel. He in fact became a friend and supporter of Gödel in the following years. Since 1929, von Neumann had been regularly visiting Princeton University (at the invitation of Oswald Veblen, probably with a view to obtaining a position there), and it was he who introduced the logicians in Princeton (in his lecture in the autumn of 1931) to Gödel's work. In 1933, von Neumann moved permanently to the newly-founded *Institute for Advanced Study* (IAS) (as did Veblen), and his positive recommendations were probably very effective in motivating the IAS to invite Gödel to visit there several times in the 1930's, as well as encouraging his permanent move there in 1940.

Hilbert in fact was present at the large joint conference which followed in Königsberg, and he gave a plenary lecture, in which he reiterated his view that no problem in mathematics would be unsolvable. Whether Gödel heard his lecture is uncertain; Gödel later stated that he had never met Hilbert, although he stayed in Königsberg for the opening of the following conferences.[18] Hilbert's reception of Gödel's proofs was recorded by Bernays,[19] who later corresponded with Gödel and received preprints and offprints of his articles. We will return to this topic later in this chapter and in Chap. 9.

This period in Gödel's career is termed the '*Moment of Impact*' by Dawson (1997), and Gödel's presentation at the Königsberg conference is called a '*bombshell*' by Hintikka (1999)—but in fact it was for the mathematical world at the time only a very small footnote. It took several years for Gödel's work to be completely appreciated and understood; for some mathematicians and philosophers, it took *many* years. Some apparently lived out their lives without really understanding it. Sadly, among those latter we must include Russell and Wittgenstein.[20]

Gödel's Proof

How did Gödel accomplish his proof? As we have indicated above, he used a 'mapping' from the *meta*-language (which speaks about the formal logic ('behind' or 'beyond') arithmetic) onto arithmetic itself, and then employed ordinary arithmetical operations to show that the system necessarily contains *undecidable* sentences: they can neither be proven, nor can their converses be proven (thus disproving the original sentence); they are thus *not decidable* (within that formal system). This means that the formal system is *incomplete*; and that was Gödel's **First Incompleteness Theorem**. He was (somewhat later) able to use the same mapping to show that a statement of the form '*this system is consistent*' could not be established *within* the formal system, i.e., consistency cannot be proven within a consistent system; this is his **Second Incompleteness Theorem**. In a modern formulation, what the First Incompleteness Theorem says is, "*For any consistent formal system capable of proving basic arithmetical truths, it is possible to construct an arithmetical statement that is true but not provable within the system*".[21] The Second Incompleteness Theorem correspondingly states, "*For any (consistent) formal system capable of proving basic arithmetical truths, the statement of its own consistency can be deduced (within the system) if and only if it is (in fact)* **in**consistent." In the next section, we examine Gödel's mapping in more detail.

Gödel Numbering

Gottfried Wilhelm Leibniz had already suggested using numbers to code concepts—or 'ideas'—employing prime numbers for primitive ideas and products of primes for complex ideas. Gödel had of course read of Leibniz's suggestion, which was never successfully put into practice, and perhaps intended to improve upon it. It is also interesting to speculate on the possible connection between Gödel's idea of using a mapping of logical symbols onto natural numbers as a code to permit logical manipulations and conclusions about arithmetic by using arithmetic itself—and his early experience with *shorthand* writing, a technique in which he acquired proficiency during his high-school years (the *Gabelsberger Schnellschrift*), and which he used throughout much of his life for taking notes and writing texts.

Shorthand is also simply a *mapping*—of invented symbols, chosen for ease of writing—onto letters, letter combinations, and often-used words in the natural language. Its goal is merely to make writing faster and more compact by replacing the many letters in typical written sentences with fewer symbols which are quickly written; and these are related in an unambiguous way to the original letters and words, so that the original sentences can be clearly and securely reconstructed from the shorthand symbols later. Of course, shorthand does not make use of a language to speak *about itself*; the shorthand symbols are a new, invented symbolic language, and represent the natural language without being a part of it. And they are generally not used to investigate *properties* of the natural language which they represent (although they *could* be, e.g., in terms of statistical analysis of word frequencies, etc.).

An interesting aspect of Gödel numbering, in contrast, is its *self-referential* quality: statements about arithmetic, and therefore about natural numbers, are represented by natural numbers, and these are manipulated using the procedures of arithmetic.

– *But now it is time for us to look at some* examples:

Gödel's first step was to designate the symbols (called the *constants* in his article of 1931) which were to be used in his formal system, and assign code numbers to them. In his (1931), he lists the *Grundzeichen* ('basic symbols' or 'elementary signs') that he will use in his proofs (on his p. 176; compare our Table 7.1).

Gödel names 7 constants: ' ~ ' ('not'; ¬ in Table 7.1), '∨' ('or'), '[ ]' ('for all'; ∀ in Table 7.1), '0' (zero), '(' and ')' (curved brackets), and 'f' (the *successor* function, which gives the next natural number following the number which is its argument; often denoted by 'S' or 's' in English texts. For example: s0 $= 1$, ss0 $= 2$ etc.). He denotes *individuals* (natural numbers 1,2,3... in this case) by x_1, y_1, z_1 ... and calls them 'variables of the first type'; variables of the second type (classes or sets of natural numbers) are denoted by x_2, y_2, z_2 ..., etc.

Nagel and Newman (1958) use a somewhat longer list of 12 elementary signs and three variables in order to exemplify how *Gödel numbering* works [cf. their Table 2 on p. 53 of their (1958)]. In addition to the signs defined by Gödel, they include ' ⊃ ' ('if... then...'), '∃' ('there exists'), ' = ' (equals), ',' (comma, punctuation), ' + ' (plus), and ' × ' (times), leaving off the universal quantifier '∀' included in Gödel's list as (x) or $[x]$. Each of these is assigned a 'Gödel number' as its code. Table 8.1 illustrates some examples of possible codes.

Peter Smith, in his very complete and clear book, '*An Introduction to Gödel's Theorems*'[22] (2007/2020), gives a somewhat longer list; in addition to the above, he includes '∧ ' ('and'), as well as '∀' [the universal quantifier, implicit in Table 8.1 as (x)]. In his scheme, all the basic symbols (*constants*) have Gödel numbers 1–27 which are *odd* integers, while the *variables* are denoted by *even* integers. Any of these sets of basic symbols and variables could be used to illustrate the principle of Gödel numbering, and we will make use of those in Table 8.1.

Table 8.1 Symbols for formal arithmetic logic (after Nagel & Newman 1958)

Elementary sign	Gödel number	Meaning
~ or ¬	1	Not
∨	2	Or
⊃	3	If... then...
∃	4	There is ...
=	5	Equals
0	6	Zero
s	7	Successor
(	8	Left bracket
)	9	Right bracket
,	10	Punctuation
+	11	Plus
×	12	Times
x	13	Individual variable
y	17	" "
z	19	" "

It should be mentioned at this point that tables like 7.1 and 8.1 are in some sense contradictory to the program of Russell and Whitehead in PM. Their symbols are supposed to be '*empty*', i.e., they have only a logical significance, but no 'meaning' in terms of the natural language. This point was also emphasized by Gödel at the beginning of the article on his thesis work (1930); it is an essential part of what is meant by 'formal' logic. On the other hand, in Tables 7.1 and 8.1 we assign *meanings* to the logical symbols.

This contradiction is discussed at some length by Nagel & Newman (1958), 2001 edition, pp. 53–55. Gödel countered it with the '*correspondence lemma*' (modern terminology), which (in modern terms) states[23] that, "*Every primitive recursive truth, when represented as a string of symbols, is a theorem of PM*". This important result is often glossed over in summaries of Gödel's proofs, and we will not try to show its proof here. It refers to 'primitive recursive truths' (PRT), which are sentences or theorems within a formal logical system (here, formal axiomatic arithmetic) that include projections (identities that pick a certain individual out of a list), and functions that are constructed by *composition of functions* (inserting the result of a function as the argument of another function). A simple example is the successor function: $s0 = 1$ is a PRT, and if inserted again as argument into the successor function, it gives $s(s0) = 2$ —i.e., the natural number '2' is obtained *recursively*—and this yields another PRT. The term 'primitive' here means that these statements are intuitively apparent, and they thus require no formal derivation or proof. In this sense, axioms and rules of inference are also *primitive* (a terminology due to Russell). They must be sought through mathematical intuition.

The logic of PM contains infinitely many PRT's about natural numbers (e.g., the multiplication table for all natural numbers). The correspondence lemma guarantees that for every formula in PM, there exists a unique Gödel number; and for every proof in PM, there is also a unique Gödel number; and, most importantly, the Gödel-number function *and its inverse* are computable. That is, the code represented by Gödel numbers is uniquely defined and reversible—the *meanings* of Gödel numbers can be read out from the numbers themselves.

We can use a simple symbolic-logical statement about arithmetic as an example of the application of Gödelization (following Smith (2020)):

$$\exists y(s0 + y) = ss0 \qquad (8.2)$$

which means, in words, 'there exists a number y such that when added to the successor of 0, it yields the successor to the successor of 0'. In other words, $1 + y = 2$, and $y = 1$. To 'Gödelize' this statement, we use the code numbers from Table 8.1:

$$4, 17, 8, 7, 6, 11, 17, 9, 5, 7, 7, 6.$$

Now, Gödel's next trick is to use these code numbers as powers of the series of the smallest prime numbers, in order $(2, 3, 5, 7, 11, 13, \ldots)$, and then to take their products to give the overall Gödel number; that is, we write the sequence

$$2^4 \cdot 3^{17} \cdot 5^8 \cdot 7^7 \cdot 11^6 \cdot 13^{11} \cdot 17^{17} \cdot 19^9 \cdot 23^5 \cdot 29^7 \cdot 31^7 \cdot 37^6 \qquad (8.3)$$

and then carry out the exponentiation and multiplication to obtain the Gödel number of the sentence (8.2). We can call it j; as can be seen, it will be a *very large* number, but still an integer. In order to *decode* the number j, reversing the above encoding process, we need to find the exponents of its prime factors. This will *always be possible*, according to a result of number theory (known as the *fundamental theorem of arithmetic*, or also the *unique factorization theorem*). It was given in Euclid's *Elements* and proved there.

The fundamental theorem states that any composite number (i.e., any number not itself a prime) can be uniquely decomposed into a product of primes (raised to given powers). Since this decomposition is unique, it gives a one-to-one correspondence of any Gödel number to a series of primes as in (8.3), and their powers indicate the basic symbols and variables in the corresponding sentence. It is quite easy to see how the first 74 composite numbers between 1 and 100 can be decomposed into products of primes raised to various powers, and the fundamental theorem guarantees that this process can be continued to arbitrarily large composite integers.

The Gödel numbers can be examined and manipulated by using ordinary arithmetic: for example, we could divide the Gödel number j given above, representing the formula (8.2), by 11^n, where n is an integer ($n = 1,2,3,\ldots$). 11 is of course the fifth small prime number, and thus represents the *fifth symbol* in the formula (8.2). We will find that $j/11^n$ divides evenly (with no remainder) for $n = 1, 2, \ldots, 6$, but for $n \geq 7$, it does *not* divide j evenly. We can thus conclude on the basis of this simple arithmetic operation that the code number for the fifth symbol in the formula is 6, i.e., the symbol that it represents is 0.

Evidently, we also need rules for forming the Gödel numbers of more complex elements, such as given *whole sentences* ('*sentential variables*'); for *sequences of sentences*; and for *predicate variables* (such as the symbolic-logic representations of statements like 'x is the successor of y' or 'z is a prime number'). These could be added to Table 8.1, but we simply give them here:

Sentential variables p, q, r (each one representing a whole sentence like (8.2)) are given as Gödel numbers by the primes greater than 12, raised to the power 2:

$$p \text{ has Gödel number } 13^2; \quad q \text{ has Gödel number } 17^2;$$
$$r \text{ has Gödel number } 19^2 \tag{8.4}$$

A *sequence of sentences* can be Gödelized by taking the Gödel numbers of the individual sentences and using them as powers of the early primes, taking their product to be the (new, larger) Gödel number for the whole sequence— if the first sentence in the sequence has the Gödel number j, and the second sentence has the Gödel number k, we can form the Gödel number m for the whole (two-line) sequence by setting:

$$m = 2^j \cdot 3^k. \tag{8.5}$$

Predicate variables P, Q, R ... are assigned Gödel numbers by taking the primes greater than 12 to the *third power*, analogous to individual variables and sentential variables (powers 1 and 2, respectively):

$$P \text{ has Gödel number } 13^3; \quad Q \text{ has Gödel number } 17^3;$$
$$R \text{ has Gödel number } 19^3 \tag{8.6}$$

With these rules, we can encode, i.e., *Gödelize, all* of the formal calculus. And, given the fundamental theorem of arithmetic, it is guaranteed that we can reverse the process and rediscover the original sentences, symbols and variables represented by a given Gödel number—that process might be tedious, but it is certain to be possible and unique; and it can be automated, that is for example performed automatically by a computer algorithm, using what Gödel, in his 1934 Princeton lectures, called β *functions*. Compare the examples in Smith (2020) and Raatikainen (2020).

Furthermore, it is simple to determine from the outset whether a given integer is a Gödel number or not: if the number is 12 or less, it is the Gödel number of one of the basic symbols (Table 8.1). Numbers greater than 12 can be decomposed into their prime factors as shown by the fundamental theorem. Numbers greater than 12 which are primes (or second or third powers of primes) are the Gödel numbers of individual variables, or of sentential variables, or of predicate variables, as defined above. If the number is a product of primes raised to certain powers, it is the Gödel number of a sentence or a sequence of sentences, which can in every case be reconstructed.

Numerical examples are given in many books, among others in Nagel & Newman (1958), and in Smith (2020). Here, we show an example from Nagel & Newman's book[24]: consider the number 243,000,000. It is moderately large, but not enormous; let us call it g. Is it a Gödel number, and if so, which statement does it encode? We can begin by noting that $1,000,000 = 10^6 = (2 \cdot 5)^6 = 2^6 \cdot 5^6$. So, a large part of the number g can be expressed as the product of small primes raised to certain powers. The remaining factor is 243, which is readily seen to be 3^5. Thus, our number g can be factorized to give $2^6 \cdot 3^5 \cdot 5^6$. We see that it *is* a Gödel number, and that it encodes a statement with three elements, whose symbols have the Gödel numbers 6, 5, 6. A quick look at Table 8.1 shows that this statement is '$0 = 0$'. It is a tautology, trivial but arithmetically correct, and is certainly a true statement about arithmetic in PM. So, we can see how the Gödel numbers for complicated statements, even for the series of statements which make up a proof, can be encoded by a single Gödel number.

We note that the above procedures for generating Gödel numbers are not the only possible ones; various others are known, but the above are similar to those used by Gödel in his (1931).

And finally, it is worthwhile to mention two typographical conventions which are often found in the modern literature[25]: for one, there is a standard notation for Gödel numbers, which consists of placing the symbol for the element that they represent between 'modified quotation marks' (often called 'corners'), ⌜ and ⌝, for example ⌜φ⌝. This stands for 'the Gödel number of the expression φ.' Secondly, it is conventional to overline (or underline) a number to indicate that the *numeral* is meant, i.e., the symbol for a number, and not a variable of the formal system. Numerals are formally defined by the successor function, e.g. $\underline{3} = \mathrm{sss}(0)$. This need only be done when a possible confusion of the symbols is to be avoided[22].

Using the Gödel Numbers

Having developed a code which was unique and reversible for encoding symbolic logic statements as natural numbers, Gödel set out to make use of it for drawing conclusions about the structure and limitations of formal-logical systems, such as PM or the Zermelo-Fraenkel (ZF) axiomatic set theory[26]; or the simpler Peano arithmetic (PA). The latter, combined with the logic of PM, forms what Gödel referred to as the 'System P', which he used to exemplify his proofs.

We first quote the introductory paragraph from his (1931) paper, to show in his own words what his goals were (English translation by the present author, with permission from its publisher, *Springer-Verlag*):

On formally undecidable Theorems in Principia Mathematica and Related Systems I[1]

By **Kurt Gödel**, in Vienna.

1.

The development of mathematics in the direction of increasing exactitude has, as is well known, led to the formalization of major areas, in such a way that proofs can be carried out by applying only a few mechanical rules. The most comprehensive of the currently-established formal systems are that of the Principia Mathematica (PM)[2] on the one hand, and the Zermelo-Fraenkel system of axioms for set theory (further extended by J. v. Neumann)[3] on the other. These two systems are so broad that all the currently-known methods of [mathematical] proofs are formalized within them; i.e. the proofs can be based on a small number of axioms and rules of inference. It would thus be reasonable to assume that these axioms and rules of inference suffice to decide all of those mathematical questions that can be formally stated within the corresponding systems. In the following, it will be shown that this is not the case; rather, in the two systems mentioned, there are relatively simple problems from the theory of natural numbers[4] whose validity cannot be decided from the axioms. This situation is due not to the special nature of the systems considered, but rather it holds for a very broad class of formal systems, to which in particular all those belong that can be generated from the two named by extending them through the addition of a finite number of axioms, presuming that no false theorems of the kind mentioned in footnote 4 become provable through the added axioms.

[Footnote: (1) See also the Notices of the Viennese Academy of Sciences (Mathematical-Scientific Class) 1930, No. 19, for the summary of the results of this work which was published there.]

[Footnote: (2) A. Whitehead and B. Russell, *Principia Mathematica*, 2nd edition, Cambridge 1925. Among the axioms of PM, we count in particular also: the axiom of infinity (in the form: there are exactly countably many individuals), the reducibility axiom, and the axiom of choice (for all types).]

[Footnote: (3) See A. Fraenkel, Ten Lectures on the Fundamentals of Set Theory, *Wissenschaft und Hypothese*, Vol. XXXI; J. von Neumann, The Axiomization of Set Theory, *Mathematische Zeitschrift* 27, (1928); *Journ. f. reine u. angew. Math.* 154 (1925), 160 (1929). We note that in addition to the set-theoretical axioms given in the literature quoted, we must include the axioms and rules of inference of the logical calculus, in order to complete the formalization.—The following considerations hold also for the formal systems considered in recent years by D. Hilbert and his coworkers (insofar as these have been published). See D. Hilbert, *Math. Ann.* 90; J. v. Neumann, *Math. Zeitschrift* 26 (1927); W. Ackermann, *Math. Ann.* 93.]

[Footnote: (4) That is, more precisely, there are undecidable theorems in which, besides the logical constants ~ (not), v (or), (x) (for all x), and = (identical to), there are no other concepts except + (addition) and · (multiplication), both referring to natural numbers, whereby also prefixes (x) may refer only to natural numbers.]

From this we can see that Gödel set out to prove the completeness of formal systems like PM or the Zermelo-Fraenkel (ZF) axiomatic set theory, or any formal logical system which includes a certain amount of elementary arithmetic; and with his genial system of coding, he in the end proved that they were *in*complete, in contrast to his own initial expectations (but that result was probably not a complete surprise to him).

– *How did he do this?*

Gödel defines a function in a formal-logical system (PM logic, or the more limited systems of Peano arithmetic, PA, or, since 1950, Robinson arithmetic, Q; or Gödel's System P, a combination of PM logic and PA) for which that system is *consistent* if and only if the function is *unprovable* within it. This procedure makes use of *self-reference*. A function can include its own Gödel number, and therefore make statements about its own status within P (or some other formal system). Note that a statement which asserts (figuratively speaking), "I am not *true* within the XY logical system" would represent a paradox, analogous to the statement, "This sentence is not true." But as Gödel recognized early on, *provability* (or *decidability*) is not equivalent to *truth*. A function which states, "I am not *provable* within the XY logical system" avoids this paradox and nevertheless carries a significant message. Also, note that the deeper meaning of the notion of *truth* is controversial, or at least complicated, within philosophy. In practical mathematics, however, it has a simple meaning: "The sentence p is true" means simply that the corresponding arithmetical statement, such as '$2 + 3 = 5$', is in fact a correct result of arithmetic. The 'truth' of such statements has no deeper significance, no metamathematical (or metaphysical or transcendental) meaning, in practice.

– *Next, we continue with an informal sketch*[27] of Gödel's actual proof:

Consider a function G which depends on one *free* variable y (not limited by some quantifier): $G(y)$. Its Gödel number (using the notation defined above) is $\ulcorner G \urcorner$. Now, substitute that Gödel number for the free variable y (using the substitution rule of inference). This makes the function *self-referential*, and the process is called 'diagonalization'. That refers to *Gödel's diagonal lemma*,[28] which in Gödel's application states that "*for any function* $G(y)$ *of a single free variable* y *in elementary number theory* (e.g., in P), *there is a unique number* $n = \ulcorner G \urcorner$ *which is the Gödel number of* $G(n)$" (n is here the *numeral*, i.e., it *symbolizes* the number n). That number is the Gödel-number code for the symbolic-logical content of G. Another formulation states that "*for some arbitrary function* $G(y)$ *with one free variable in the language of a formal system* T, *a sentence D can be (mechanically) constructed such that.* $T \vdash D \Leftrightarrow G(\ulcorner D \urcorner)$".

Note that Gödel in his 1931 proof did not mention the diagonal lemma[29] explicitly, but it is implicit in his proof; he demonstrated it only for the specific case of the *provability predicate* (see below).

[A brief aside concerning the meaning of *self-referential*: As various authors [e.g., Hintikka (1999)] have pointed out, 'self-referential' is not a straightforward concept in this connection. This is because of the double role played by the natural numbers in Gödel's scheme: they are on the one hand simply 'everyday numbers' which can be used for counting and for expressing simple arithmetic results, sums and products; but at the same time, they (as 'Gödel numbers') represent terms, sentences, and proofs in a formal logical system (e.g., in P). So, in defining 'self-referential', one can rightly ask which *self* is meant. In the case of simple but paradoxical statements like "This sentence is false", the identity of the sentence—its *self*—is evident and unique. The case of Gödel numbers is not so clear—they are like the actor who may be speaking as a natural person, as 'himself', or may be speaking his lines onstage, as a fictitious person, invented by a playwright. Nevertheless, a kind of self-reference is present in a function like G(n), and it is essential for the success of Gödel's proofs. Some authors have objected that the term 'self-referential' is too anthropomorphic. See the article by Saul Kripke (2021)[30] and references therein.]

Now we consider a formal *provability predicate Pr*[x], that is a formula whose 'meaning' is, "the sentence whose Gödel number is my argument x is provable" (of course within the given formal-logical system, e.g., P). Every formal system contains (by definition) the possibility of determining whether a given sequence of formulas is indeed a proof of a particular sentence. Therefore, it is possible to define a provability predicate within a formal system.

Gödel denotes such a predicate as $\overline{\text{Bew}}$ [x] (for *Beweis* = proof); the over-line means *negation*, in his notation. In other words, this predicate means "*the statement whose Gödel number is my argument x is not provable in the System P*". In our notation, that predicate[31] is denoted by $\neg Pr[x]$. Applying the diagonal lemma to this formula, we find that there is a number n which is the Gödel number of $\neg Pr[n]$ within the formal system considered; that is, the formula with the Gödel number n is not provable (within that system)—but that formula is simply $\neg Pr[n]$ itself. If this statement were provable, then it is *not* provable—a classic contradiction, meaning that the formal system would not be consistent. But we have taken as a precondition that the system *is* consistent. The only way to resolve the contradiction is then to assume that *neither* $\neg Pr[n]$, *nor* its converse, $Pr[n]$, is provable within the system; it is therefore *not decidable*. But a formal system containing a non-decidable sentence is *incomplete*, since completeness requires that all the sentences derivable from its axioms can either be proven or disproven. Gödel would have termed this situation '*nicht entscheidungsdefinit*'.

In fact, the statement '$\neg Pr[n]$ is not provable within this system' is *true*, in the semantic sense that it is indeed not provable. So, summarizing, we can say that the formal System P (or whichever formal system with its partic-ular axioms is being considered) is *consistent* but contains a sentence which is *undecidable*, although true. This is precisely the situation that Gödel foresaw in the introduction to his (1930), a year earlier. (Gödel later gave a detailed proof that the statement is indeed *true*, in the semantic sense mentioned above, which we need not repeat here; see below and backnote [42]).

Finally, we must consider the meaning of the notion *consistent*. We dealt with it in the previous chapter, and defined it there to mean that *no contra-dictions can be found by any finite sequence of logical statements within the system considered*. This is termed 'simple consistency'. For his incompleteness proofs, Gödel used a special (stronger) form of consistency, which is called ω-*consistency*. It is defined as follows: "*a system S is ω-inconsistent if, for some property P, S proves P(n) (where n is a natural number, the argument of P), i.e. S proves P(0), P(1), P(2) … up to some* [transfinite] *number ω, but then fails*". The number for which it fails is a 'nonstandard' number, i.e., it is larger than any standard number *n*. (The concept of nonstandard numbers was introduced by Skolem in 1934). If the system is not ω-*in*consistent, it is ω-consistent.

In proving his (1st) incompleteness theorem, Gödel assumed that the system considered was ω-consistent. This limitation was later eliminated by *J. Barkley Rosser* (1936), an American logician who was a graduate student of *Alonzo Church* in Princeton when Gödel gave his first lectures there. He was given the duty of preparing lecture notes from Gödel's 1934 talks on incompleteness[2], along with *Stephen C. Kleene*, another Church graduate student. Rosser (1936) gave a proof of the incompleteness theorem which presupposes simple consistency rather than ω-consistency. He makes use of what is now called 'Rosser's Trick': Instead of a sentence asserting '*I am not provable within* P' (the '*Gödel sentence*'), he uses a sentence which asserts: '*If I am provable within* P, *then there is a shorter proof of my negation*' (the '*Rosser sentence*'). This eliminates the need to assume ω-consistency.

Some Amusing Comments

Douglas Hofstadter, in his (2007), presents Gödel's proof in his usual elegant and amusing manner, beginning around page 180. He emphasizes two aspects that are seldom mentioned in more formal treatments: *First*, the existence of Gödel numbers that denote *provable sentences*. He calls them 'Prim numbers'. He initially discusses the Gödel numbers of all 'well-formed formulas' (wff), terming them 'wff numbers'. Considering the examples given above, our Gödel number *j* (the Gödel number of formula (8.2) above), or the numerical example 243,000,000, encoding the formula $0 = 0$, are both wff numbers.[32] The number 125,000,000, obtained erroneously in Nagel & Newman (1956), encodes only a fragment (see backnote [23]), and thus is *not* a wff number. Hofstadter's wff numbers are a legitimate aspect of formalized number theory which can be used to investigate properties of formal systems from *within* those systems themselves. He then considers what he calls '*Prim numbers*', defined as those Gödel (wff) numbers that denote formulas which are *provable* within the formal system considered; that is, they would be the Gödel numbers of our predicate $Pr[x]$ with various values of the argument x (denoting various provable formulas). This is simply the arithmetization of the proof sequences given in PM (represented there using typographical symbols as in Table 8.1). Thus, for example, the Gödel number j given above is not only a wff number, it is also a Prim number (since it denotes the formula (8.2), provable in PM). Unfortunately, there is no algorithm which allows us to 'mechanically' pick out the 'Prim numbers' from among all the wff numbers, nor to identify the *non*-Prim numbers (which Hofstadter calls 'frivolous numbers'), denoting wff's that are *not* provable within PM. But we see that the wff number n used above belongs to the 'frivolous' numbers, since it denotes a formula that is *not provable* within PM.

Second, Hofstadter takes up a topic which was discussed by the Harvard philosopher *Willard van Orman Quine* [see Quine (1962) and (1987)]; namely, *self-referential sentences*, and how to put compact information about itself into a sentence. He calls this 'putting an elephant into a matchbox'. That is exactly what is done in Gödel's proof predicate—and it is possible because n in its argument firstly is a numeral, the *symbol* of the (very large) Gödel number referred to, and secondly, even that very large number can be reduced to a reasonable-sized chain of numbers, and thus logical symbols, by reversing the Gödelization process, i.e., factoring it into its prime factors with their exponents. He summarizes this section by giving a Quine sentence about self-referential sentences:

'Put in front of itself in quotation marks yields a complete sentence' put in front of itself in quotation marks yields a complete sentence.

This is, as Hofstadter points out, analogous to the simple self-referential sentence " 'is white' is white". And he concludes with another analogous sentence, a *metaphor* for Gödel's proof predicate:

'Yields, by inserting its own Gödel number as argument, a non-Prim number' yields, by inserting its own Gödel number as argument, a non-Prim number.

This is the kind of self-referentiality which Hofstadter terms 'Gödel's strangeness'.

The Second Incompleteness Theorem

The so-called 'Second Incompleteness Theorem' follows (although by no means automatically or immediately) from the First Theorem, and it was already formulated (without giving a detailed proof) in Gödel's (1931). It had also already been anticipated by von Neumann, some time after his discussion with Gödel in Königsberg; but it was not known to either Gödel or von Neumann immediately after the meetings there.[33] It concerns directly the question of *consistency*, the main topic of Hilbert's 2nd Problem. Using a predicate of the same form as the provability predicate $Pr[x]$, we can take an inconsistent formula in arithmetic—e.g., '$(1 = 0)$', which is evidently false (call it $\perp$, the symbol for 'contradiction'). Then the consistency of the system T (e.g., P) can be expressed as $\neg\, Pr\,[\ulcorner\perp\urcorner]$ (which we can abbreviate as *Cons* $[T]$, for '*the consistency of the system T*'). Its provability within the system T is negated by the same logic as used in the first theorem, so if the first theorem is correct, then *Cons* $[T]$ is also not provable within the system, i.e. the system *cannot prove its own consistency*. This is the basic content of the second theorem. In symbolic form, it can be stated as: "*If T is a consistent formal logical system which contains elementary arithmetic, then $T \nvdash Cons[T]$*"; or, as we quoted it above, '*if $T \vdash Cons[T]$, then it is inconsistent*'. This is of course only a sketch of an actual proof; Gödel intended originally to follow his first paper (1931) with a second, containing a more detailed treatment of the second theorem. But the reception of his (1931) was such that he never wrote its planned sequel. Perhaps he simply didn't find the time to write it, or felt that it was unnecessary; or, as Dawson suggests, the various negative reactions to the first paper discouraged him from writing a sequel (although he was aware that those reactions were unfounded).

A detailed proof of the second theorem was given by Hilbert & Bernays (1939) (evidently written by Bernays). But it was still dependent on the structure of the specific formal-logical system considered; it must satisfy the 'derivability conditions' of that specific system. A general proof for all formal-logical systems that contain elementary arithmetic was given by Feferman (1960); he discussed his route to that proof later in a memoir (Feferman 1997). See Myhill (1952), Kleene (1976) or Franzén (2005) for an overview of the consequences and later developments resulting from Gödel's incompleteness theorems, and his later related work.

Response to the Incompleteness Theorems

A complete treatment of the reception of Gödel's incompleteness theorems would fill whole volumes, and it depends most especially on answering the question, "Reception by *whom*?" before trying to address the topic seriously. The reaction of *logicians* was relatively rapid, and Gödel's achievement was recognized and appreciated within that small, elite group, although approval was not universal, in particular by those who failed to understand it in detail (such as Zermelo, who however did not consider himself to be a logician and indeed rather looked down upon that group).

The reaction of general mathematicians and of philosophers was naturally slower in coming, while the reaction of a wider public occurred only after Kurt Gödel's death, and it has been mixed. However, among mathematicians, his work on incompleteness was generally recognized as important and even epochal; for example, consider von Neumann's remark when Gödel received the first Einstein Award in 1951 (together with Julian Schwinger)[34]: "*Kurt Gödel's achievement in modern logic is singular and monumental—indeed it is more than a monument, it is a landmark which will remain visible far in space and time.*"

At home in the *Josefstädter Straße* with his brother Rudolf and his mother Marianne, Kurt apparently remained silent about his newfound fame, although he must have announced the achievement of his various career milestones, such as receiving his doctorate and, almost exactly three years later, his *Habilitation*. But even 20 years later, Rudolf and Marianne were practically unaware of Kurt's reputation among his mathematical and scientific peers. His worldwide fame, following the publication of Hofstadter's '*Gödel, Escher, Bach*' (1979), came too late for Marianne and, indeed, for Kurt himself. His brother Rudolf lived to be almost 90; he died in January 1992, shortly before his 90th birthday in February of that year, and he was informed of his brother's international reputation—and wondered why they had not been more aware of it during Kurt's lifetime.

The book by Franzén (2005) gives a particularly detailed treatment of the reception of Gödel's incompleteness results, but it is also considered in other books, notably Dawson (1997), Smith (2020), Goldstein (2005), and the *Collected Works*, Vol. I.[35] We will defer the discussion of the reception of Gödel's work to Chap. 10 and the following chapters, mentioning specific applications and reactions as they occurred in the years following the publication of the theorems in March of 1931; but we summarize here his interactions with other logicians immediately after the announcement of his incompleteness results at the Königsberg conference. We have already described von Neumann's interest and the von Neumann-Gödel correspondence in November, 1930. Von Neumann had apparently returned from Königsberg without mentioning Gödel's results to Hilbert or to others at the conferences there.

The next logician to inquire about the new work was Bernays, by now a nontenured professor in Hilbert's institute in Göttingen. He wrote to Gödel in late December of 1930[36] to thank him for sending a copy of his 1930 completeness paper, and he mentioned that he had heard (through the mathematical grapevine, probably originating with von Neumann) that Gödel had new and interesting results, asking for a preprint (in those days, a typed copy of the manuscript, or a photocopy of the galley proofs of the article in question). Gödel sent one, and that was the beginning of a correspondence that continued for some months, until Bernays had completely understood the paper. Gödel's interactions with both von Neumann and with Bernays were cordial and productive—von Neumann later introduced Gödel's work to the logic community in Princeton, and Bernays' reaction was positive; he subsequently provided the detailed proof of the second incompleteness theorem (in Volume II of Hilbert and Bernays' *Grundlagen der Mathematik*, published in 1939).

But not all of Gödel's colleagues were so polite and cooperative. In September of 1931 (after meeting Gödel at the *Bad Elster* conference, see below), *Ernst Zermelo* wrote to Gödel, saying that he had found an error in the latter's proof of incompleteness. A lengthy correspondence ensued, in which Gödel went to considerable effort to make his proof clear to Zermelo. The correspondence remained civil, but Zermelo was unconvinced; when Gödel showed his reply of Oct. 29th to Carnap, both agreed that Zermelo had simply not understood the proof.

Ernst Zermelo was born in 1871. He spent his youth in Berlin, where he graduated from a *Gymnasium* in 1889. He then studied philosophy, physics and mathematics in Berlin, Halle and Freiburg, obtaining his doctorate in mathematics from the University of Berlin in 1894. He served as assistant to Max Planck in Berlin, where he remained until 1897, and he then went to Göttingen, obtaining the *Habilitation* there in 1899. After more than 10 years in Göttingen, he was offered a professorship in mathematics in Zurich in 1910, which he occupied for only six years, resigning in 1916. He held an honorary professorship at the University of Freiburg from 1926 until 1935. At the time of his meeting with Gödel, he had just turned 60, and it seems that he was somewhat of a curmudgeon.

Gödel after Publication of the Incompleteness Theorems

In an interview with Werner DePauli-Schimanovich and Peter Weibel for their television documentary in 1986,[37] Kurt Gödel's brother Rudolf, by then 84, recalled that Kurt had suffered a serious depression "shortly after the publication of his famous work" (or in late 1931; see Wang 1987), and had to be committed to a sanatorium (in *Purkersdorf*, near Vienna) for fear that he would become suicidal. Dawson[38] has carefully researched the available

archival documents (Gödel's correspondence, Carnap's diaries, Gödel's attendance at meetings etc.), and came to the conclusion that there was no gap in
his schedule in the period from March, 1931 (when his paper on incompleteness appeared, presumably the 'famous work' referred to by Rudolf)
until the end of that year which could be attributed to a stay at Purkersdorf. Gödel was, however, indeed committed there in 1934 after returning
from his first stay in the USA. It seems probable that Rudolf simply confused
the dates (not unlikely after 55 years!). Dawson [37] suggests that the 'famous
work' mentioned by Rudolf referred instead to the lecture notes from Gödel's
talks on his Incompleteness Theorems, delivered at Princeton University in
the spring of 1934. However, those notes[2] were not published immediately;
instead, mimeographed copies were distributed privately, and the notes were
published only in 1965 (and Rudolf was probably not aware of them). Thus,
it seems that the report by Rudolf is specious, and, while Kurt may have
experienced some stress as a result of disagreements with colleagues about
his proofs, in particular with Zermelo, it was apparently not sufficient to
incapacitate him or require hospitalization before 1934.

We again take up the thread of Kurt Gödel's life in Vienna in the years
1931 and 1932: After the Königsberg conference in September 1930, he was
busy formulating and writing up his incompleteness results for publication.
There followed a period of intensive correspondence with other mathematicians, as we saw above. He also gave several lectures on his new results,
notably at the Vienna (Schlick) Circle in mid-January of 1931, and at the
Menger Colloquium a week later.

Furthermore, he worked on other problems, including one suggested by
Karl Menger in early 1931 (Menger was at the time on sabbatical and was
working at the Rice Institute in Houston/TX, USA—now Rice University).
This was an extension of '*Lindenbaum's Lemma*' to uncountable sets of propositional formulas, which Gödel announced to the Colloquium in June[38].
He also corresponded with the French logician *Jacques Herbrand*, who was
visiting in Germany (he was an 'external member' of Hilbert's group) and who
had heard of Gödel's incompleteness work from von Neumann in Berlin. This
correspondence had important consequences, as we shall see in Chap. 12.

When Menger, still in the USA, received a copy of Gödel's article (1931)
on incompleteness, he was so enthusiastic that he interrupted the lecture
series that he was giving and spoke on Gödel's new results—the first
announcement of them in the western hemisphere. Similarly, 6 months later,
von Neumann used a planned lecture in Princeton to introduce Gödel's
(1931), the first time that most of his listeners there had heard the name
'Gödel'.

Indeed, Gödel was busy in the spring and summer of 1931. Herbrand made a suggestion about recursive functions in a letter to Gödel, but he never received the latter's reply—two days after it was sent, he was killed in a mountain-climbing accident, and the only reply subsequently received by Gödel was a death notice. Gödel however took up the problem and included it in his 1934 lectures in Princeton, attributing it to Herbrand (see Chap. 12). Gödel also published (formally or in talks and notes) a number of shorter papers and reviews, as can be seen from the *Collected Works*, Volumes I and III. During this period, he also read intensively on philosophy—post-Kantian metaphysics as well as Leibniz, according to Menger.[39]

An important event for Gödel later in 1931 was a conference sponsored by the *Deutsche Mathematiker-Vereinigung* (DMV) which took place in Bad Elster, Germany, in September of that year. He gave a talk[40] there on September 15th, entitled '*On the existence of undecidable arithmetical propositions in the formal systems of mathematics*'.

Bad Elster is a small town in the southernmost region of Saxony, the *Vogtland*. It is very near the Czech border, and, like the nearby *Karlsbad*, where the Gödels had often spent their summer vacations during Kurt's youth, it is a spa, popular during the *Belle Époque* among better-situated citizens of the German and Austro-Hungarian Empires. One of its chief attractions is the *Kurhaus* (Spa house) built in 1890 and sponsored by the Saxon royal family (it is thus still called the 'Royal Spa House'; see Fig. 8.3). Whether some of the sessions of the DMV annual conference (which, as in Königsberg the previous year, was held together with the annual conferences of the GDNÄ and the DPG, for the last time in 1931) in fact took place in the *Kurhaus* is not known, but it is certainly possible.

Among other things, the conference offered Gödel the possibility of meeting colleagues from outside Vienna. Ernst Zermelo attended it, and he and Gödel interacted there. Both of them gave talks, but Zermelo was not impressed by Gödel's lecture on his incompleteness theorems, and he made some critical remarks about it during a discussion. He was in fact reluctant to meet Gödel personally, but was persuaded by other colleagues, and the two of them spoke animatedly during an excursion.

Zermelo's views on metamathematics were orthogonal to those of Gödel,[42] and he apparently considered the latter to be an upstart whose results were probably incorrect, or if not, unimportant (as emerged in their later correspondence). Nevertheless, they behaved politely towards each other at the conference. A week later, on September 21st, Zermelo, who had in the meantime written a summary of his own talk including a critique of Gödel's

Fig. 8.3   The *'Königliches Kurhaus'* at the *Sächsisches Staatsbad*, the Royal Spa House at the Saxon State hot springs in Bad Elster. Photo: Wiki Media, public domain[41]

lecture, wrote to Gödel, claiming that he had found an error in the incompleteness proof. Gödel responded, not immediately but very thoroughly, with a long letter in which he patiently explained his proof and pointed out its correctness. [That letter (dated Oct. 12th, 1931) in fact deserved to be published at the time: In it, Gödel gave a new proof showing that 'truth' in a formal language is not definable within that language (later also published by Tarski (1936), who arrived at the same conclusion independently; it is now called the 'Gödel-Tarski theorem on the undefinability of truth'); and he presented a new, alternative proof for his 1st Incompleteness Theorem].

Nevertheless, Zermelo was not impressed; in his reply on Oct. 29th, he essentially admits the correctness of Gödel's proof, but fails to acknowledge its significance (although he had done so in his earlier letter, when he still thought it to be erroneous). This ended their correspondence.[43] Zermelo indeed later referred to Gödel as "the most advanced" among contemporary logicians[44] (of whom collectively he apparently held a low opinion, however). The episode with Zermelo must at the very least have been discouraging to Gödel, who was shy of confrontational interactions with other people, and it could indeed have contributed to a depression in late 1931; but that, if it occurred at all, was apparently not serious enough to require hospitalization or prevent his working normally.

How did Hilbert react to Gödel's 'famous paper' when he became aware of it? *Hilbert* and *Russell* were the logicians whose programs were most immediately called into question by Gödel's results; the latter however had left the field of mathematical logic soon after the publication of the last volume of PM, and he did not concern himself much (or immediately) with Gödel's article when it appeared. Hilbert apparently became aware of it after Bernays' correspondence with Gödel in January 1931, when Bernays received a copy of the galley proofs of the paper. He later recounted Hilbert's reaction (in a letter in 1966 to Constance Reid).[45] Dawson[46] quotes his letter: Bernays himself had expressed doubts about the completeness of formal systems sometime before Gödel's proofs. Hilbert was angered by that suggestion, and likewise by Gödel's results; but after a short time, he accepted that they were correct and tried to adapt his own program accordingly. Gödel however found that some of those attempts failed to take his results properly into account,[47] and they were also not consistent with Hilbert's own principles, in his (private) opinion.

Gödel's Habilitation

As we saw at the beginning of this chapter, the next step in the progress of Kurt Gödel's career would be for him to obtain the *Habilitation* and become a *Privatdozent* at the University of Vienna. He would have to request the opening of a *Habilitationsverfahren*, the process leading—if successful—to the conferral of that advanced degree; and to submit the necessary documents to support his application. They would of course include his *Habilitationsschrift*, the thesis on the basis of which his research would be judged. Each university in the German-speaking world has its own rules and regulations governing this process (called a *Habilitationsordnung*), and those differ somewhat in detail although their general character is unified. A common rule is the *two-year interval* normally required between the conferral of a doctoral degree and the opening of a *Habilitationsverfahren*. The rationale behind this rule is that the research on which the *Habilitation* is based is supposed to be new, original and independent of earlier research, e.g., the research done for the doctorate. The two years are considered to be the minimum time for performing and writing up that new research. On the average, a time of 4–6 years in fact elapses between the doctoral dissertation and the initiation of a *Habilitationsverfahren*, which itself typically takes 4–8 months.

Dawson[48] wonders why Gödel did not submit his (1931) as a thesis and initiate the *Habilitation* process at the beginning of the Winter Semester in

October of 1931. The 'two-year rule', if it were enforced at the University of Vienna, would have automatically prevented that; he would have been allowed to submit his application only after February 1932, two years after the formal conferral of his doctoral degree.[49] Of course, that rule is not practiced strictly everywhere, and exceptions can be made. As we saw in Chap. 7, Paul Bernays was granted the *Habilitation* in Zurich in 1912, the same year that he obtained the doctorate in Berlin (at age 24). Five years later, when he became Hilbert's assistant in Göttingen, he however went through a second *Habilitation* process there, probably on Hilbert's advice and/or because the University refused to recognize his 'rapid' *Habilitation* from Zurich. Another exceptional case is that of von Neumann, who was made *Privatdozent* (the youngest in the whole university; this implies the *Habilitation*) at the University of Berlin in 1927, just one year after his doctorate in mathematics in Budapest. His reputation as a mathematical genius probably preceded him to Berlin and may have made the exception possible.

Gödel may also simply have felt that he needed more time to prepare for the *Habilitation* process, which involves much more than simply submitting a thesis. And Hahn, who was to be secretary of the *Habilitationskommission* (*Habilitation* Commission) and Gödel's advocate during the process, may have suggested the later submission date for his own reasons. However this may be, Gödel submitted his application for the *Habilitation* to the Philosophical Faculty at the University of Vienna on June 25th, 1932.[50] He was 26 years old at the time, still rather young for becoming a *Privatdozent*. No document containing his *Habilitationsschrift* seems to have survived. In contrast to doctoral dissertations, this thesis is usually not reproduced in a large edition, since it is mainly of interest only to the Commission.

There are two forms in which the thesis may be submitted: One is for the candidate to perform original research, not published elsewhere, and to write it up as a *Habilitationsschrift*, similar to a doctoral dissertation, but usually more detailed. It may then be published later as a review article or as a series of research articles. This is what Bernays did for his second *Habilitation* in Göttingen in 1918. (He however waited too long to publish, around 8 years, so that he lost priority for some of the work in the thesis). The other possibility is a *cumulative* thesis, in which already-published articles are collected, usually with a general introduction and summary of the research field at the beginning, and short introductions for each article showing its significance and its relationship to that field. Gödel apparently chose that route, with his (1931), certainly a significant and even momentous publication, as the single article demonstrating his research ability.

In the event, he included in his application, as required in addition to the thesis itself, a *curriculum vitae* and a list of topics on which he was prepared to give a public lecture (*Probevortrag*) demonstrating his potential teaching ability. He was also required to defend his thesis before the Faculty. All of this was a daunting series of examinations, and that may have been part of the reason for his hesitation in submitting his thesis and initiating the process.

Wang[51] and Dawson[52] both document many details of Gödel's *Habilitationsverfahren*, which was formally completed on February 11th, 1933, and led to his being instated as *Privatdozent* one month later by the Dean of the Faculty. Of interest is Gödel's *curriculum vitae*, his own private history and personal evaluation of his achievements; and also, the recommendation to the Faculty (via the Commission), written by Hahn in support of Gödel's application. Both are quoted as the German originals by Wang (1987)[53] and by Christian (1980); we give their English translations (by the present author, with permission from their publisher, *Springer-Verlag*) here:

Gödel's *curriculum vitae* in his application for the *Habilitation* (dated June 25th, 1932)

I was born in Brünn in 1906 as the son of German parents, and I attended four classes of elementary school and eight classes at the German State Gymnasium, where I passed the *Reifeprüfung* [matura, *school leaving examination*] in the year 1924. In the autumn of that same year, I moved to Vienna, where I have resided without major interruptions ever since, and where I obtained Austrian citizenship in 1929. In the Winter Semester of 1924, I matriculated as a regular student in the Philosophical Faculty and dedicated myself initially to the study of physics, later to mainly mathematical studies. In addition, encouraged by Prof. Schlick, in whose philosophical circle I often participated, I engaged in studies of modern works on the Theory of Knowledge. My own scientific activities were concentrated mainly on the field of fundamentals of mathematics and symbolic logic. In the year 1929, I submitted a work from this field entitled '*On the Completeness of the Logical Calculus*' as my doctoral dissertation, and I received the doctorate in February 1930. In the same year, I gave a lecture at the Königsberg conference on that topic, and another at the 1931 Annual Conference of the German Association of Mathematicians in Bad Elster, on the work submitted as my *Habilitationsschrift*. In Vienna, I took part in the Colloquium led by Prof. Menger and I was active in editing and publishing its Reports, which appear annually. I also contributed to the Seminar on Mathematical Logic given by Prof. Hahn in the academic year 1931/32, assisting in choosing the topics and in preparing the students for their talks.

In the year 1931, I was asked by the Editors of the *Zentralblatt für Mathematik*, to which I am a regular contributor, to write a report on research in fundamentals of mathematics, together with A. Heyting, and I am currently working on preparing it.

[Signed] Vienna, June 1932, Dr. Kurt Gödel.

Hahn, secretary to the *Habilitationskomission*, prepared a report on Gödel's *Habilitation* thesis and his previous scientific record, which he submitted to the Commission to prepare them to make a decision on Gödel's suitability to receive the *Habilitation*. Its text (from the German as quoted by Wang (1987) and Christian (1980) and translated by the present author) was as follows:

Report of the *Habilitationskommission* on Gödel's application (December 1st, 1932)

The doctoral dissertation *[of K. Gödel]* already demonstrated its significant scientific value (on 'The Completeness of the Logical Functional Calculus'). It solved the important and difficult problem posed by Hilbert as to whether the axioms of the restricted logical functional calculus form a complete system, by showing that every generally valid formula of the restricted functional calculus is provable. *[Gödel's] Habilitationsschrift 'On Formally Undecidable Theorems of Principia Mathematica and Related Systems'* is an achievement of the first rank, which has aroused considerable attention in all the circles of experts in the field, and – as can be foreseen with confidence – it will take its place in the history of mathematics. Mr. Gödel was able to show in that work that within the logical system of Whitehead and Russell's *Principia Mathematica*, problems can be derived which are undecidable with the methods of that system, and that the same holds for every formal-logical system which is capable of expressing the arithmetic of the natural numbers; thus it has also been demonstrated that the program advanced by Hilbert of proving the consistency of mathematics cannot be carried out.

Apart from some additional works by Gödel which deal with the field of symbolic logic, his note *'On the Intuitionistic Propositional Calculus'* deserves to be particularly singled out; in it, the theorem is proved that: There is no realization of Heyting's axiomatic system of the intuitionistic propositional calculus by means of a finite number of truth values for which *[all]* the, and only the provable formulas are found by arbitrary insertion of particular values. There are infinitely many systems between that of Heyting and the system of ordinary propositional calculus.

The works presented by Dr. Gödel tower far above the level that is usually required for a Habilitation. Dr. Gödel can already be considered to be a first-rate authority in the field of symbolic logic and research into the fundamentals of mathematics. In close scientific cooperation with the present reviewer and with Prof. Menger, he has also proved his worth to a great extent in other areas of mathematics.

Dawson[54] gives a detailed chronic of Gödel's *Habilitationsverfahren*. The Commission had 10 members, and it was chaired by the Dean of the Faculty (Prof. *Heinrich Srbik*). It included all of the senior faculty members of the Mathematics Institute, and several others from neighboring institutes (physics, philosophy). Its first meeting was held on November 25th, 1932, and it used Hahn's report as the basis for its decision (above, shown in its final form as submitted to the faculty; the date 'Dec. 1st' given on the report is the date of its submission to the faculty). The Commission had basically a pre-screening function, to decide whether to pass the application on to the faculty, which it in fact approved unanimously. The first Faculty Meeting to deal with Gödel's *Habilitation* process was held on December 3rd, and it was attended by 52 professors from all the institutes within the faculty. At this first hearing, the faculty had to decide two questions: (i) Does the candidate have the professional and scientific qualifications to justify the *Habilitation*?—and, if so, (ii) Should the process be continued? (These two questions are not independent, but they are also not identical—there might be formal, administrative or personal reasons for not pursuing a *Habilitationsverfahren* at a particular university, even though the candidate was technically qualified. Such 'derailments' occasionally occur, and the candidate then has the option of moving to a different university, and can present the positive answer to (i) as the basis for a *Habilitation* there). In Gödel's case, the qualification was clear, and the vote on (i) was 51 'yes', no abstentions, and one 'no'. In the vote on (ii), two members had apparently left, and the result was 49 'yes', no abstentions, and one 'no'.

Just who was responsible for the single 'no' vote is not known, but Dawson, based on a report by Olga Taussky-Todd, suspects that it might have been Prof. Wirtinger, the oldest faculty member in Mathematics, who had been overshadowed by the successes of Furtwängler and Hahn and was nearing retirement, apparently somewhat embittered. He had evidently voted in favor of Gödel at the Commission's meeting, where the vote was probably taken by hand sign; a 'no' vote there would have been obvious and would have resulted in protracted discussions. But in the anonymity of the much larger Faculty Meeting, where the votes were probably collected on slips of paper, he could vent his distaste for the new generation. This of course remains speculation, but it is quite plausible. Professors from other institutes would have no motive for opposing Gödel's *Habilitation*, especially in view of the convincing arguments in Hahn's report.

The next step in the process was Gödel's defense of his *Habilitationsschrift*, which was scheduled for a colloquium on January 13th, 1933 (Friday the 13th !). The result had to be accepted by a full Faculty Meeting, which took place on January 21st, where Gödel's defense was approved by a majority of the members.

The final step was Gödel's *Probevortrag*, a public 'test lecture' in which he was required to present a topic outside his thesis to a general audience. He had submitted a list of suggested subjects for the lecture with his application materials, but the Commission chose a different topic, namely the area noted by Hahn in his report, termed by the Commission '*Über den intuitionistischen Aussagenkalkül*' ('*On the Intuitionistic Propositional Calculus*'). This is the same as the title of a talk that Gödel had given on February 25th, 1932, in Menger's Colloquium, initiated by a question from Hahn. That talk was communicated by Hahn to the Vienna Academy, and also appeared in the Reports of the Colloquium. On June 28th, 1932, Gödel gave another Colloquium lecture about the translatability of intuitionistic arithmetic into classical arithmetic, which appeared in the Reports the following year ([*1933a*] in the Collected Works), and he also gave a talk (undated, but noted in the Collected Reports of the Colloquium for 1931/32, [*1933b*] in the *Collected Works*) on adding a 'proof predicate' Bp to propositional logic and thus relating it to intuitionistic propositional logic and modal logic. All of these could serve as material for his *Probevortrag*, so he was indeed well prepared. His lecture was scheduled for February 3rd, 1933 (again a Friday). Its precise content is not known, although a summary must have been included in the proceedings of the Commission or the minutes of the final Faculty Meeting, which took place on February 11th. Gödel's *Probevortrag* was approved by the faculty with a majority vote, so that he had surmounted all the hurdles; and he was then granted the *Venia legendi* (or *Venia docendi*, as it typically was called in Austria; in German also termed the '*Lehrbefähigung*'), giving him the right to deliver lectures at the university level. His appointment as *Privatdozent* (the '*Lehrbefugnis*') was announced by the Dean's office on March 11th, 1933.

The title *Privatdozent*, as the name suggests, was not an official position, but rather a kind of freelance lectureship or unpaid instructorship. The professors in German-speaking countries were '*Beamten*', civil servants who worked for the State and usually had to take a loyalty oath. *Privatdozenten*, in contrast, had no obligation to the government and were thus a sort of guarantee of academic freedom. It is therefore not surprising that this title was quickly annulled by the Nazis after the German annexation of Austria in 1938; they replaced it soon after by the '*Dozent neuer Ordnung*', an office of a similar rank but under State control, salaried like the professorships (but of course at a lower level).

An important event for Gödel in 1932—presumably in the early autumn; the date is unknown, but the fact is mentioned by Dawson (1997), Wang (1987) and Christian (1980)—was his visit to Göttingen, where he met several well-known contemporary mathematicians, including Emmy Noether, Carl Ludwig Siegel, and possibly Ernst Zermelo (whom he of course already knew from the Bad Elster conference the previous year and their ensuing correspondence); but *not* David Hilbert. Gödel later affirmed to Hao Wang that he had never met Hilbert. Whether he gave a seminar or colloquium talk in Göttingen, and on what topic, is not recorded.

Gödel's talk to the Menger Colloquium on June 28th, 1932 had other repercussions for his later career: Menger had invited Oswald Veblen, from Princeton, to attend. He presumably met Veblen while on sabbatical in the USA, and the latter was visiting Europe in the summer of 1932; and Veblen was very impressed by Gödel's lecture, so that he invited him to visit Princeton the following year. Von Neumann had already introduced Gödel's work to the growing logic group in Princeton in his talk in the autumn of 1931. There, something new was hatching—the founding of the *Institute for Advanced Study*, which would play an important role in the second half of Gödel's life. It was the reason for Veblen's European trip that summer. We will come back to this important topic in later chapters. We will also return to Gödel's project with Heyting, the review article for the *Zentralblatt für Mathematik* on the fundamentals of mathematics which he mentioned in his *c.v.* for the *Habilitation*. But in the next chapter, we first examine a different topic which dominated most of Gödel's life, namely matters of *health*.

Notes

1. Carnap (1963), 1997 edition, p. 29.
2. Rosser & Kleene, Princeton interview, in: *The Princeton Mathematics Community in the 1930s,* Transcript Number 23 (PMC23). Interview by William Aspray, 1985. Online at: https://web.math.princeton.edu/oral-history/c21.pdf.
3. Dawson (1997), 2005 edition, p. 63 and Endnote 147.
4. Photo: by Maria Kokoszyńska-Lutmanowa, 1935. Reused from *calisphere* (Bancroft Library, UC Berkeley, original photo cropped). Copyright status unknown, fair use. See: https://calisphere.org/item/ark:/28722/bk0016t8g2w/.

5. The term 'arithmetization' dates from the later nineteenth century (Kronecker, constructionist school) and is more general.
6. Wang (1987), p. 84.
7. Dawson, *op. cit.*, pp. 58, 59 and 61.
8. Referring to *'The Concept of Truth in Formalized Languages'* by A. Tarski, first published in Polish in 1933, then as a preprint in German in 1935, then as *'Der Wahrheitsbegriff in den formalisierten Sprachen'* in *Studia Philosophica*, Vol. I (1936), pp. 261–405, and finally in English in Tarski's book, *Logic, Semantics, Metamathematics* (Oxford, 1956).
9. Dawson, *op. cit.*, p. 67, footnote 5.
10. This is indeed a compact rendering, but because of the several other connotations of the word 'decisive' in everyday English, it might be preferable to use a more specific term. Panu Raatikainen (in his article on *'Gödel's Incompleteness Theorems'* in the *Stanford Encyclopedia of Philosophy* (2020)) calls it *'decidable in a formal system'*; Davis (1965) calls it simply *decidable*.
11. As quoted in Dawson, *op. cit.*, p. 61.
12. Quoting an earlier remark by DuBois-Reymond; see Franzén (2005), p. 16.
13. See e.g. the *Stanford Encyclopedia of Philosophy*, *'The Continuum Hypothesis'*, by Peter Koellner (2013). Online at: https://plato.stanford.edu/entries/contin uum-hypothesis/.
14. Photo from *picture-alliance.de*. Photographer unknown; licensed for re-use; source: picture-alliance/IMAGNO/*Wiener Stadt- und Landesbibliothek*.
15. Nagel & Newman (1958), 2001 edition, Chapter VI.
16. See Verena Huber-Dyson, Interview in *The Edge*, entitled *'Gödel and the Nature of Mathematical Truth'*, recorded on 25.07.2005: Huber-Dyson (2005). Consulted Feb. 2022.
17. *ibid.*, online at: https://www.edge.org/3rd_culture/vhd05/vhd05_index. html.
18. Wang 1987, p. 85, states that Gödel *did* attend Hilbert's lecture. Gödel evidently believed (correctly) that hearing a lecture by someone was not the same as 'meeting' that person, as we also saw regarding Brouwer in Chapter 7.
19. See Wang, *op. cit.*, p. 87.
20. This is clear for Russell, who said as much in his autobiography. Wittgenstein's case is more controversial; the 'notorious paragraph' in his posthumously published (1956) has been widely interpreted as showing that he failed to understand the 1st incompleteness theorem, but some recent commentators disagree. See Victor Rodych on *'Wittgenstein's Philosophy of Mathematics'* in the *Stanford Encyclopedia of Philosophy* (2018), and references therein.
21. Sebastian Bader, *'Gödel's Incompleteness Theorems'*, contribution to the *Knowledge Representation and Reasoning Seminar*, Dresden, April 25th, 2006 (GK334, DFG). Online at http://logic.amu.edu.pl/images/f/f2/Sebastian bader.pdf
22. Smith (2007/2020).

23. Quoted from [21], *op. cit.*

24. This example was also used in Nagel & Newman's *Scientific American* article in June, 1956, which was essentially the seed of their later book (1958). There, an amusing error occurred: the three prime factors were given correctly, but in the calculation of their numerical values, the middle factor was treated as 5^3 instead of 3^5, giving 125 instead of 243. The resulting number is $g = 125,000,000$. It is indeed also a Gödel number, but it encodes only a fragment, '0)'. Whether this error was noticed by the readers is not known; it was corrected by an *erratum* in the August 1956 issue, presumably by the authors. There were some letters from readers in the September issue, but they dealt with typography (distinguishing meta-language statements and object-language statements) and an analogy from a chess game.

25. Smith, *op cit.*, pp. 138–142.

26. See e.g.: Zermelo, E., 1908, '*Untersuchungen über die Grundlagen der Mengenlehre, I* ', *Mathematische Annalen*, Vol. 65, pp. 261–281; and Levy, A., 1979, '*Basic Set Theory*', Springer, New York/Heidelberg 1979.

27. For a more rigorous but still very compact exposition of Gödel's proof, see e.g., Nagel & Newman (1956); or Kleene (1976); or Davis (2006).

28. Named by analogy to the lemma introduced to number theory by Georg Cantor in 1891.

29. Also known as the *Fixed-point theorem* (similar to Kleene's recursion theorem), or the *Self-reference lemma*. See Enderton (1972), p. 227, or Smith (2020), p. 181.

30. Kripke 2021: '*Gödel's Theorem and Direct Self-Reference*'. Saul A. Kripke, preprint in arXiv:2010.11979v2 [math.LO] 7 (June 2021). Online at https://arxiv.org/pdf/2010.11979.pdf.

31. We are simplifying the argument used by Gödel in his (1931) in order to make it intuitively clear. In the original, he defined an 'ordering relation' $R(n)$ which orders sequences of symbols representing formulas (with one free variable) in the formal system considered (e.g., Gödel's system P) – Gödel calls these 'class-signs' ('*Klassenzeichen*'). For any arbitrary class-sign α, $[\alpha;n]$ is the formula corresponding to α with the argument n. A triple relation $x = [y;z]$ [or, in more modern notation, $x \Leftrightarrow y(z)$] can also be defined in the system. He then defines a class (or set) K of numbers n by: $n \in K \equiv \overline{Bew}[R(n); n]$ (in modern notation: $K = \{n \in \mathbb{N} | \neg Pr[R(n);n]\}$, i.e., K is the set of all the numbers n for which the formula $R(n)$ is *not provable* when n is used as its own argument. The function $[R(q);q]$ for some particular q then becomes an *undecidable sentence* in the system P. This is an *indirectly* self-referential sentence, which says, in effect, '*I am not provable within P*'. Compare Kripke (2021) (cf. backnote [29]). Gödel is at pains to show that he is not using a circular argument in defining this sentence [cf. footnote [15] in his (1931)].

32. Note that Hofstadter uses a different coding scheme; in his scheme, code 2 represents '0' and code 6 represents ' $=$ '. Thus the simple equation $0 = 0$

has the Gödel number 72900 in his notation, more compact than Nagel & Newman's.

33. See Sieg (2005), p. 175.
34. Quoted by Kleene (1976), taken there from the *New York Times* article of March 15th, 1951, p. 31, on the Einstein Award to Gödel and Schwinger that year.
35. Kurt Gödel: *Collected Works*, Volume I (1986), eds. Feferman, Dawson et al. Item 5 in Appendix B.
36. Wang, *op. cit.*, p. 87.
37. See Schimanovich & Weibel (1997).
38. Dawson, *op. cit.*, p. 77.
39. Quoted by Wang, *op. cit.*, p. 88.
40. Dawson, *op. cit.*, p. 75.
41. Photo from Wiki Media, photographer unknown, public domain; cropped. Reused from: https://commons.wikimedia.org/wiki/File:Bad_Elster_Kurhaus_1900.jpg
42. See Wang, *op. cit.*, p. 89 ff., and Dawson, *op. cit.*, p. 76.
43. This correspondence was rediscovered about the time of Gödel's death, and was reported, with copies of Gödel's Oct. 12th letter and Zermelo's Oct. 29th reply, by I. Grattan-Guinness in 1979, including the complete text (English translation) of the letters. The original letter from Zermelo (Sept. 21st) was believed to be lost, but was later discovered in Gödel's papers, and its text (with English translation) was published by J.W. Dawson in 1985. See Gratan-Guinness (1979) and Dawson (1985).
44. See Wang, *op. cit.*, p. 91, for Zermelo on Gödel.
45. Constance Bowman Reid (1918–2010) was a writer who specialized in introductory texts on mathematics, as well as biographies and historical writings about mathematicians, among them David Hilbert.
46. Dawson, *op. cit.*, p. 72.
47. *ibid.*, p. 73 and endnote [162].
48. *ibid.*, p. 81.
49. Budiansky (2021) claims that the University of Vienna in fact had a *four-year* rule (p. 137) and "made an exception for Gödel", but he gives no sources.
50. Wang, *op. cit.*, Dawson, *op. cit.*; see also backnotes [43] and [45], this chapter.
51. Wang, *ibid.*, pp. 92–94.
52. Dawson, *op. cit.*, pp. 86–89.
53. Wang, *op. cit.*; quoted there from Christian (1980), pp. 261 and 263.
54. Dawson, *op. cit.*, pp. 86–89.

9

Matters of Health

For Kurt Gödel, his *health*—both mental and physical—was an important topic during most of his life. And it was important not only to him personally, but also to his family and friends, over many years. At various times, repeatedly, it was the predominant influence on his daily existence. After three chapters devoted to mathematical logic, let us now take a step back and look at a more general aspect of Kurt Gödel's life—and here, we depart from the strictly chronological order that we have thus far followed, in order to gain an overview of the role that health matters played throughout his life.

Early Years

As we saw in Chap. 2, Kurt Gödel's first health issue occurred early in his childhood, when he was around 5 years old. Whether he had any detailed recollections of that period is uncertain; however, his brother Rudolf, four years older, remembered it clearly. It may have been a topic of conversation for their parents, who were no doubt somewhat disturbed at the time. Young Kurt underwent a period of what would now be called an anxiety neurosis, and he was very insecure and easily upset for a time. He also developed an unusually strong and anxious attachment to their mother, so that he cried whenever she left their apartment. This is unusual for a child of 5. In those days, psychotherapy was in its infancy, and a practical man like Gödel's father

© The Author(s), under exclusive license to Springer Nature
Switzerland AG 2022, corrected publication 2023
W. D. Brewer, *Kurt Gödel*, Springer Biographies,
https://doi.org/10.1007/978-3-031-11309-3_9

Rudolf August would probably not have considered it in any case, so there is no medically-reliable record of Kurt's difficulties.

This 'phase' of his behavior passed by with no apparent aftereffects, and for the next several years, he was evidently normally healthy and enjoyed outdoor activities as well as more intellectual pursuits. However, when he was about 8, he suffered an attack of rheumatic fever, with swelling and painful joints and at times a high fever, which caused him to miss school for several months. This would be a serious setback for any child, although Kurt apparently recovered with no lasting physical aftereffects, as diagnosed by several doctors after his recovery from the fever. But Kurt was sensitive, and unusually inquisitive—he wanted (and *needed*) to understand the world as deeply as he could, so that he read about the possible aftereffects of rheumatic fever in various medical books. As a result, he became convinced that he had suffered heart damage, which is indeed a possible complication of a rheumatic fever episode. He clung to that belief for the rest of his life, regardless of medical opinions.

This shows another aspect of his personality, also mentioned in his brother's recollections: Kurt maintained very strong opinions and beliefs, once he had established them, and he was completely resistant to being convinced otherwise, even by much older and better-informed people (such as the doctors who examined him after his illness). One might be tempted to attribute this to his superior intelligence—since he had often observed that he understood many things better than other people, he became so sure of himself that he rejected differing opinions and conclusions, even when he was evidently wrong. But such an explanation is too facile; in fact, his resistance to differing opinions bordered on obsessive/compulsive behavior, and it was probably rooted in a deeper personality dysfunction. In any case, after the illness, he became more reclusive and less willing to undertake physical activity or play outdoors, and he also worried more about his health—what his brother termed '*his hypochondria*'. This later also became a full-blown obsession, coupled with worries about his diet and eating disturbances.

Nevertheless, Kurt's school days passed rather normally, although he had few close friends, as we noted in Chap. 3. He had wide interests and was very successful in his schoolwork, and was no doubt admired by many of his schoolmates. His health seems to have been normal during the rest of his school career, although he avoided strenuous activity. Likewise, his student years, 1924–1929, were apparently a period when he had few health problems, and his preoccupation with his diet and other aspects of his health was not problematic for his daily life. We might suppose that the 15-year period from around 1915–1930 was the healthiest and most worry-free in his entire life.

After 1930

As we have seen in Chaps. 7 and 8, life became more stressful for Kurt Gödel after he completed his doctoral dissertation and began working as an independent researcher. He was more in demand and experienced more confrontational interactions with others as he became better known in his field and took on more professional and teaching responsibilities; and this was evidently reflected in his health. While the sanatrium stay in 1931 mentioned by his brother probably was simply a confusion with events in 1934, there is some evidence that he was indeed suffering under the increased professional stress in his life by the year 1931 and thereafter.

One example of this is the review article on the foundations of mathematics which Gödel promised to write, together with Arend Heyting, for the *Zentralblatt für Mathematik und Ihre Grenzgebiete*, a journal to which he often contributed reviews in the period 1931–36 [cf. Dawson (1983)]. Heyting had originally been asked (in June of 1931) to write the article, but he felt that he lacked competence for some of its chapters, and he requested that Gödel take over responsibility for those. Gödel readily agreed, in a letter dated September 3rd, 1931, and made some suggestions for improving the plan of the article. A detailed chronology of this project is given by Dawson (1997), pp. 83–86, based on correspondence found in Gödel's and Heyting's personal archives, and it is also mentioned by Wang (1987), p. 89. Gödel cited the project as evidence of his professional standing in the *c.v.* that he wrote for his *Habilitation* application in June of 1932 (see Chap. 8).

Unfortunately, every time the editor reminded him of an approaching deadline, he asked for more time. In May of 1933, a few months after completing his *Habilitation*, he wrote to the editor and asked for another extension until September, just before his planned departure for the USA, citing his *Habilitationsverfahren* and other obligations as reasons for the delay, and mentioning that he had been ill for a time [see Dawson (1997), p. 85, and footnote 3]. In the footnote, Dawson wonders if that illness was real or just evidence of his increasing hypochondria, and he mentions that Gödel had read books on mental disturbances and on drug treatments for psychiatric problems during the previous year, suggesting that he was now worried about his *mental* as well as his physical health. This is the first indication in the written record that he might have been questioning his own mental state.

In Princeton, the new *Institute for Advanced Study* (IAS), which had been in the planning stage since 1930, was scheduled to open in early October of 1933, in time for the fall semester. It was associated with, but not part of Princeton University (and its School of Mathematics used office space in the

University's *Fine Hall* for a few years until its own building was completed). Gödel planned to be there for the opening, and informed his contact in Princeton, Oswald Veblen, of that plan. What actually occurred is prescient of his later psychiatric problems and is also typical of his difficulties in dealing with everyday life. He was seen off at the rail station in Vienna by a group of his colleagues, as Olga Taußky-Todd later recalled[1]:

> On Gödel's first visit to Princeton in 1933, several colleagues, including myself, went to see him off at the train station, where he boarded the Orient Express *[at the* Westbahnhof *in Vienna, on Sept. 23rd]*. Later he confided to me that this was not his actual departure from Vienna. He was taken ill before reaching his boat; he took his temperature and decided to return home. His family persuaded him to try again, however.

Gödel in the end sailed from Southampton on the *Aquitania*, leaving September 30th and arriving in New York on October 6th. He had informed Veblen by cable on September 25th that his arrival would be delayed due to illness.[2] As to his planned review article with Heyting: Finally, while Gödel was in Princeton in early 1934, he sent a letter to the editor of the journal, saying that he could not deliver the manuscript as promised (by then for the end of 1933) and would send it in July 1934. The editor lost patience at that point, and published Heyting's part of the article separately. This episode was rather embarrassing for Gödel, and unfortunately, it was the beginning of a pattern which repeated itself all too frequently in his later life. Thereafter, he hardly published any collaborative work, and he failed to deliver several promised manuscripts, even after repeated extensions of their deadlines, sometimes over several years. What Wang calls "a fragmentary draft" of his part of the article with Heyting was found in his papers after his death, and it shows that he had in fact made little progress during the 2-½ years in which he had ostensibly been working on it.

One may see this failed project as simply a youthful overestimation of his own ability to formulate the article in a short time, coupled with the press of other work (such as reviewing draft chapters for books by Menger and Carnap, which he may have worked on with a higher priority, and his duties at the University), along with the pressure of his *Habilitation*. But it was no doubt also related to his great attention to details and his perfectionism (which some people have called *legalism*), that kept him from ever being satisfied with his writings even when they had progressed much further than he had on the planned article with Heyting. This episode might be seen as the first small crack in the otherwise smooth façade of an aspiring and successful scholar.

Kurt Gödel also suffered from a certain loneliness while in Princeton. In his obituary for the Royal Society, of which Gödel was a Foreign Member, *Georg Kreisel* (1980) mentions that Gödel was subject to two major frustrations that contributed to his breakdown after returning to Vienna from Princeton in June of 1934. He led a strictly monastic life during his more than 8 months in Princeton, and this was frustrating to him; Kreisel writes,[3] "... *there were two frustrations, each perhaps sufficient to trigger a breakdown in someone of Godel's personality. More than twenty years later he still spoke of the frustrations of—tacitly—his bachelor life in Princeton where he had just spent a year*". Stephen Budiansky, for his (2021) biography of Gödel, examined the latter's personal notes (his *Protokolle*) from the first phase of his psychiatric problems, and found evidence that he was to a certain extent sexually dominated by Adele. Quoting Oskar Morgenstern's diary,[4] "*the wife is appalling. A riddle. One can hardly speak with him when she is present... He is doubtless under her spell; incomprehensible, how he can stand it*". Gödel apparently took a sober view of sexual intimacy: "*'A visit to Adele', he observed, has a wonderful way of clearing his mind—'a walk has a similar effect', he added*".[5] Another frustration was his family's disapproval of his relationship with Adele, which caused him to essentially hide it over almost a decade.

First Breakdown

Following his return to Vienna in June of 1934 from his stay at Princeton, Gödel did indeed begin to experience genuine psychiatric difficulties, quite possibly related to built-up stress resulting from his stay in an unfamiliar and to some extent challenging environment. He had not eaten well while in Princeton, probably in part because of the unfamiliar food (and the early closing hours of restaurants in Princeton),[6] and in part because he suffered from an inflammation of the gums after having a tooth filled.

According to Wang (1987), Gödel left Princeton in late May and crossed the Atlantic on the Italian liner *S.S. Rex*, arriving in Genoa on June 3rd. He disembarked there, but he then stayed overnight in Milan and spent three days in Venice before returning home on June 7th. A few weeks later, on July 24th, 1934, his mentor Hans Hahn died after surgery for a cancer which had been diagnosed only a short time before. He was just 54, and his death came suddenly and unexpectedly for his friends and colleagues.

Soon thereafter, Gödel suffered a 'nervous breakdown', an episode of depression and anxiety attacks, perhaps triggered by Hahn's unexpected death, and/or the frustrations mentioned above. He was initially treated by Dr. *Julius Wagner (von) Juaregg*, a well-known psychiatrist (and Nobel prizewinner, in 1927 for Physiology or Medicine, for his introduction of malaria inoculation to treat *dementia paralytica*). In 1934, Wagner was practicing at the *Clinic for Psychiatry and Nervous Diseases* in Vienna, where he was officially retired, but still professionally active. In the autumn, Gödel's condition had become so acute that he was sent in October to the sanatorium in Purkersdorf, about 15 km due west of Vienna. That sanatorium (Fig. 9.1) was not a psychiatric clinic in the modern sense, but rather a sort of care home where those patients who could afford it could lead stress-free lives and restore their mental and physical health by resting, eating a healthy diet and taking moderate exercise. It had been built in 1904/05, originally intended to be a 'mineral spa and cure park'. Gödel was sent there because his doctors and his family felt that he was so depressed that he might commit suicide. Just what sort of therapy, if any, Gödel received while there is unknown, but the prognosis was good and he was in fact released by early November, as evidenced by the fact that he attended the Menger Colloquium on November 6th. This was however only the first of his stays in various sanatoria in the coming years. Wang (1987) quotes Menger [(1994), p. 211] from around this time, who remarked,[8]

> All in all, Gödel was more withdrawn after his return from America than before, but he still spoke with visitors to the Colloquium. To all members of the Colloquium, Gödel was generous with opinions and advice in mathematical and logical questions. He consistently perceived problematic points quickly and thoroughly, and made replies with greatest precision in a minimum of words, often opening up novel aspects for the inquirer. He expressed all this as if it were completely a matter of course, but often with a certain shyness whose charm awoke warm personal feelings for him in many a listener.

Gödel had planned to return to the IAS in Princeton for the winter semester 1934/35, but postponed the visit for a year due to his health problems. The years 1933/34 saw drastic political changes, particularly in Germany but also in Austria. Menger reports that Gödel was aware of those events (which we will discuss in more detail in Chap. 10) and spoke about them, but was strangely '*impassive*', as though they did not concern him in any real way. His brother Rudolf mentioned that their whole family was '*not political*', but in Kurt's case, this seems almost pathological, and his lack of responsiveness about current events and political disasters was later the cause of a certain estrangement between Gödel and Menger, who could not accept Gödel's apparent lack of interest or empathy for his fellow humans in the face of the political and social catastrophes of the later 1930s.

Fig. 9.1 The sanatorium *Westend* in Purkersdorf, where Kurt Gödel spent some weeks in October of 1934. *Photo* from Wiki Commons, by Roman Klementschitz, Vienna[7]

In the spring of the following year, 1935, Gödel gave his second lecture course in Vienna, on '*Ausgewählte Kapitel der mathematischen Logic*' ('Selected Topics in Mathematical Logic'), beginning on May 4th. (His first lecture course had been held in the summer semester of 1933, soon after he completed his *Habilitation*, and was on the topic '*Grundlagen der Mathematik*' ('Fundamentals of Mathematics').)

He also was active in Menger's Colloquium and published several short papers (one on the length of proofs, on June 19th, showing that proofs which are long and tedious in first-order logic may be made much briefer and more compact when carried out in second-order logic). This can be seen as the beginning of 'speed-up theorems' in computational (complexity) theory. He also began working on set theory, and on Hilbert's first problem, the *continuum hypothesis*, and he discovered the usefulness of *constructible sets* in that connection. He informed von Neumann of this work, but he did not publish it at the time.

But he did all this in spite of the state of his (mental) health, which necessitated several shorter stays at the sanatorium in Purkersdorf, as Gödel later recalled. In August, 1935, he also spent some time in another sanatorium, in *Breitenstein am Semmering*, a village located in the foothills at the eastern end of the Austrian Alps, about 80 km (50 mi.) to the southwest of Vienna,

Fig. 9.2 The Sanatorium and Recuperation Home in *Breitenstein am Semmering*, where Kurt Gödel stayed in August 1935. The building, by then in disrepair, was razed in 2006. *Photo* From an old postcard, public domain

easily reachable by rail. That institution, called '*Sanatorium und Erholungsheim Breitenstein am Semmering*' ('Sanatorium and Recuperation Home in *Breitenstein* on the *Semmering* '; see Fig. 9.2), had been built in the years following 1903 on the initiative of Henriette Weiss, a social activist and supporter of the education of women for nursing careers. It was conceived as a '*Volkssanatorium*', i.e., a rest home at a higher elevation (880 m. above sea level), where people of all means could be treated in particular for tuberculosis. The '*Semmering*' refers to a small mountain in the pre-alps, around 1500 m. (5000′) high. The village of *Breitenstein* is located on its slopes, at above 800 m. altitude. By the time of Gödel's stay there in 1935, the Sanatorium was a general sort of rest home, not specifically for tubercular nor for psychiatric patients. Gödel probably went there voluntarily for a vacation and relaxation in preparation for his presumably strenuous second visit to Princeton, planned for the coming months.

Princeton *Redux*

In the autumn of 1935, Gödel again set out for Princeton, to begin the stay originally planned for the fall semester of 1934 and postponed for a year due to his health problems. He left Vienna in mid-September and crossed the Atlantic on the Cunard ship *MV Georgic*,[9] arriving in New York on September 28th. As Wang points out,[10] Paul Bernays was on the same ship, but whether they met each other on board is in question; Wang (1987) presumes that they probably didn't, since Bernays was traveling in tourist class. Goldstein (2005) suggests[11] that they did, as does Dawson (1997), and that Gödel informed Bernays in detail about the proof of his Second Incompleteness Theorem while on board, which Bernays later published [with Hilbert (1939)]. Wolfgang Pauli was also on that ship, and he sent Gödel a polite note and certainly *did* meet him during the voyage.

Gödel's stay in Princeton was much briefer than planned; he arrived there at the beginning of October, but in mid-November, he suddenly resigned his position as a visiting scholar due to health problems and exhaustion, and, with Veblen's help, returned to Europe on the steamer *Champlain*, leaving New York on November 30th and arriving at Le Havre on December 7th. His family had been warned of his impending return, and he hoped that his brother would meet him in Paris and accompany him back to Vienna; but Rudolf was occupied with his medical duties and probably did not take Kurt's request very seriously. Kurt managed to travel to Paris and spent several days in a hotel there, making a long and expensive telephone call to Rudolf in Vienna, after which he was able to make his way back alone.

Kurt's brother Rudolf in fact later told two different stories about Kurt's arrival back in Vienna.[12] In an interview given to John Dawson in July 1983, he said that Kurt had called him after arriving in Paris, but that he was unable to go there to accompany him home, and Kurt came alone after spending 3 days in a hotel. Later, in an interview with Eckehart Köhler in 1986, Rudolf said that he had in fact gone to Paris to pick up his brother.

Whatever the details of the last stage of his journey home, Kurt was in a poor state when he arrived, as Menger recalled: "*Gödel returned to Vienna in December in quite a bad state of health and mind*". He required considerable time and repeated stays at Purkersdorf and various other sanatoria to recover. Menger was at the time carrying on a correspondence with Veblen in Princeton, hoping to find a position in the USA, since he had decided to emigrate due to the increasingly difficult conditions in Austria, in particular the growing numbers of Nazi sympathizers there, especially in the universities. In a letter dated December 17th, he wrote, "*It is too bad that he* [Gödel] *overworked himself to a degree that he needed soporific things, but perhaps still worse that he actually started easing up on them*"[13] Menger also reported somewhat later, "*When I saw him again in 1936, he told me he was striving to prove the consistency of the continuum hypothesis in set theory*".

The Disastrous Year 1936

In January of 1936, Gödel's condition worsened, and his mentor Moritz Schlick, worried that he might not recover, sent an almost desperate letter to the psychiatrist Otto Pötzl, who was Wagner-Juaregg's successor at the Neuropsychiatric Clinic, describing Gödel's contributions to mathematics and begging Pötzl to do his best to restore Gödel to health.[14]

Gödel postponed his planned lecture courses in 1936 more than once, and he spent "*several months*" in sanatoria and health spas, on and off. He wrote to Veblen (but crossed out the sentence in his draft letter) that he had stayed "*in a sanatorium for nervous diseases for several months in 1936*".[15]

That was the *Sanatorium Rekawinkel*, in the valley of the River Wien, in the Vienna Woods to the southwest of the city (Fig. 9.3). It is in the same direction as Purkersdorf, but about 12 km further west. In contrast to the *Sanatorium Purkersdorf*, which was primarily a rest home, *Rekawinkel* was specialized as a clinic for psychiatric disorders. Gödel was committed there in part because his symptoms had worsened and even his mother was afraid that he might become violent or commit suicide. He was later convinced that he had been given drugs there, against his will, at night. He thought that he had been dosed with a heart stimulant[16] (he probably was indeed suffering from cardiac arrhythmia and tachycardia—a racing, uneven heartbeat; these are common symptoms of anxiety attacks and are often psychosomatic in origin). He repeated those suspicions to his psychoanalyst, Dr. Philip Erlich, many years later in Princeton.[17] He also told Dr. Erlich of an abortion which Adele had sometime before 1936, and for which he had guilt feelings and fears of being arrested, which may have contributed to his breakdown early that year.

Apparently, he was at least spared the dangerous and drastic treatments popular at the time, electro-shock therapy and insulin-shock therapy, both intended to cause a deep unconsciousness in the patient that was believed to induce a sort of 'reset' of the brain, allowing a 'fresh start' without the psychosis afterwards. Gödel was however worried about being poisoned, and refused to eat, losing considerable weight. Adele visited him and encouraged him to eat, feeding him like a baby and testing the food herself so that he could see that it was not poisoned.

According to receipts found in his legacy, Gödel also spent some time in June of 1936 in *Golling*, near Salzburg, a resort region on the *Salzach* River (probably at a spa hotel there). Later, in August, October, and November, he spent three successive stays at the spa town of *Aflenz*, a mountain village in the *Steiermark* (Austrian State of Styria, to the southwest of Lower Austria and

Fig. 9.3 The Sanatorium at *Rekawinkel. Photo* from an old postcard, public domain

Vienna), where his family had vacationed earlier (see Fig. 9.4). Interestingly, he was accompanied by Adele on the second of those stays,[18] from October 2nd to 24th. (It later became known that she was also serving as his 'food tester' there, since he was still worried about eating spoiled—or poisoned—food, an early sign of his increasing paranoia).

Adding to Gödel's depressed mental state, on June 22nd, 1936, his mentor and friend Moritz Schlick was assassinated on a stairway in the University's main building while on the way to his last lecture of the semester. This unexpected disaster was doubtless a great shock to Gödel and probably set back his recovery considerably. We will examine it in more detail in a later chapter; here, we are primarily interested in its effects on Gödel's state of mental health; and those who knew him at the time saw them as very detrimental.[19] His stay at *Golling*, far from Vienna, may have been an effort to escape for a time from the city where this tragedy had occurred.

1936 was also, understandably, not productive for Gödel's career as a mathematician. Consulting the *Collected Works*, Vols. I and III (and also Dawson 1983), we can see that he had only two publications in that year, both published in the *Reports* of the Menger Colloquium. The first was a very brief note quoting a comment that he made on economics theory during the Colloquium of November 6th, 1934, replying to his friend *Abraham Wald*; and the second was an (important, but initially neglected) paper from his

Fig. 9.4 The spa town of *Aflenz Kurort* (*Aflenz* Health Spa), on the slopes of the Austrian Alps. *Photo* Wiki Commons.[20] Karl Menger had also spent nearly a year there around 1923/24, in a successful effort to cure his tuberculosis

talk at the Colloquium of June 19th, 1935, on the Length of Proofs. He also had no publications at all in 1937, reflecting his lack of activity during the previous year. In the list of his *Unpublished Essays*, there is also no entry between 1933 and 1938. Nevertheless, in private and between sanatorium stays, he was working on set theory, as we can see from Menger's remark, quoted above.

Back to Work: 1937

Gradually, Gödel's condition improved, and by early 1937, he was back at work. He had announced a lecture course based on his 1935 results, titled '*Axiomatik der Mengenlehre*' (Axiomatization of Set Theory), the third course that he gave as *Privatdozent*. Wang was able to find only one registration slip for the course (from his own teacher, *Wang Xian-jun*); but he was informed by the latter that there were in fact 5 or 6 students participating, among them *Andrzej Mostowski* (1913–1975), a Polish mathematician who later worked in foundations of mathematics.

Later in that summer, Gödel corresponded with von Neumann, who subsequently visited in Vienna and met with Gödel on July 17th. He also corresponded with Karl Menger in July and in December. Menger had accepted a position at Notre Dame University, near South Bend, Indiana/USA, where he went (initially as Visiting Professor) in January of 1937. He invited Gödel to visit Notre Dame during his next planned stay in Princeton, scheduled to begin in autumn 1938. In his December letter, Gödel wrote,[21] "*As you know, I had bad experience in America with my health and hence do not want to bind myself for a longer period in advance*". However, he accepted the invitation to the IAS and indeed also went to Notre Dame in early 1939.

That acceptance may have been one of the factors which decided Gödel's mother Marianne to return to Brünn. His brother Rudolf, quoted in Chap. 5, attributed her decision mainly to economic factors (her house was in Brünn, as well as the investments and her widow's pension which constituted the rest of her inheritance from Rudolf August; and life in Vienna was expensive). But he also mentions Kurt's decision to visit America for a third time—Marianne was probably opposed to that visit, since she, like Kurt himself, would have noticed that his breakdowns seemed to occur in connection with his visits there. But she could not control him, neither in terms of his travel plans nor concerning his relationship with Adele, and this may have made her feel that she had nothing more to do in Vienna. Rudolf could certainly take care of himself without her help. In any case, she went back to her villa in Brünn in the autumn and remained there for the next seven years.

An immediate result of her departure, evidently, was that Kurt and Adele finally moved together and lived essentially as a married couple, although their formal marriage ceremony took place nearly a year later. In his interview with Werner DePauli-Schimanovich and Peter Weibel for their television documentary,[22] Rudolf Gödel mentions that Kurt lived in "*a few dwellings*" which he himself had never visited, and he "*didn't know*" if Kurt had lived together with Adele in any of them. But as we have seen, Kurt and Adele went together to Aflenz in October 1936, and registered in the hotel there as husband and wife; and that may not have been their only trip together. Rudolf's remark was perhaps mainly directed toward their apartment in Vienna's 19th District (*Döbling*, where Karl Menger had grown up and attended *Gymnasium*). They moved there, to a house at *Himmelstraße* 41–45 ('Street of Heaven', a highly symbolic name), on November 11, 1937, and lived there for almost exactly two years, departing on November 9th, 1939. We shall see some photos of the house, in the village of *Grinzing*, part of the 19th District, and learn of some more details in Chap. 11. This was Kurt's first apartment outside the inner ring of Viennese districts, and it probably was a certain relief for him to be almost in the countryside, away from the University area with its constant demands. Budiansky (2021) suggests that he chose this relatively far-away address to avoid letting other people know that he was living with Adele. In any case, his health was much better than in the previous year, and he withstood not only the stress of his next America visit but also that of numerous other problems which arose for him in the following two years.

Life in Vienna however must have become somewhat lonely for Kurt Gödel by 1937. Many of his former friends and colleagues had departed that city (and indeed Austria) in one manner or another, beginning with *Herbert Feigl* in 1931. Feigl (1902–1988) was a fellow student and friend of Gödel's, and we have encountered him previously as an early member and supporter of the Vienna Circle. Feigl spent a fellowship year at Harvard, and then after his marriage in 1930, he emigrated to the USA, first settling in Iowa and later moving to Minnesota, where he had a distinguished career at the University of Minnesota, retiring in 1971. He remained loyal to logical positivism (he preferred the name *logical empiricism*) until the end of his life.

The next to go was *Rudolf Carnap*, one of Gödel's closest associates (and teachers) in the years 1927–30. In 1931, Carnap accepted a position at the *Karlsuniversität* in Prague, and left Vienna, although he often returned to attend colloquia and the meetings of the Circle. But the rise of Nazism put him at risk by the mid-1930s, and in 1935, he emigrated to the USA (probably with help from W.v.O. Quine, from Harvard University. They had met in 1933 when Quine visited Prague); cf. Chap. 4.

Hans Hahn, Gödel's teacher and his mentor for his doctorate and his *Habilitation*, died unexpectedly at only 54 years of age in the summer of 1934, as we have seen (at the same age and of similar causes as Kurt Gödel's father, just over five years earlier). *Marcel Natkin* (1904–1963) was the third member of the threesome (Gödel, Feigl, Natkin) who spent many hours together in coffeehouse discussions on philosophy, logic and other themes.[23] Natkin was born in Łodz, Poland and worked as a graduate student in Schlick's group in Vienna, receiving his doctorate with a thesis on '*Causality, Simplicity and Induction*' in 1928. He was an admirer of Mach's philosophy. He however gave up philosophy and left Vienna, moving to Paris in the early 1930s; there he started a successful second career as a photographer.

As we have seen, *Moritz Schlick* was murdered by a deranged former student at the University in June, 1936, effectively marking the end of the Vienna Circle. The remaining members of the Circle rapidly dispersed soon after: *Franz Alt*, whose interest lay in economics, left Vienna in 1938 for New

York, and two other economists, *Abraham Wald*, with whom Gödel had been friends, and *Oskar Morgenstern*, both also went that same year to the USA (Wald to Colorado, and Morgenstern to Princeton, where he and later also his wife became good friends with the Gödels). *Friedrich Waismann* moved to Cambridge University, joining his idol Wittgenstein there in 1937. He later moved to Oxford, where he remained until his death in 1959. His book of conversations with Wittgenstein, '*Ludwig Wittgenstein und der Wiener Kreis*' was published posthumously by B. F. McGuinness in 1979. *Olga Taußky*, after a year in the USA, also went to Cambridge on a fellowship for two years, later working in London. She visited Vienna in September 1937 and spoke to Gödel about possible positions abroad. He was still keeping notes (his '*Protokolle*') about his daily life, and he was by that time evidently considering seeking a position abroad,[24] in part because he had little income in Vienna and needed some financial security. In his notes from this period, Kurt also expresses uncertainty about his personality and the impressions that others might have of him.[25]

Karl Menger's departure in early 1937 was in some sense very sad for Menger himself, especially since his Colloquium was now orphaned; and he was also particularly worried about Gödel, who he felt needed guidance. In a letter to *Franz Alt*, also a participant in the Menger Colloquium and friend of Gödel's, he wrote.[26]

Tief betrübt bin ich darüber, so wenig für den so schönen und mir so lieben Wiener Mathematikerkreis tun zu können. Ich glaube, ihr sollt alle von Zeit zu Zeit zusammenkommen und insbesondere bewirken, dass Gödel am Kolloquium teilnimmt. Das ist nicht nur für alle anderen Teilnehmer von größtem Nutzen, sondern obwohl er das vielleicht nicht wahrhat, auch für ihn selbst. Der Himmel weiß, in was er sich einspinnen könnte, wenn er nicht von Zeit zu Zeit dich und die anderen Wiener Freunde spricht. Sei deshalb auf meine Verantwortung wenn nötig auch zudringlich.

(English translation, by the present author):

I am extremely sorry to be able to do so little for the brilliant circle of Viennese mathematicians whom I hold so dear. I believe you should all get together from time to time, and especially, you should ensure that Gödel takes part in the colloquium. That would be of great advantage not only for all the other participants but also for Gödel himself, although he may not realize that. Heaven knows what he might get involved in if he doesn't talk to you and his other friends in Vienna now and then. Be insistent, if necessary, on my responsibility.

Menger's almost touching concern stands in stark contrast to his estrangement from Gödel sometime later, while Gödel was visiting at Notre Dame and teaching a course together with Menger. It was triggered by Gödel's lack of empathy for refugees from Europe. Menger apparently could not see that this was most likely due to a personality disturbance, and not to callousness or selfishness—nor to simple naïveté—on Gödel's part.

Gödel was in fact presented with an alternative to the Menger Colloquium in October, 1937, when *Edgar Zilsel*, a student of the philosopher Heinrich Gomperz, set up a seminar group. Gomperz had been dismissed from the University during the brief '*Austrofascism*' period in 1934, but stayed on in Vienna until 1935, and had led various discussion groups during that interval. His assistants, first Viktor Kraft, then Edgar Zilsel, hosted the groups privately, and Gödel joined in the founding of Zilsel's group in 1937. Dawson[27] gives a detailed summary of the members of that group and Gödel's participation in it, which culminated in a report that he gave to a meeting of the group on January 29th, 1938, on consistency problems in logic. Its significance is that it was apparently his last lecture in Vienna. The manuscript was found in his papers after his death and reconstructed as [Gödel (*1938a*)].

Just how Gödel reacted to Menger's absence and that of his other friends is not clear; it is however certain that having Adele as a full-time partner at this point, even if reluctantly, was good for his health, both mental and physical. Kurt could relax in her presence, and he didn't need to 'keep up a front' as he did with many other people, probably including his mother and brother. And she kept him supplied with food that he was not afraid to eat. So, the two of them finished out the year 1937 in splendid isolation in Vienna–*Döbling/Grinzing*.[28]

The *Ostmark*—1938/39

The year 1938 was a decisive one for Austria. Prussia, which had been an annoying but small rival to Austria in the time of Maria Theresia in the eighteenth century, had become the driving force of the *Deutsches Reich*, and now, since the Nazi takeover in early 1933, the menacing '*Third Reich*'—and a very much bigger brother to little Austria. In March of 1938, the *Anschluss* took place: Germany annexed Austria, making it the 'Eastern province' (*Ostmark*) of the *Reich*, and Hitler arrived in Vienna on March 15th to take part in a triumphal parade to celebrate the event. Budiansky (2021) gives a striking description of the unleashed Nazi frenzy which occurred that night in Vienna, quoting *Carl Zuckmayer*, who witnessed it. [Zuckmayer was a very good contemporary witness of much of the history of the first half of the twentieth century, and his autobiographical book, '*Als wär's ein Stück von mir*' (Zuckmayer 1966) is well worth reading; an English translation ('*A Part of Myself*', 1993) is available].

Gödel, in his correspondence at the time, made no mention of all this, although he had predicted it in earlier conversations with Menger (see Chap. 11). This was part of his personality, reflecting an underlying disorder—on the one hand, his single-mindedness, which kept him focused on his 'inner life', and on the other, a notable lack of (felt) empathy, which was later the cause for a break with Menger. That was not due to any callousness in his thinking—on the contrary, he was rather oversensitive; but he suffered from a personality disturbance which could not be overcome simply by rational thought, however much he wished to do so.

However, in his correspondence from spring 1938 were also official letters from the University, informing him that his *Lehrbefugnis* was cancelled as of April 23rd. The Nazis wasted no time in eliminating any position, such as the *Privatdozentur*, which might harbor political dissent. They wanted to put everyone under the control of the State as quickly as possible. Gödel, in turn, felt that annulling his *Lehrbefugnis* was a violation of his *rights*, a notion that Karl Menger found laughable (in view of the fact that so many people had no rights at all under the Nazi regime).

According to Dawson (1997), Gödel was working in late 1937 and the first half of 1938 on the *Generalized Continuum Hypothesis* (GCH), originally proposed by Cantor and extended by Jourdain, which he had been considering since at least 1935; but he had not yet published any results. In January of 1938, von Neumann again visited Vienna, and met with Gödel in the city. They apparently discussed plans for Gödel's next visit to Princeton and for publication of his results on the GCH (which he had divulged only to von Neumann and to Menger, asking each of them to keep it to himself). He continued working on that topic throughout early and mid-1938, apparently oblivious to the new regime in Austria (his Austrian citizenship was automatically converted into citizenship of the *Deutsches Reich* following the *Anschluss*). After some negotiations with Veblen, it was agreed that Gödel would visit the IAS/Princeton in November/December, lecturing on his set-theoretical results, and then go early in the following year, 1939, to Notre Dame, where Menger had invited him (via the President of the University) to give a joint lecture course on mathematical logic.

The *momentous event* in Kurt Gödel's life later in 1938 was his marriage to Adele Porkert on September 20th. We will describe that event in more detail in Chap. 11; here, again, we are concerned with his health, which was generally stable in 1937/38. His departure for Princeton a few days after his marriage was chaotic, as usual. Surprisingly, he was able to obtain a visa for the USA and an exit visa from the German *Reich* without problems; but booking passage on a ship to New York turned out to be difficult, since every available berth was already sold out. He finally was able to book a ticket on the *S.S. Hamburg*, leaving from that city and scheduled to arrive in New York on October 7th. Dawson[29] offers some interesting speculations on why he was not able to use the ticket that he had booked, which we will consider later. In any case, he was finally able to rebook on its sister ship, the *S.S. New York*, sailing from Cuxhaven, and arrived in New York on October 15th, a week later than planned.

Princeton, 1938—Notre Dame, 1939

On arrival in Princeton, Gödel chose not to live in a rented apartment or house as he had on his earlier visits, but instead stayed at the '*Peacock Inn*', a colonial-style hotel dating from the later eighteenth century. It was originally located two blocks southeast of its present site, at the corner of Nassau St. and University Place (see the map in Fig. 10.3). In 1875, it moved to the current location on Bayard Lane, and was owned at that time by a Princeton professor, a Mr. Libbey. The present hotel was founded in the same building in 1911 by Joseph and Helen O'Connor, who renamed it after an inn in the British Midlands, adopting the peacock as its emblem, a sign of 'royalty, good food, and good luck'. Einstein lived there for some time in 1933 while waiting for his (first, rented) house at 2 Library Place to be readied. (He bought his later, last house at 112 Mercer St. in 1935).

The Inn is not far from the (former) Fine Hall, a building belonging to Princeton University, a 'pet project' of Oswald Veblen, and at that time still relatively new, housing the Mathematics Department. The IAS School of Mathematics was also still located there, pending the completion of its own building, Fuld Hall (this occurred only later, in the autumn of 1939). Gödel's stay in a hotel probably reflected his feeling that this visit would be short. His lecture series was scheduled to begin on November 1st, and he evidently worked hard after arriving to prepare his lectures and also a first publication on his consistency results in set theory, which he sent to the *Proceedings of the National Academy of Sciences* (PNAS) on November 9th [cf. Gödel (*1938*)].

However, he also found time in late October to travel to New York City [to a conference of the *American Mathematical Society* (AMS); see Gödel (1939)], where he conferred with Menger on the details of their planned joint lecture course at Notre Dame, and also met Emil Post, who worked at City College (CCNY). We heard about that meeting in Chap. 7 (in the section on '*Tales of Bernays, Post and Henkin*'). The meeting and the subsequent correspondence between Post and Gödel were rather sad, perhaps for both Gödel and Post, but certainly for the latter. Both of them were reserved and shy, and not aggressive about trumpeting out their results. Post, who suffered from chronic depression and bipolar syndrome, died young (at age 57, in 1954), from heart failure or a burst artery during electroshock therapy, a common but drastic remedy for psychiatric disorders in those times.

Gödel's health held up, and he finished his lecture series on set theory at the IAS and went to Notre Dame in early 1939. Karl Menger (1994) describes Gödel's visit there in his autobiography[30]:

> During his stay at Notre Dame, Gödel appeared to be in fairly good health, but not particularly happy. He lived on campus—I do not recall whether for the whole period, but certainly for a large part of the semester. But he had quarrels with the Prefect of his building....[31]

In other words, his *physical* health was good, but his *mental* state less so. Kurt Gödel's disputes with the Prefect were typical of his insistence on his *rights*, while the Prefect—an old priest—had a strict sense of order and how things were to be done. Gödel had similar conflicts with the new Nazi authorities in Vienna, over more important topics, to be sure; and there, he also insisted on his *rights*, which, as Menger had reminded him, was rather futile given the situation. In Chap. 11, we will learn more about his activities while at Notre Dame.

Before leaving the US in early June, he passed by Princeton again and conferred with Flexner about returning in the autumn. His problem was to obtain a long-term visa, and this preoccupied him in the coming months, along with several other 'administrative' problems, as we shall see. Gödel sailed on June 14th, 1939 on the *Bremen*, arriving in Germany a week later.

A Brief and Hectic Interlude in Vienna

Back in Vienna, Kurt was reunited with Adele, but was soon confronted with numerous (non-scientific) problems. His original immigration visa for the USA (dating from 1933) had expired and he had entered the country in October 1938 on a visitor's visa (which would allow him to remain only until the following June). His *Lehrbefugnis* in Vienna had been revoked the year before, and he would have to apply for a new position (the *Dozentur neuer Ordnung*, 'Docent of the new Order'), a salaried position under the control of the Nazi authorities—by October 31st in order not to be permanently excluded from university employment. He had trouble transferring funds from his Princeton bank account to Vienna, and his savings were running low (and were to a considerable extent tied up in the villa in Brünn, in which he had inherited a 1/3 interest; but of course his mother was living there, and selling it was not an option).

Czechoslovakia—that is Bohemia and Moravia, since Slovakia had declared itself independent—had in the meantime been completely taken over by the *Reich* as a 'Protectorate', in violation of the Munich Agreement of September 1938, where Hitler had limited his claims to the *Sudetenland* and Moravian *Schlesien*, mostly populated by ethnic Germans. War was becoming more and more imminent, although Gödel seems not to have realized this.

And—the final threat—he was called up for a medical examination for military service, where, to his great surprise, he was declared fit for service. He could have mentioned his stays in psychiatric clinics in 1935/36, but that would have been exceedingly dangerous, as Nazi policy included 'euthanasia' ('mercy killing' of people judged to be mentally ill or incompetent). He also concealed his medical history from the American authorities, fearing (probably quite rightly) that it would endanger his chances of obtaining a long-term visa.

For a time, Gödel tried to keep all his options open, applying in September for the *Dozentur*, and negotiating at the same time for a non-quota entry visa to the USA. This in turn was held up by bureaucratic problems, involving his status as a 'professor' in Vienna and the terms of his possible employment at the IAS (which was not a teaching institution in the traditional sense). The annulment of his *Lehrbefügnis* in Vienna[32] of course endangered his status as 'professor', but the recommendations submitted from the IAS to the State Department avoided mentioning that.

What was perhaps the final straw, convincing him to leave Vienna permanently, was an attack on Gödel by a gang of young Nazi thugs on the *Strudelhofstiege*, a stone stairway leading down from the end of the *Strudelhofgasse*, the street with the buildings where the mathematics and philosophy lectures were held, to a small street at a lower level, the *Pasteurgasse* (Fig. 9.5).

(This public stairway is the centerpiece of the eponymous novel by *Heimito von Doderer*, which describes the years just before and soon after WWI in Vienna. See Doderer (1951)—English edition 2021. It has thus become iconic for many readers, long after Gödel's unpleasant experience there).

Fig. 9.5 The *Strudelhofstiege* in Vienna. Above is the *Strudelhofgasse*, a short street leading from the *Währingerstr.* to the *Pasteurgasse* (foreground), the latter about 10 m lower. The staircase connects the two streets for pedestrian traffic; compare the map in Fig. 4.2. *Photo* Wiki Commons[33]

Just why Gödel was attacked is not entirely clear; he may have been mistakenly thought to be Jewish, or he simply seemed 'different' or 'too intellectual' to his young attackers. His glasses were knocked off, but he was not injured, and Adele famously drove off the attackers with her umbrella. But the incident no doubt left Kurt somewhat shaken and uncertain about his future safety in Vienna.

Gödel thus continued pursuing his plan to return to Princeton, aided by Veblen and von Neumann, but most energetically by IAS Director Flexner. After the latter retired as Director, due to a disagreement with the IAS faculty about new appointments, his successor *Frank Aydelotte* continued the correspondence about Gödel with the State Department and the German Consulate in Washington, and finally succeeded in obtaining 'non-quota' visas for Kurt and Adele.

Kurt also mentioned, in his correspondence with the education authorities in Vienna[34] (and, in greater detail, in a statement found in his papers, dated December 11th, 1939) that his financial assets were dwindling and that he needed to begin earning a salary. In view of that situation, it is rather surprising that Kurt and Adele purchased an apartment in the Central District of Vienna (*Hegelgasse* 5, Wien-1) in November, and moved there, leaving their bucolic dwelling in *Grinzing* on November 9th. In the event, they spent only two months living in the city, since they left Vienna (forever) in early 1940, the visa problems having been solved in their favor in the meantime. We shall hear more about their new apartment in Central Vienna in Chap. 11.

Back to Princeton—For Good!

The trip to Princeton was no longer so simple as in earlier years. The North Atlantic was unsafe for travel by ship, since the German *U-Boot* fleet was actively trying to suppress trade between the Americas and Europe and the British were intercepting ships and interning German citizens. The alternative was the eastern route, around ¾ of the Earth's circumference, across Asia and the Pacific and finally by transcontinental rail across America to its East Coast; and that route was in fact stipulated in their German exit visas. With the help of the IAS, they had been able to book railway tickets, passing through the *Reich* and occupied Poland to *Vilnius* (Lithuania) and Latvia, crossing the Soviet border at *Bigosovo* (now in Belarus) and on to Moscow, where they boarded the famous *Trans-Siberian Railway*, crossing Central Asia to *Vladivostok*, the eastern outpost of the Soviet Union. From there, their trip continued by ship and train to Yokohama, Japan, and then across the Pacific to San Francisco (on the *President Cleveland*, a ship belonging to the American President Lines, a US shipping line which served the Pacific area. Their original booking on the *President Taft* was missed due to their late arrival in Yokohama). This adventuresome voyage will be the subject of Chap. 13, and we note here only that it was an exhausting and potentially very stressful journey that lasted nearly two months; and given his past travel history, it would probably have been impossible for Kurt if he had been traveling alone. With Adele's competent, practical, and effective help, he survived the trip with no major aftereffects.

The Gödels finally arrived in Princeton on March 10th, 1940. Dawson (1997), Hao Wang (1987/1997), and Budiansky (2021) all quote extensively from *Oskar Morgenstern*'s diaries and memoirs about the Gödels' first months in Princeton, and later on. When Gödel arrived in Princeton in March 1940, Morgenstern renewed his acquaintance (initially out of interest in the state of affairs in Vienna). Morgenstern, in his diary, recorded his conversation with

Gödel[35]: "*Gödel has come from Vienna. Via Siberia. This time with his wife. When asked about Vienna, he replied, 'The coffee is abominable'. He is very droll in his mix of profundity and unworldliness. It was new to me that he is interested in ghosts* ". Morgenstern was evidently taken aback by Gödel's professed interest in ghosts. He had not previously met Adele, and he was initially not at all impressed by her, "... *a Viennese washerwoman type: verbose, uncultured, feisty...*". Later, he discovered her more human qualities, and admitted that she had "*probably saved his life*". Morgenstern (and later his wife Dorothy) are important sources of information on the Gödels' Princeton years, as they became good friends with the Gödels, and remained so until Morgenstern's death in 1977, a half-year before Gödel's.

In Kurt Gödel's own estimation, the 1940s were a generally good time for him, health-wise. He had no health crises, neither physical nor mental. However, things were not entirely rosy. He became a fresh-air addict, especially after their vacation in Maine in the summer of 1941 (see Chap. 3, section on '*Mathematics and Philosophy*'), and he was subject to chronic fears that he was being poisoned by gases from the heating system and/or the refrigerator in his various dwellings. He and Adele at first lived in a rented apartment at 245 Nassau Street (see Fig. 10.3). However, Kurt began to suspect the heating system there, and they moved to another apartment nearby, at 3 Chambers Terrace—where he again soon complained of "*the bad air from the heating*" and removed all the radiators (making it impractical to live there in winter).

This strange behavior became noticeable to his colleagues, especially after the summer of 1941, and caused them some worry. IAS director Aydelotte even wrote to Gödel's personal physician,[36] inquiring as to whether he might become violent (expressing similar fears to those that plagued Gödel's family during the difficult time in 1936). Kurt and Adele moved several more times before finally purchasing a house in 1949 (at 145 Linden Lane; see the map in Fig. 10.3, and also Fig. 14.5. The house number was 129 when the Gödels bought it, but it was changed to 145 in 1960[37]). Kurt's fears of poisoning by 'bad air' gradually subsided, although his general paranoia recurred, especially as he grew older. These episodes of 'strangeness' caused the IAS administration to be cautious about hiring him on a permanent basis, and it was only in 1946, after an initiative by von Neumann, supported by Veblen, that he was given a long-term contract. He remained merely a 'permanent member' until 1953, when he was finally offered a professorship.

A Safe Haven at the IAS

Both Oskar Morgenstern and Alfred Tarski referred to Gödel at various times as 'crazy', and only half jokingly. Karl Menger, in contrast, expected him to behave 'normally', and was disappointed and shocked when he failed to meet those standards. While Gödel could perform professionally and intellectually, sometimes at a very high level, his personal relations were often difficult and limited to a few chosen friends. He was shy to an almost pathological extent, and it was noted by the students who attended his lectures at Notre Dame (and elsewhere) that he always faced the blackboard, even when he was not writing on it. On the other hand, the lectures were clear and well structured, and he made serious attempts to explain difficult concepts to the uniniti-ated. Obtaining a position at the IAS, where he had no regular teaching or lecturing duties and, until he was named professor in 1953 (at age 47), also no administrative responsibilities, was a great advantage for him. When he *did* have administrative duties (for example in helping to choose candi-dates from his field for membership at the IAS), he took them *very* seriously, and his 'legalism' often delayed such processes painfully. Rebecca Goldstein recounts some of those difficulties, based on interviews with some of Gödel's former colleagues in the School of Mathematics at the IAS, which resulted in his increasing isolation ('Logic' was finally declared to be a department of its own, at times with only one member, who was thus no longer invited to the faculty meetings of the other mathematicians).

In addition to Morgenstern, Gödel found another new friend in Princeton, an unlikely one in several respects: *Albert Einstein*. Einstein had been recruited as a Founding Member of the IAS in 1932 by Flexner and Veblen; he was a professor at the Institute, and his presence on the faculty greatly increased its prestige in its early years. While Einstein had suffered under the increasing antisemitism and polarization in Germany throughout the 1920s, and he was well aware of the unstable political situation by the early 1930s, his departure from Berlin on a lecture tour to the USA in late 1932 was regarded as another routine trip by his associates there. He himself was partic-ularly prescient, however, and when he and his wife Elsa were closing their little summer house ("*unsere Villa*") in *Caputh*, south of Potsdam (Fig. 9.6), for the winter in early December of 1932, he turned to her and said, "*Schau' sie Dir sehr gut an… Du wirst sie niemals wiedersehen.*" ("Take a good look at it. You'll never see it again").[38]

And events proved him right. For Einstein, as well as for John von Neumann and Hermann Weyl, also founding members who had come from Germany, the IAS was a place of calm in the storm, a refuge which appeared

Fig. 9.6 The Einstein summer house in Caputh, near Potsdam, 2006. *Photo* by Stephan M. Höhne, from WikiMedia[39]

'just in time' to take them in when their previous lives were rendered impossible by the rise of fascism.

The other two founding members of the mathematical faculty, Oswald Veblen and James Alexander, were Americans by birth and not under a similar threat. In Chap. 14, we will take a closer look at Gödel's friendships with his colleagues in Princeton, and especially with Einstein.

Professionally, Gödel initially continued his work on set theory. In 1940, he finished his monograph on consistency problems, and it was published in the *Annals of Mathematics Studies* [Princeton University Press; Gödel (1940)].

After his summer vacation with Adele in Maine in 1940, which seems to have intensified his obsession with being poisoned by 'bad air' from heating systems or leaking gas from refrigerators, he gave no more public lectures until the autumn of 1946. But he continued his intensive efforts to prove the independence of the Continuum Hypothesis and the Axiom of Choice from the other axioms of (Zermelo-Fraenkel) set theory.

The rest of the decade of the 1940s seems to have passed uneventfully, as far as Kurt's health was concerned. In 1941, he again vacationed with Adele in Maine, where he worked on a proof of the independence of the Axiom of Choice (AC). However, he never completed or published his attempts to prove the independence of the AC and the Continuum Hypothesis (CH) [that was accomplished by Paul J. Cohen in 1963; see Cohen (1963/64), and his later book, Cohen (1966)].

Gödel became a close friend to Einstein around 1942, and soon the two of them could be seen walking together to and from the Institute. Gödel's work on mathematical logic wound down in the first half of the decade, and apart from mentions in review articles and short notes on particular aspects, he made no more important contributions to that field. He did publish a significant article in 1944, a contribution to a volume in honor of Bertrand Russell, which was however more philosophical in nature—his first publication of a predominantly philosophical character.

In 1946–50, Gödel made a brief excursion into physics, specifically into General Relativity, and found a new solution to Einstein's field equations which has busied relativity theoreticians on and off ever since. However, his motivation for this departure from his previous areas of work was, as we shall see, in the main *philosophical*. It resulted in three major publications in the years 1949–1952, and a talk at the IAS in May 1949, as well as a lecture to the ICM-1950 in Cambridge, MA at the end of August, 1950—almost exactly 20 years after his first major public lecture, in Königsberg. We will return to this topic, '*Gödel's Universe*', in Chap. 15.

The 1950s and 60s

In the following decade of the 1950s, Gödel indeed had some health issues which were not just due to his hypochondria or to other psychiatric disturbances. In February of 1951, he suffered from a hemorrhaging duodenal ulcer. There had been no indication of his illness in the months before, although, due to his paranoid fears of poisoning and his poor eating habits, he was seriously underweight. In addition, he was taking a variety of self-prescribed medications for imaginary illnesses, some of which may have had serious side-effects or conflicting interactions. His ulcer was life-threatening, and although he was given blood transfusions and fed intravenously, there was some worry that he might not survive during his first days in the hospital. He himself was convinced that he would die, and he suffered from a depression which did not improve his health outlook.

Oskar Morgenstern wrote of this episode in his diary, and there he mentioned that Gödel had dictated his testament to him when he visited him in the hospital. However, after several days, Gödel's condition improved and he was able to return home, where he rapidly recovered, both physically and mentally. He was not quite 45 at this point, and basically healthy, apart from his compulsions and other personality disturbances. His rapid recovery was no doubt to a considerable extent due to Adele's no-nonsense attitude and to her cooking.

But in addition, Morgenstern and IAS Director J.R. Oppenheimer hatched a plan to give him a psychological boost, in the form of the *Einstein Award*, which was to be presented on Albert Einstein's birthday, March 14th, for the first time in 1951 and every three years thereafter. We will take a closer look at that event in Chap. 16, where we recall Gödel's life and work during the 1950s.

Later in 1951, Gödel was awarded his first honorary doctorate, a D.Litt. from Yale University. Near the end of that year, he gave the prestigious *Josiah Willard Gibbs Lecture*, sponsored by the American Mathematical Society (AMS), at Brown University in Providence, RI. His topic was '*Some basic theorems on the foundations of mathematics and their implications*'. The text is reprinted in the *Collected Works*, Vol. III, pp. 304–323, with the citation [Gödel (*1951*)]. In 1952, he was awarded an honorary doctorate (D.Sc.) from Harvard University (proposed by W.v.O. Quine). His reputation was growing, and he received some attention from the national press in the USA, which led to correspondence with interested people of all sorts. He seems to have taken that very seriously, and he answered their queries when it was at all reasonable, sometimes spending an inordinate amount of time on them.

Once he had recovered from his ulcer crisis, Gödel had no more major health issues during the 1950s, although his psychiatric problems continued and may have even worsened; his hypochondria was gradually becoming a full-blown paranoia, and it sought out new threats to focus upon, after his fear of 'bad air' declined following the Gödels' move to their house on Linden Lane in 1949.

More than once during the 1950s, Kurt promised his mother that he would visit her in Vienna, but each time he found some reason to postpone his trip. In the autumn of 1951, he was probably working on his Gibbs Lecture, scheduled for December 26th of that year, and had no time to travel. [See Wang (1987)]. Later, he reported to her that he had been advised not to travel to Europe. In fact, he was simply afraid to go back to Austria—his memories of the catastrophic year 1936, and of his dealings with the Austrian bureaucracy in 1938/39, and what he had learned after the war, made him increasingly unwilling to even visit there, leading to nightmares about being back in Austria and unable to leave again, etc. He finally admitted this to his mother, and in the meantime, she and his brother Rudolf began traveling to the US to visit Kurt, and evidently had some enjoyable stays there (more about their visits in Chap. 16).

In 1953, he was (finally) promoted to a professorship at the IAS (a status which in the opinion of many of his supporters, including von Neumann and Oppenheimer, was long overdue; but there had been opposition from the faculty, apparently mainly due to fears that his 'perfectionism' and 'strange ideas' would complicate discussions and decisions about faculty policy, for example the hiring of new faculty members). This promotion no doubt improved his own self-image, and certainly his finances. He had doubts about his true worth and the impressions he made on others, as recorded by Dawson and by Budiansky on the basis of Gödel's notes and memos to himself, found in his legacy after his death. On the other hand, his new position increased his demands upon himself and the mental stress of 'living up to what was expected' of him. And, like some of his other colleagues at the IAS, he suffered from 'imposter syndrome', feeling that his work was not important enough to justify his position there.

Adele went on her second trip to Vienna in 1953, and Kurt was apparently able to function well in her absence. In October of that year, he spent some time with L.E.J. Brouwer, who was visiting the IAS from the Netherlands. Gödel was also invited that spring by Paul Schilpp to contribute to a Carnap volume in his series on *Living Philosophers*. Carnap was in fact working at the IAS in 1952–54, but there seems to be little information about his interactions with Gödel; that may emerge when their respective unpublished writings are more completely analyzed. Schilpp encouraged Gödel to write a chapter for the book, to be entitled '*Carnap and the Ontology of Mathematics*'—but Gödel, in a letter to Schilpp in July 1953, agreed only to write a shorter article on the topic '*Some Observations on the Nominalistic View of the Nature of Mathematics*', intending to prove that mathematics is not (only) syntax. He worked for some time on writing it, and as was his habit, wrote several unfinished manuscripts; but he finally begged off (in early 1959!) on the grounds that he had no well-founded comments to make. In fact, his manuscripts were critical of Carnap's ideas, and he was probably unsure about how his old friend would react. The book was finally published in 1963, so that Carnap would have had ample time to respond to Gödel's article; but the latter was not published, although the various texts were later found among his papers. [Two versions, III and V, are reconstructed in the *Collected Works*, Volume III (1995); see (*1953/9*) there]. In his later conversations with Hao Wang, he regretted having spent so much time on it.

Oskar Morgenstern noted in his diary that Gödel was working on the Carnap paper in the spring and summer of 1954 and again in spring 1955. He also mentions that Gödel complained of 'feelings of unreality' in the autumn of that year. Morgenstern also reported a brief relapse on the part of Gödel in November/December of 1954, when he suffered an outbreak of his hypochondria and believed that he was suffering from heart failure. That episode was apparently acute, but subsided rather quickly.

In April of 1955, *Albert Einstein* died after a brief illness. This apparently came as a surprise to Gödel, who knew that Einstein was ill, but was not aware of the seriousness of his final illness (an aortic aneurism, which had been provisionally treated several years before but had begun leaking again in the spring of 1955. Einstein refused another operation or any other measures to prolong his life, feeling that 'his time had come'). Gödel's immediate reaction was not recorded, but it is certain that he was less disturbed than he had been 20 years earlier by the premature deaths of his mentors Hans Hahn and Moritz Schlick. He seems to have sublimated the loss and turned to other things, but he must have missed the almost daily conversations with his wise old friend (as he indeed wrote to his mother in Vienna). Gödel, with Bruria Kaufman, Einstein's last assistant, helped to order Einstein's papers in his office at the IAS.

In 1955, Gödel was elected to membership in the U.S. *National Academy of Sciences*. In the summer, he again planned to visit his family in Vienna, but was 'advised by his doctor not to make the trip'. Adele, however, did go to Vienna in February of 1956, and brought her mother back with her when she returned in March. Adele's mother lived with the Gödels in Princeton until her death three years later.

Also in 1956, Kurt received an invitation to contribute to a *Festschrift* for Bernays, which led to his 1958 article [his '*Dialectica*' paper; see Gödel (1958)]. The following year, he was elected to membership in the *American Academy of Arts and Sciences*. His colleague and vigorous supporter *John von Neumann* died in February, 1957, at the age of 53, after a long illness with cancer. He and Oswald Veblen had been the most active 'local supporters' of Gödel in Princeton, and von Neumann had not spared his praise and encouragement for Gödel. His *laudatio* of Gödel when the latter received the Einstein Award is legendary (see Chaps. 8 and 16). Veblen had retired from the IAS in 1950, and spent his final decade in Maine, near where Kurt and Adele had vacationed several times, so that when he died in 1960, it was less of an immediate loss and shock to Gödel.

In 1957, Gödel was working on his 1958 paper and on preparing some other articles for reprinting. In May of 1958, Kurt's mother Marianne and his brother Rudolf paid their first visit to the U.S. and to Princeton, seeing Kurt again for the first time since his departure from Vienna in January of 1940. Paul Bernays celebrated his seventieth birthday in 1958, and he visited the IAS for several months early in that year. Gödel sent off his contribution to the Bernays *Festschrift* in July; we will hear more about it in Chap. 16.

Adele's mother died in March of 1959. Adele herself went for a vacation to the White Mountains for most of July, leaving Kurt to work 'again, with great concentration' while alone in Princeton, where he enjoyed the unusually hot summer. In October, Adele made another voyage to Europe, spending two months abroad. Around this time, Kurt began his study of the writings of *Edmond Husserl*, which would occupy him for the next 18 years. He was interested only in Husserl's later work, after 1909 (the year of Husserl's *transformation*; cf. Chap. 18).

1960 was a calm year for the Gödels—Marianne and Rudolf again visited Princeton, this time for a longer stay (March to May). Adele also went to Vienna for a month in November. When she returned in December, she found that Kurt had eaten practically only eggs during her absence. This clearly presaged another downturn in his (mental) health, which made itself felt the following year, 1961, later mentioned by Kurt to have been one of the three worst during his life. Wang reports this (1987, p. 122) but did not know exactly the nature of Gödel's health problems in that year. There is some evidence that his hypochondria had once again taken on threatening proportions, and he was dosing himself with a variety of (unnecessary) medicines. Apparently, his depression and anxiety had returned, exacerbated by his increasing tendency to paranoia, a self-reinforcing spiral. Nevertheless, he was able to carry on a more or less normal life that year, and was not hospitalized or put in therapy, as he had been during his previous crises in the 1930s.

Gödel noted around this time that he had begun to go to bed earlier and get up earlier, and that this schedule was an improvement (in a letter to Marianne, dated March 18th). He was elected to membership in the *American Philosophical Society* in 1961, an indication that his increasing concentration on philosophy had been taken notice of, even though he had published little. In the summer of 1961, Adele spent a longer stay in Italy (July–September). His letters to his mother that summer and autumn concerned more philosophical topics (his 'theological' letters, and his personal wish to have a good philosophical library). We will consider them in more detail in Chap. 17.

During the remainder of the 1960s, Gödel worked mostly on philosophical topics, continuing to study the later writings of Husserl. He produced few manuscripts during that decade: one is an unpublished draft for a planned lecture to the American Philosophical Society, which he was invited to give in 1963, as a new member. It was to be on the foundations of mathematics in the twentieth century and their place in a modern *Weltauschauung*. He apparently never replied to the invitation, although he prepared a draft of his proposed text [Gödel (*1961)]. In 1962, he wrote a postscript to an article [Spector (1962)] published posthumously for a young logician, Clifford Spector, who had developed a proof of the consistency of analysis using recursive functionals; and in 1964, a revision of his 1947 paper for the AMS, 'What is Cantor's Continuum Problem?' [Gödel (1964)]. His health, although not perfect, was not a major problem for Gödel himself nor for Adele and his friends in Princeton during this period.

Gödel's mother, Marianne, died in July of 1966. She had hoped to visit Kurt again to celebrate his 60th birthday in April of that year, but she was too weak to travel. She died just 6 weeks before her 87th birthday.

The Beginning of the End

Starting in 1970, Kurt was in therapy with his psychiatrist, Dr. Philip Erlich. Steven Budiansky has evaluated some of the records of his visits to Dr. Erlich, who noted initially that Gödel had come for therapy due to 'pressure from his wife and brother'. His main complaints were feelings of inadequacy and the 'imposter syndrome' noted above. His other, much more serious problems (his hypochondria, his incipient paranoia, his attacks of anxiety and depression) emerged only gradually in the course of the therapy. The year 1970, according to his later conversations with Hao Wang, was the third of his 'three worst years', and he had another psychiatric crisis that year, compounded by his advancing age and the long-term effects of the many drugs with which he had been dosing himself for the past 40 years. It began early in the year, according to Oskar Morgenstern's diaries, and became rather serious by April, when Gödel's weight had become dangerously low and he appeared to be weak and gaunt, in addition to his usual psychiatric symptoms. Morgenstern's record breaks off at that point, as he was traveling and had no contact with Gödel until August, when he returned to Princeton. Gödel by then had completely recovered—miraculously, as Morgenstern found. The agency of his rapid recovery is not recorded, but there is evidence that he had been given psychopharmaceuticals which were unexpectedly effective. (In his unsent letter to Alfred Tarski, regarding a manuscript that he had sent to Tarski for reviewing before its planned submission to PNAS, he mentions that the manuscript was chaotic because he had been taking 'drugs impairing the mental functions' [cf. Gödel (*1970c)]). He stopped seeing Dr. Erlich the following year, but returned to therapy in 1977, when his condition had once more become rather desperate.

All in all, his last decade (in fact only 8 years including 1970) was a story of accelerating decline, accompanied by the loss of those who had been his guardians in the past, in particular Adele, who—6½ years older than Kurt— was herself having increasingly serious health problems; and his loyal friend Oskar Morgenstern, also 4 years older, who suffered from cancer for more than eight years before his death in the summer of 1977, preceding Kurt by 5½ months. His various biographers describe this period in uniformly but diversely pessimistic tones. Dawson (1997) gives a matter-of-fact description, as does Hao Wang, who himself was in fairly intensive contact with Gödel, mainly by telephone, in 1971–72 and again in 1975–76, due to his interest in Gödel's philosophy and mathematics. This contact continued right up to the end of Gödel's life. Wang's documentation based on those numerous conversations, along with Oskar Morgenstern's diaries, provide the most complete and accurate record of Gödel's final years, although both contain occasional mistakes concerning dates etc.

After his recovery from his psychiatric crisis in the first half of 1970, Gödel enjoyed a period of around 3½ years of relatively good health.[40] In April of 1974, his enlarged prostate reached a point at which he had to seek medical help (which he had been resisting, in spite of Adele's urgings, for several years). He showed his usual antagonism toward medical advice and treatment, and willfully removed a catheter inserted to open his urinary tract. He adamantly refused an operation (this may be related to the fact that his father had died as a result of a similar operation, to be sure 45 years earlier, and his closest friend—almost his *only* local friend by that time, Oskar Morgenstern—was suffering from prostate cancer). Eventually, after being transferred to another hospital, he acquiesced and continued to use the catheter for the rest of his life, in spite of the accompanying inconvenience and danger of infection.

Nevertheless, Gödel remained extremely interested in philosophy and mathematics, having manic episodes in which he was very excited about new proofs and insights. He maintained contacts to various old friends, in particular to Paul Bernays (up to about 1975) and to Hao Wang, the latter especially during Gödel's last academic year as an active member of the IAS, when Wang was at the Institute as an academic visitor and spoke frequently with him (but mainly by telephone).

Gödel published nothing during those last 8 years, but in his 'unpublished writings' (*Collected Works*, Volume III), there are four items from 1970, his '*ontological proof* ' (which has attained a certain fame in the intervening years), two manuscripts on the continuum hypothesis, and the (unsent) letter to Alfred Tarski, mentioned above.

After his official retirement at age 70, on July 1st, 1976, Kurt Gödel lived only another 18 months. His wife Adele had become seriously ill in November of the previous year, and, after being cared for by a neighbor who was a nurse, Elizabeth Glinka, for a few months in 1976, she suffered several more illnesses which required increasingly long hospitalizations. Kurt was also admitted to Princeton Hospital in March/April of 1976. He 'checked out' without permission and returned home, where Adele and Elizabeth were shocked at his condition. His health and that of Adele were now his chief preoccupations, but attempts (e.g., by Hao Wang) to convince him to go to the University of Pennsylvania hospital for examinations and treatment fell on deaf ears. Adele was no longer able to help him, being too weak and intermittently ill herself.

In the following year, 1977, he was still able to carry on a number of telephone conversations with Wang, in which he described his work on constructible sets. Adele required an emergency operation in early July and remained in the hospital and in recuperation homes for the next 6 months. His steadfast friend Oskar Morgenstern succumbed to his cancer on July 26th, apparently unexpectedly for Gödel, who had not faced the truth of his friend's condition, too preoccupied with his own health problems. All this proved to be too much for Kurt, left to his own devices. He refused offers of help and even a visit from Dr. Erlich, who had ordered an ambulance to take him to the hospital, which he would not accept. He apparently was unaware of the death of Paul Bernays that September, with whom he had contact until the mid-1970s. He also broke off contact with his brother Rudolf during his last two years. The last of his friends to see Kurt was Hao Wang; he had been traveling for two months and visited Gödel on December 17, 1977, after his return. Wang found Kurt 'not apparently very sick' and still 'nimble of mind'. But Kurt was very pessimistic, saying that he *could no longer make positive decisions, only negative ones*. He was beginning to distance himself from the world.

Adele left the hospital on her own responsibility and returned home around December 20th, and found Kurt emaciated and weak. She finally convinced him, with the help of Hassler Whitney, to go to the hospital, which he entered on December 29th. But he was too far gone—his self-starvation could not be reversed. (Whitney was a mathematician colleague from the IAS, about the same age as Gödel, who had no very close personal contacts to him, but nevertheless tried to help him in his last years).

Hao Wang spoke to Kurt Gödel in the hospital by telephone on January 11th, finding that he was *polite but sounded remote*.[41] He had evidently begun withdrawing from the world for the final time, and he died three days later, curled up on a chair in the fetal position, at 1 pm on Saturday, January 14th, 1978. His death certificate lists the cause of death as '*malnutrition and inanition due to a personality disturbance*'.

Whether his death could have been avoided at that time remains an open question—he had entered a downward spiral of his own creation, with fear fueling his paranoia which made it impossible for him to accept the help that he desperately needed, combined with the absence at a critical time of those who were still close to him—Adele, Morgenstern, Wang. His final decline, at a relatively early age (he would have been 72 three months after his death) seems somehow fated, given his personality.

Adele, in poor health, survived him by three years and died in 1981, three months after her 81st birthday, in a residence home near Princeton.

Conclusions: Gödel's Health

What conclusions can we draw about the origins of Kurt Gödel's health problems from the above chronology?

Kurt himself often spoke of his 'desire for precision', which Hao Wang called 'his need for security', both probably euphemisms. In fact, his behavior strongly resembles the intense need for *predictability* which is often observed in children (and adults) who are somewhere on the autistic spectrum. Owing to an overactive processing of sensory inputs, or to some deeper psychological or neurological problem, such people are often inordinately upset whenever any (even very minor) aspect of their environment does not meet their expectations or what they have previously observed. Kurt Gödel was obviously not anywhere near the extreme, severe end of the spectrum, but he may well have been somewhere near its other extreme, the 'high performance' end. This possibility was not explored during his lifetime, and there is no way to make a 'distant diagnosis' with any certainty, but the record of his psychological problems speaks for itself.

Rebecca Goldstein, in her (2005) biographical work about Gödel,[42] gives a rather rationalistic (or rationalized) explanation of the 'anxiety neurosis phase' in Kurt's childhood: she supposes that around the age of 5, he became aware of his own superior intelligence, and, by contrast, the relative mental short-comings of the other members of his family—and responded with anxiety since he, as a child, was dependent on his supposedly mentally-incompetent parents. This seems rather bizarre; a child of 5, no matter how intelligent, would not likely objectivize his own parents to such an extent. Furthermore, the day-to-day experience in his family would have completely contradicted such a conclusion. Both of his parents, each in their own way, were indeed very competent in dealing with life and in providing Kurt with all that he needed. In fact, they had a degree of competence which he, at 5, could hardly imitate, and he was reminded daily of that. His anxiety, in contrast, is typical of those in need of extreme predictability—for example, if Kurt's mother left their apartment, how was he to be sure that she would ever return? This type of anxiety is a well-known symptom of personality disorders at the 'high performance' end of the autistic spectrum. In more serious cases, it can result in uncontrollable tantrums. He was also unusually attached to his mother and had a comparatively cool relationship with his father.

This type of 'nearly normal' autism was first described by *Hans Asperger* in 1944. He called it 'autistic psychopathy' and noted that it was often accompanied by high intelligence, not necessarily limited to a very narrow area of experience (as in the '*idiot-savant*' phenomenon). He thought that it occurred only in boys and men, and he wrote that it was "*an extreme variant of male intelligence*". The general phenomenon of infantile autism was studied about the same time by the Austrian/American psychologist *Leo Kanner*, but many authors note a real difference between his observations and those of Asperger (see below).

The history of 'Asperger's Syndrome' begins around 1938, in—where else? Vienna. Hans Asperger carried out studies in the period 1938–1944, at the University Children's Clinic in Vienna, of 'unusual' children, when he observed and described the new syndrome, *autistic psychopathy*.

The story of the discovery of autism is fascinating (and it has been summarized in a very thorough manner in a recent book by Steve Silberman[43]); the following quote is from a review of that book by Simon Baron Cohen in '*The Lancet*' (2015)[44]:

> … child psychiatrist Leo Kanner at Johns Hopkins University School of Medicine in Baltimore, MD, USA, wrote a seminal article in 1943 in which he described—'for the first time'—11 children in his clinic without the social instinct to orient towards other people, who were mostly focused or even obsessed with objects, and who had a 'need for sameness' or a 'resistance to (unexpected) change'. To give a name to this new psychiatric condition, Kanner coined the term 'infantile autism'. Kanner's article made medical history, as befits someone who discovers a new medical condition. But just one year later, paediatrician Hans Asperger, at the University of Vienna in Austria, wrote an article describing a group of children in his clinic who shared many of the same features. Kanner's paper became highly cited and high profile, whereas Asperger's article went almost unnoticed …

Hans Asperger (1906–1980) was born two months before Gödel in Haus-brunn, Austria. During his youth, he himself exhibited some symptoms of mild autism. He studied medicine at the University of Vienna, parallel to Gödel's studies there, receiving his medical degree in 1931, and soon became director of the Special Education department at the University Children's Clinic. He was a supporter (and a member) of Dollfuss's '*Fatherland Front*' (and therefore probably not an opponent, if perhaps not an enthusiastic supporter, of the Nazi regime after the *Anschluss* in 1938). His work was practically forgotten later, having been published in German at a time when very little medical research was being done due to the war. His work was rediscovered by *Lorna Wing*, a child psychiatrist at the Institute of Psychiatry in London, who published an article in 1981 calling attention to what she called 'Asperger's Syndrome', which showed many similarities to—but also important differences from—the 'infantile autism' described a year earlier by Leo Kanner.

Leo Kanner (1894–1981) was born in Klekotów, then in the Austro-Hungarian province of Galicia (now Klekotiv, Ukraine). At age 12, he was sent to Berlin, where he initially lived with his uncle. After finishing the *Gymnasium* there in 1913, he began studying medicine at the University of Berlin; his studies were however interrupted by WWI, during which he served as a medical assistant in the Austro-Hungarian army. He was finally able to graduate in 1921. After working as a cardiologist at the *Charité* (university hospital in Berlin), he emigrated to the USA for economic reasons in 1924. There, he was introduced to pediatric psychiatry at the Yankton State Hospital in South Dakota, where he did his residency. He went with a fellow-ship to the Johns Hopkins University in 1928, and he helped to establish the Children's Psychiatry Service there. From 1938, he studied a series of patients and published a paper in 1943 on what he called 'infantile autism', which established his reputation as the discoverer of autism as a personality disorder.

In recent years, there have been two new developments which—on the one hand—strengthen Asperger's claim to having been the real discoverer of a previously unknown personality disorder; and on the other, calling his general reputation into question. Silberman's (2015) book, quoted above, reveals that Asperger had given lectures in Vienna as early as 1938 on 'autistic psychopathy', and that *Georg Frankl*, a diagnostician in Asperger's clinic, left Austria and went to work in Kanner's group at Johns Hopkins later that same year, undoubtedly bringing the news about Asperger's discovery with him. Kanner later claimed to have been unaware of it.

In 2018, an article was published by *Herwig Czech* in the journal '*Molec-ular Autism*' [cf. Czech (2018)] which shed doubt on Asperger's resistance to

Nazi euthanasia programs and suggested that, on the contrary, he cooperated actively with them, which he had denied after WWII. That article suggests that his name should be removed from the syndrome which he most likely discovered, due to his probable involvement in Crimes against Humanity during the Nazi era.

In fact, after Lorna Wing's publications, 'Asperger's Syndrome' was added in 1994 to the *Diagnostic and Statistical Manual of Mental Disorders* (DSM IV), but in 2013, it was removed (from DSM V), not due to Asperger's dubious past, but rather to the fact that the Syndrome is now considered to be part of the Autistic Spectrum of disorders (near to its 'high-performance' end) and therefore not to require a separate name. It remains as a sub-category of Autism Spectrum Disorder (ASD) in the *International Classification of Diseases* (ICD-11) compiled by the WHO.

Gödel and Asperger

In 2004, Michael Fitzgerald and Viktoria Lyons, of Trinity College Dublin, published an article in '*Autism—Asperger's Digest*' entitled '*Kurt Gödel (1906– 1978). The Mathematical Genius who had Asperger Syndrome*' [see Lyons & Fitzgerald (2004), and also their subsequent book (2005), Item 25 in Appendix B]. In their brief article, they put forth the thesis that Kurt Gödel was subject to Asperger's Syndrome, and give evidence based on his symptoms and comorbidities as found in the written record.

Such a 'distant diagnosis'—distant in both time and space, since their work occurred more than 25 years after Gödel's death and far from the places where he had lived—is of course only speculative and perhaps not even ethical, by modern psychiatric-medical standards. On the other hand, the evidence is highly suggestive, if not conclusive. They give a brief biography of Gödel and list six items which support their suggestion: (i) Gödel's *social behavior* (his extreme shyness, avoidance of personal confrontations or even any face-to-face meetings, his 'hermit-like' or 'social outcast' behavior); (ii) his *lack of empathy* (documented for example by Menger's recollections, in particular his remarks about Gödel's seeming indifference to the fates of his colleagues in Europe during the Nazi regime); (iii) his *naïveté/childishness* (again cited by Menger and also by Morgenstern, who said that Gödel was often like a boy of 11 or 12 in his views and behavior); (iv) his *narrow interests and obsessiveness* (comparing him to Newton, who was so obsessed with his work that he often forgot to eat); (v) his *routines and control issues* (citing Dawson about how Gödel collected '*every scrap of paper that crossed his desk, including library request slips, luggage tags, crank correspondence and letters from autograph seekers and mathematical amateurs'*); and (vi) his *comorbidities* (the technical term for other personality disturbances that may accompany Asperger's/autism, such as Gödel's extreme hypochondria, his outbreaks of anxiety and depression, and his increasing paranoia, all known to be comorbidities of Asperger's syndrome). In their conclusion, they remark that,

> Based upon a retrospective analysis, it appears that Kurt Gödel meets the criteria for Asperger syndrome (APA 1994). His life is an example that Asperger syndrome does not necessarily have to be an obstacle for major intellectual achievements, maybe quite the contrary....

Whether this diagnosis is indeed correct will necessarily remain a matter of speculation. Michael Fitzgerald has made similar claims about many well-known figures in the sciences [see his (2005), with Viktoria Lyons], and one might suspect that he would include anyone in Einstein's category of '*those who have found favor with the angel*' in that diagnosis [cf. Chap. 14, Einstein quote at the end of the section '*Friendship with Albert Einstein*']—i.e., those who have dedicated their lives to working exclusively and exhaustively 'in the temple of science'.

Gödel's earlier biographers have not discussed this possibility. Hao Wang and John Dawson, as we have noted, chronicle Gödel's 'eccentricities' in a sober manner, not attempting to explain their possible root causes. Palle Yourgrau considers the etiology of *paranoia* and its manifestations, while Steven Budiansky[45] does the same for *hypochondria*, relating it to other personality disorders which Gödel exhibited. In his (2021), Budiansky writes:

> At the same time *[1950s]*, his hypochondria began undeniably to loom ominously larger. The etiology of hypochondria is still not well understood, but as with other anxiety disorders like obsessive-compulsive disorder it follows a self-reinforcing logic of its own, not necessarily tied to any definite external events. It is hard not to see at least a connection between his *[Gödel's]* insistence on finding hidden causes in everyday affairs, his belief in the existence of absolute truth in his mathematical investigations, and his steadfast rejection of empirical medical evidence about his own condition. An intolerance of uncertainty is characteristic of most hypochondriacs, as is a tendency to perceive innocuous bodily symptoms as evidence of serious illness.

Earlier in his book, Budiansky suggests that Gödel suffered from obsessive–compulsive disorder.[46]

Both Rebecca Goldstein and Verena Huber-Dyson tend to rationalize Gödel's paranoia as exaggerated by other observers and as simply a strong

form of normal anxieties and fears (but each of them experienced him personally only once: Goldstein at a reception hosted by the director of the IAS in
1973, a garden party (cf. Chap. 18, Fig. 18.2) where Gödel famously 'held
court' for his younger admirers[47]; and Huber-Dyson at a private tea at the
Gödels' house in 1957, which she attended with Georg Kreisel, where Kurt
made a brief and elegant appearance).

Huber-Dyson describes her impressions of that occasion,[48] noting that
Kreisel, who was on familiar terms with the Gödels at the time, had graciously
invited her and her two young children (her children with Freeman Dyson) to
accompany him on a visit to Adele and Kurt. She recounts the conversation,
which was mostly 'small talk', and the attention received by her children, as
well as her feelings of sympathy for Adele, who was on the fringe of Princeton
society, like Huber-Dyson herself; albeit for different reasons. She mentions
Adele's sharp-witted jabs at that buttoned-up society, and describes Gödel's
appearance, coming down from his study to join them for a half-hour and
receiving them in a natural and friendly manner. This underscores the observations of other witnesses from that time, who pointed out that Gödel often
appeared relaxed and casual in the company of Adele.

Rebecca Goldstein[49] emphasizes Gödel's 'legalism' and his tendency to
have an overblown respect for authority, almost regardless of what the authorities demanded; these caused him problems with the other faculty members in
the School of Mathematics at the IAS, especially after he was named professor
in 1953, and led to his effective isolation in a 'department of one member'.
John Dawson recounts Gödel's meeting with *Gustav Bergmann*[50], a refugee
from Nazi terror, when he arrived in the USA. Gödel greeted him with the
words, *"And what brings you to America, Herr Bergmann?"*, an example of
Gödel's lack of (*felt*) empathy. In fact, Gödel was always very polite and seemingly sympathetic to others, but this may have been *learned* behavior rather
than an expression of genuine feelings, as such anecdotes suggest.

Solomon Feferman,[51] echoed by Palle Yourgrau[52] and Jaakko Hintikka,[53]
regrets the loss to mathematics caused by Gödel's failure to follow up on
his innovative work and '*develop his own insights systematically*' (quoted from
Hintikka, who attributes this failure on the part of Gödel to his philosophy
of mathematics). Yourgrau imputes it to his paranoia (which Feferman does
not mention in this connection). Feferman[54] also implies that it was due to
Gödel's *philosophy*:

In general, Godel shied away from new concepts as *objects of study*, as opposed
to new concepts as *tools* for obtaining results ... *[He]* was understandably
cautious about making public his Platonist ideas, contrary as they were to the
'dominant philosophical prejudices' of the time ... Godel did the major part

of his logical work in isolation … As much as anything, Gödel's achievement lay in arriving at a very clear understanding of which problems in logic could be treated in a definite mathematical way. Along with others of his generation, but always leading the way, he succeeded in establishing the subject of mathematical logic as one that could be pursued with results as decisive and significant as those in the more traditional branches of mathematics [emphasis added].

While there is no doubt much truth in both of these suggestions, it may also be the case that his hesitation to continue and systematize his results was due to a strong inhibition against publishing *anything* that he had not 'perfected'. This would be in line with aspects of autism spectrum disorder (ASD). The question is, was his philosophy the result of his personality, or vice versa?

We conclude this survey of Kurt Gödel's health issues by quoting from Lyons and Fitzgerald once more, from the end of their (2004) [and again at the end of their chapter on Gödel in their (2005)]:

Despite the difficulties associated with his condition, Kurt Gödel was able to use his enormous mathematical talents to achieve major scientific successes with the help of a few friends and the total dedication of his wife Adele.

Whatever 'his condition' may have been, they hit the mark with their observation.

Notes

1. Taussky-Todd (1988), p. 28.
2. Dawson (1997), 2005 edition, p. 97.
3. Kreisel (1980), p. 154. Quoted with permission from the Royal Society, here and in Chap. 16.
4. Oskar Morgenstern, diary entries from July 12th, 1940 and December 11th, 1947. Translated from the diary entries in [Morgenstern: Diaries].
5. Quoted by Budiansky (2021), pp. 174/175, from [Engelen, ed. (2019, 2020)] and from Gödel's '*Protokolle*'.
6. Dawson, *op. cit.*, p. 97.
7. Image from Creative Commons, 10.04.2007, by Roman Klementschitz, Vienna; used under GNU Free Documentation license 1.2. Reused from: https://commons.wikimedia.org/wiki/File:Sanatoriumpurkersdorf1.JPG.
8. Wang (1987), p. 97.

9. The *MV Georgic* was the last ship built for the White Star line, and after the White Star/Cunard merger in 1934, it sailed under the White Star-Cunard flag. It was the sister ship of the *Britannic*, and was used as a troop transport during WW II; it was partially destroyed by a bomb attack in 1941, but was reconstructed and remained in service until 1955.

10. Wang, *op. cit.*, p. 99.

11. Goldstein (2005), e-book version, p. 212.

12. Dawson, *op. cit.*, p. 110, Endnote 276.

13. *ibid.*, quoted on p. 111.

14. Letter found in Schlick's archives, quoted by Budiansky (2021), p. 172 and Note 68.

15. Dawson, *op. cit.*, quoted on p. 111 and Endnote 279, from the draft of a letter to Veblen dated November 27th, 1939.

16. Budiansky (2021), p. 178. The drug was called *Strophanthin*; it is also known as '*Ouabain*', and is based on an arrow poison used in Africa and derived from plant extracts.

17. Budiansky, *ibid.*, p. 173.

18. Dawson, *op. cit.*, see Endnote 280 on p. 111.

19. See the quote from Kurt Gödel 's brother Rudolf in Chap. 5 (from Schimanovich & Weibel; cf. backnote [19] in Chap. 5); he says, "*...the murder of the philosopher Prof. Schlick, in whose circle my brother had been active. This latter event was in the end the reason that my brother suffered a severe nervous breakdown and had to spend some time in a sanatorium, which was naturally a cause for much concern on the part of our mother*".

20. Image: Wiki commons, by IKAI, CC A-SA 3.0 license. Compare: https://creativecommons.org/licenses/by-sa/3.0.

21. Letter to Menger, December 1938, quoted by Dawson (1997).

22. Quoted in footnote 20, p. 111 in Dawson (1997), 2005 edition.

23. Goldstein, *op. cit.*, p. 106.

24. Budiansky (2021), p. 185.

25. *ibid.*, p. 168.

26. Sigmund (2015–17), p. 292 (e-book version): Letter from Karl Menger to Franz Alt, dated December 31st, 1937.

27. Dawson (1997), 2005 edition, pp. 124 ff.

28. Budiansky (2021, p. 190) suggests on the basis of Gödel's private notes ('*Protokolle*') that he may have been reluctant to move together with Adele and gave in only to put an end to her complaints. But the move was in fact clearly beneficial to him, even if he failed to see that (not for the first time did he fail to recognize his own best interests).

29. Dawson, *op. cit.*, p. 128.

30. Menger (1994), *on Gödel*, Sect. 11.

31. Dawson (1997), 2005 edition, p. 134, mentions that there was a receipt in Gödel's papers for a room in the *Morningside Hotel* at Notre Dame from January 27th through May 31st; but Menger's recollection and that of a

graduate student at the time, who recalled that Gödel lived in the Lyons Hall dormitory annex, are clear. He may have rented the hotel room as a 'backup'. Wang (1987) also cites that hotel as his residence while at Notre Dame.

32. The *Habilitation* (*Lehrbefähigung*) is conferred for life, unless revoked due to some serious academic misconduct. It was in fact revoked in the cases of certain refugees or anti-Nazis, mainly Jewish, for political reasons, but this did not apply to Gödel. But the *Lehrbefugnis* (status as *Privatdozent* or some other teaching position) is granted by a particular institution for a limited time, and it may be revoked on comparatively short notice.

33. Image from Wiki Commons, by Welleshik, 2008. Licensed under a GNU free documentation license, Creative Commons. Reused from: https://commons.wikimedia.org/wiki/File:Strudlhofstiege4.JPG.

34. His applications for an extension of his leave from Vienna, and for the *Dozentur*, were finally approved, much too late to be relevant; the latter in the summer of 1940.

35. From Morgenstern's Diaries, March 11th, 1940 and July 4th, 1940.

36. Dawson, *op. cit.*, p. 158.

37. Wang (1987), p. 107.

38. Quoted in Philipp Frank's Einstein biography; see for example https://www.einstein-website.de/z_biography/caputh.html.

39. Image: Wiki Commons, by Stephan M. Höhne, from WikiMedia. Used under Creative Commons Attribution-Share License 3.0.; reused from: https://commons.wikimedia.org/wiki/File:Albert_Einstein_home_Berlin_100.JPG.

40. Dawson, *op. cit.*, p. 238.

41. Wang, *op. cit.*, p. 133.

42. Goldstein (2005), e-book version, pp. 56 ff.

43. See'*Neuro Tribes…* ', by Steve Silberman [Silberman (2015)].

44. Review of Silberman (2015) by Simon Baron-Cohen, in *The Lancet*, Volume 386 (Oct. 3rd, 2015), pp. 1329–1330.

45. Budiansky, *op. cit.*, p. 255.

46. *ibid.*, p. 169.

47. Goldstein (2005), p. 211.

48. Paraphrased from Huber-Dyson (2005).

49. Goldstein, *op. cit.*, p. 236 ff.

50. As told to Dawson in a letter from Bergmann dated March 4th, 1983.

51. Feferman (1998).

52. Yourgrau (2006), p. 156 ff.

53. Hintikka (1999), p. 46.

54. Feferman, *op. cit.*, end of the section on '*Character, Impact, and the Influence of the Work*'.

10

A Sojourn Abroad, 1933/34

After having investigated Kurt Gödel's health problems throughout his whole life in the previous chapter, we now return to our chronological story of his life, rejoining him in Vienna in early 1933, soon after he became a *Privatdozent* at the University of Vienna. With his *Lehrbefugnis* in hand, he immediately embarked upon his first lecture course, entitled '*Grundlagen der Mathematik*', a survey course on Fundamentals of Mathematics, which was reportedly attended by around 20 students—a large number for such a specialized lecture course. He had announced the course too late for it to be included in the official course catalog, and a slip of paper with the belated announcement was inserted into copies of that volume.

But by then, he was also planning to ask for leave in the winter semester from Vienna and its University, since he had been invited (by Abraham Flexner, on the recommendations of Oswald Veblen and John von Neumann, both of whom had been impressed by Gödel's work in 1931/32) to visit the newly-founded *Institute for Advanced Study* in Princeton, NJ/USA, beginning in the academic year 1933/34. We saw in Chap. 9 how this first visit was successful, in spite of its also being stressful for Gödel, and we will come back to the academic and professional work that he accomplished while he was there. It seems clear that his own intention at that point was to divide his time between Princeton and Vienna, perhaps spending a successive half-year in each place in the coming years.

© The Author(s), under exclusive license to Springer Nature
Switzerland AG 2022, corrected publication 2023
W. D. Brewer, *Kurt Gödel*, Springer Biographies,
https://doi.org/10.1007/978-3-031-11309-3_10

Princeton and the Founding of the IAS

Princeton University has a long history (by American standards). It was founded in 1746 as the *College of New Jersey*, a Presbyterian college in the town of *Elizabeth*, and the fourth college in what was then *British North America*. Ten years later, it moved to its present location in the town of *Princeton* (settled in 1696; see the town map in Fig. 10.3), about 75 km (47 mi.) to the southwest of New York City. The College was renamed *Princeton University* in 1896.

The idea for the *Institute of Advanced Study* (the IAS), a proposed semi-independent institute to be loosely associated with the University but privately funded, was the brainchild of *Abraham Flexner*, an educator who was interested in new models for education and research, and who wanted to establish a research institution in the USA which would be comparable to the research universities in Europe, such as Oxford University, the Collège de France, or the University of Heidelberg. Flexner had spoken and written publicly about his idea, and he was contacted in late 1929 by the *Bamberger* family, who had sold their department store in Newark, NJ and wanted to use the proceeds to establish an educational institution in the region. Flexner (Fig. 10.1) convinced them to support his idea of an independent institute for high-level research, where recognized experts in mathematics and physics (and later in history, natural and social sciences) could carry on their work without the pressure and distraction of teaching and administrative duties which are omnipresent in university positions. Flexner, in his autobiography,[2] later recalled his first contact with the Bamberger family:

> I was working quietly one day when the telephone rang and I was asked to see two gentlemen who wished to discuss with me the possible uses to which a considerable sum of money might be placed. At our interview, I informed them that my competency was limited to the education field and that in this field it seemed to me that the time was ripe for the creation in America of an institute in the field of general scholarship and science, resembling the Rockefeller Institute in the field of medicine—developed by my brother Simon—not a graduate school, training men in the known and to some extent in methods of research, but an institute where everyone—faculty and members—took for granted what was known and published, and in their individual ways, endeavored to advance the frontiers of knowledge.

Flexner's first academic ally was *Oswald Veblen*, whom we have met in previous chapters as a mathematician and logician. His chief interest was topology, and he was one of the most eminent professors in the Mathematics

Fig. 10.1 Abraham Flexner, 1910. Portrait. *Photo*, by W. M. Hollinger, public domain[1]

Department at Princeton University by the late 1920s. He joined Flexner's project and became the first professor at the IAS, and he was instrumental in searching for other faculty and in obtaining land for the new Institute. The story of Flexner's recruitment of Albert Einstein (with Veblen's help) is legendary, and is recounted for example in Abraham Pais' Einstein biography (1982).[3] The IAS was officially founded in 1930, with Flexner as its Founding Director, and besides Veblen, its first professors were Albert Einstein, John von Neumann, Hermann Weyl, and James W. Alexander II. Its School of Mathematics (which included physics), the first School to be put into operation, opened in early October, 1933.

Oswald Veblen (1880–1960) (Fig. 10.2) was born in Iowa, where his father was professor of physics at the University of Iowa. His uncle was the famous economist and sociologist *Thorstein Veblen*. After school in Iowa City, he attended the University of Iowa, finishing with a Bachelor's degree in 1898. He then went to Harvard, obtaining a second BA in 1900. He went on to the University of Chicago for graduate work, finishing his Ph.D. with a thesis on '*A System of Axioms for Geometry*' in 1903 (see *categoricity* in Chap. 7, and backnote [29] there). He was a member of the mathematics faculty at Princeton University from 1905–1932, interrupted by military service during WWI, when he was a Captain/Major in the US Army. He

Fig. 10.2 Oswald Veblen, portrait photo, ca. 1915. From Wiki Commons, public domain[4]

gave a plenary lecture at the ICM conference in Bologna in 1928 (where the Hilbert-Brouwer dispute came to a head). He later assisted in planning the construction of Fine Hall as a home for the Princeton Mathematics Department, and, as we have seen, he was strongly involved in the planning and realization of the IAS beginning around 1930, and helped to recruit its early faculty members, including Albert Einstein and—later—Kurt Gödel. Professionally, he made important contributions to topology, and to projective and differential geometry; and during WWII, he was involved in the development of an early electronic computer.

In 1932, Vebeln resigned from the Princeton faculty to become the first professor (in the School of Mathematics) at the IAS. There, he was a long-term supporter of Kurt Gödel, together with John von Neumann, as we have seen. Their mathematician colleague at Princeton University, James Alexander, also became a Founding Professor of the IAS. Veblen remained on the IAS faculty until his retirement as Professor *emeritus* in 1950, and he died 10 years later in Brooklin, Maine. He was honored posthumously by the creation of the *Oswald Veblen Prize in Geometry*, established by the AMS.

The Institute for Advanced Study (IAS)

The IAS Princeton has enjoyed a very positive international reputation since its founding, and has been the model for many other 'institutes for advanced study' in various locations around the world. It has an impressive record of prizes won by its faculty and members, including 34 Nobel Prizes and 42 Fields Medals (the 'Nobel Prize for mathematics') as of 2021, as well as numerous other honors and awards. It can make a solid claim to being a birthplace of the modern digital computer and has been a breeding ground for many modern developments in mathematics (including fundamentals), theoretical physics, and other areas. But most importantly for the first generation of faculty and members, it provided a refuge for many talented scientists from the vicissitudes of life in academia and from political and social persecution, especially in the Fascist era of the 1930s and 1940s. How Kurt Gödel's life after 1940 would have continued if he had not found support in Princeton is not a pleasant topic to contemplate.

The Institute is not completely uncontroversial, however, and it has been the object of sarcastic remarks by various people. The famous theoretical physicist *Richard P. Feynman*, for example, who earned his PhD at Princeton University with John Wheeler in 1942, remarked[5] that,

> When I was at Princeton in the 1940s I could see what happened to those great minds at the Institute for Advanced Study, who had been specially selected for their tremendous brains and were now given this opportunity to sit in this lovely house by the woods there, with no classes to teach, with no obligations whatsoever... Nothing happens because there's not enough real activity and challenge: You're not in contact with the experimental guys. You don't have to think how to answer questions from the students. Nothing!

When Feynman was offered a position at the IAS, he refused, even though the salary and other conditions were much better than the usual academic posts. He felt on the one hand that it would stifle creativity (for that reason, the Institute offered him a half-time position, combined with a professorship at the University), and on the other, he felt unworthy of such a position.

Feynman's colleague *David Bohm*, who was assistant professor of physics at Princeton from 1947–51, ironically referred to the IAS as the '*Princetitute*', and his opinion of it was similar to Feynman's.

Even Adele Gödel made jokes about it[6]: She called it an '*Altersversorgungsheim*' ('Old-Age Care Home') and entertained fantasies that it was full of pretty young co-eds who were waiting in line to seduce the aging professors.[7]

Buildings and Grounds

A map of Princeton in 1938 is shown below as Fig. 10.3, indicating the mathematics building (former *Fine Hall*) at the University and the first IAS building, *Fuld Hall*, as well as various other locations of interest. The new mathematics building at Princeton University in 1930 was named for the department's founding chairman, Dean *Henry B. Fine*. Previously, the department had shared space with physics in Palmer Hall.[9] The new (today the 'former') *Fine Hall* (Fig. 10.4) was luxurious, with wood paneling, fireplaces in some offices, a common room and a locker room with showers in the basement. When the IAS began operation in 1933, Oswald Veblen had arranged for it to rent space in Fine Hall until its own building, *Fuld Hall*, could be completed (Fig. 10.5).

The IAS campus is located to the southwest of the University campus on what was once *Olden Farm*, purchased by the Bambergers at the behest of Veblen for use by the new Institute (see map, Fig. 10.3). The Institute now owns hundreds of acres of land, much of it woods. Built in a neo-Georgian style, Fuld Hall, like Fine Hall, was luxurious by the academic standards of the day. It is named for one of its donors, Mrs. Felix Fuld (née Bamberger).

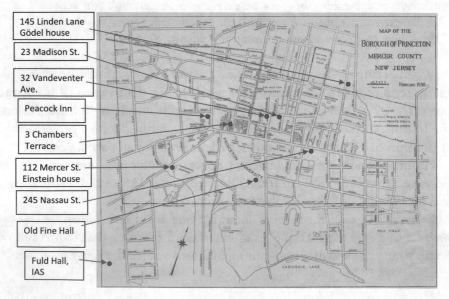

Fig. 10.3 Map of Princeton, 1938. The arrows point to places where Gödel lived and worked, as well as to Einstein's house. Courtesy of the Princeton University Library[8]

Fig. 10.4 The former *Fine Hall* (now Jones Hall) at Princeton, home of the Mathematics Department from 1930 until 1968. The IAS also rented office space here from 1933–1939. Photo courtesy of Princeton University Archives, Mudd Library[10]

Fig. 10.5 *Fuld Hall*, IAS, photographed from the back side. The Common Room is on the ground floor in the central section; above it (with tall arched windows) is the library. Einstein's office was at the far right at the end of the building, Oppenheimer's (while he was Director, 1947–1966) was at the left end, with its own outside entrance. *Photo:* Wiki Commons, 2005[11]

Gödel's First Stay at the IAS

After Kurt Gödel arrived in October 1933 for his first stay in the USA, at Princeton's IAS, he lived in a rented apartment, no doubt arranged for him by the Institute and located at 32 Vendeventer Ave. (compare the map in Fig. 10.3). The IAS at that time resided in Fine Hall, on the University campus, and was conveniently close to his apartment, about 500 m. (ca. 4 city blocks) away. Just what he did during his first semester at the IAS is not clear, since he left no notes, receipts, book slips or other scraps of information from that time.[12] His declared intention was "*to attend the Quantum Mechanics seminar, consult with* [Alonzo] *Church, and improve my English*".[13]

There is also no direct information about his conversations with Alonzo Church, but there was in any case no serious later collaboration between them, although their work had many points of contact. Dawson (1997, 2005 edition, p. 98) includes an interesting discussion of an earlier exchange between Gödel and Church about the latter's new formal-logical system, carried out by correspondence in the summer of 1932. Church had hoped that his proposed system would be 'immune' to Gödel's incompleteness results, which he thought to be dependent on 'peculiarities of type theory' [the latter was introduced by Russell in his (1903) to avoid the Russell Paradox, and included in Whitehead & Russell (1910–1913, 'PM')].

Church's system in fact proved to be *inconsistent*, as Gödel may have suspected, but a subsystem of it, the 'λ-calculus', remained valid and was investigated further by Church's students *Stephen Cole Kleene* and *J. Barkley Rosser* (Fig. 10.6). Their work led to 'Church's Thesis', in which he hypothesized that all effectively calculable functions are 'λ-calculable'. All of this occurred previous to and in the spring of 1934, while Gödel was giving his first Princeton lectures, and there was an interesting interaction involving Gödel, Church, and 'Church's Thesis', of which Gödel was initially skeptical. Church challenged him to provide a satisfactory definition of 'effective calculability', and Gödel realized after some time that Herbrand's definition of recursiveness could be modified to provide the required definition. He included that at the end of his lectures [see Item (3) below, in the quote from the notes on Gödel's 1934 lectures; also Davis (1982)], and Church indeed used it to demonstrate his 'λ-definability'. We will come back to this topic when we consider *Computability* in Chap. 12.

At the end of his first semester in Princeton, Gödel attended the joint annual meeting of the *Mathematical Association of America* (MAA) and the *American Mathematical Society* (AMS), held at the end of December 1933 in Cambridge, MA. He gave a talk there, on '*The Present Situation in the*

Fig. 10.6 *Left*: *Stephen Cole Kleene*, around 1950. Photo: University of Wisconsin, Madison archives. Used with permission.[14] *Right*: *John Barkley Rosser*, 1965. Photo used with permission, taken at the AMS Meeting in Denver, Winter 1965[15]

Foundations of Mathematics (reconstructed as [*1933o] in Volume III of the *Collected Works*). It contains what is probably his first public remark on mathematical Platonism, quoted by Dawson (1997, 2005 edition, p. 100)—speaking of the axioms for a simple theory of types and for set theory: "... *if interpreted as meaningful statements,* [those axioms] *necessarily presuppose a kind of Platonism, which cannot satisfy any critical mind and which does not even produce the conviction that they are consistent*". As Dawson points out in a footnote, the placement of the comma following '*Platonism*' is critical to interpreting this sentence, i.e., whether the following phrase applies to 'a kind of Platonism' or to 'the axioms as meaningful statements presupposing ...'. Since Gödel was still in an early phase of using English in talks and papers, and given the differences in comma placement in English and German (which have duped many a more experienced non-native speaker in both directions), the precise interpretation of that sentence has to be taken with caution. It would seem to contradict his long-term championing of Platonism, which he himself dated as beginning in 1925 (early in his undergraduate career in Vienna).

Gödel seems to have adjusted to life in the—for him—unfamiliar environment in the USA without too many problems (but see the remarks in Chap. 9, at the end of the section '*After 1930*'). Of course, he could devote himself full-time to his mathematical logic, which was his main intellectual pursuit at the time. One of his principal difficulties during his first Princeton stay seems to have been with *eating*.

In Vienna, he had lived with his mother for the past 4 years, and she no doubt saw to it that he was supplied with food that he preferred, even when he worked late hours. In addition, Adele was there to 'take care of him' when he was not at home with his mother and brother. [In Princeton, the wife of Oswald Veblen took over that 'feeding' function to a limited extent, as recounted by Dawson (1997)]. Furthermore, in Vienna there were restaurants and even cafés that remained open until 10 pm or later, so that he had many alternatives. In Princeton, the evening meal would have typically been taken between 6 and 7 pm, and most restaurants, of which there were probably not very many in 1933, would have closed by 8 pm at the latest. Gödel was not used to cooking for himself and was often so engrossed in his work that he neglected meals, and he apparently did not eat well while in Princeton, a problem which recurred, sometimes with serious consequences, throughout the rest of his life.

Lectures on Incompleteness

During the second semester of his visit, from February to May of 1934, Gödel gave a series of lectures on his incompleteness proofs, and that period is well documented in memoirs and interviews, for example by *Stephen C. Kleene*, a logician who was a graduate student of Alonzo Church (and, together with *J.B. Rosser*, another Church graduate student, was recruited by Veblen to write up the notes from Gödel's lectures). See their joint interview: Kleene & Rosser (1984). Cf. also Kleene (1981) and Davis (1982) on the history of 'Church's Thesis' and the λ-calculus.

Those lecture notes, entitled '*On Undecidable Propositions of Formal Mathematical Systems*', were circulated among interested parties as mimeographed copies for over 30 years, until they were finally published in 1965 in Martin Davis's book, '*The Undecidable*' [see Davis (1965), pp. 39–74], and later in Gödel's *Collected Works*, Vol. I (1986), pp. 346–372.

They contain essentially the results published in his (1931), but with some additions, innovations and generalizations that are worth noting. The introduction to the notes as published by Davis[16] mentions the following points:

(1) The specific formal system chosen as example (although again based on the theory of types) permits variables ranging over numerical-valued functions; this makes it possible to give a much simpler proof than before, in that recursive (i.e., primitive recursive) functions are representable in the system.

(2) There is a discussion of the relation between the existence of undecidable propositions and the possibility of defining 'true sentences' of a given language in the language itself. (This matter has been treated more fully by Tarski…) [now known as the 'Gödel-Tarski Theorem'].

(3) In addition to the specific formal system chosen for consideration, there is given an undecidability theorem for 'formal mathematical systems' in general. Since Gödel's characterization of 'formal mathematical system' uses the notion of a rule's being *constructive*, that is, of there existing a 'finite procedure' for carrying out the rule, an exact characterization of what constitutes a finite procedure becomes a prime requirement for a completely adequate development. Church's thesis … identifies the functions which can be computed by a finite procedure with the class of recursive functions (general recursive functions). In the present article, Gödel shows how an idea of Herbrand's can be modified so as to give a general notion of recursive function…

The notes themselves are of course very detailed and contain some additional explanations and aids to understanding not in the original paper (1931). Gödel's interactions seem to have been mainly with the Church graduate students Kleene and Rosser. In their (1984) interview, Kleene says of Gödel's presence at Princeton in 1933/34, "*He* [Gödel] *affected research very much in that he gave lectures on his undecidability stuff. In the course of that, he produced a notion of general recursive function. This became the subject of my research, which was a sequel to the things I'd done with Church*".

Rosser, on the other hand, emphasizes Gödel's unapproachability: "*Gödel was not the kind of man who spoke to people on his own. It was hard to get to talk to him at all.*" He goes on to describe Gödel's politeness and willingness to explain any unclear points; but "*… he would talk to you, answering whatever questions you had to ask him. Very seldom would he volunteer any himself. You'd ask him questions, and unless you saw him again, that was the last you heard from him.*"

About the notes from Gödel's lectures, Kleene reports, "*… He was about to leave for Europe. I gave him a set of the notes, as we had them ready just before he left. He was going to take it to his hotel room in New York, prior to flying. Or rather sailing; it was probably a ship in those years. I told him, 'We have a copy* *at Princeton too. If you don't send us word before you embark, we will bring them out from the copy we have'. When he gave his Gibbs lecture at Yale somewhat later, somebody should have played the same kind of trick on him: recorded it and said, 'if you don't give us the manuscript, we'll bring it out from our copy', because they never did get a copy of that"*.

Regarding Gödel as a lecturer, Kleene says, "*He didn't have that extremely careful, slow, deliberate style of Church, but I think he was pretty clear. We took notes, sometimes we would have to check them with him a little bit.*" He describes a visit of the Gödels to his Maine farm in the summer of 1941 (when Kurt and Adele also stayed at an inn in Brooklin, ME): "*... We exchanged some mathematical thoughts. I told him about my realizability interpretation of intuitionistic number theory, and he said he had an interpretation also using partial recursive functions ... That came out, I guess, in an article or two later*".

During that semester, and in between his lectures, Gödel also traveled to New York and Washington to give more popular talks on his incompleteness theorems. The first, to the *Philosophical Society* of New York University, on April 18th, was on *decidability*, while the second, at the Washington *Academy of Sciences* on April 20th, was on *consistency proofs*.

At the end of April, he was in New York again for several days, probably to visit the U.S. and Austrian consulates and discuss the question of visas for his return to Princeton in 1934/35 (for which Flexner had offered him a grant-in-aid). As we saw in Chap. 9, however, that visit never came to fruition (instead, it was delayed by a year, and then interrupted after only a short stay)—due to Gödel's (mental) health crises, which began in earnest in late July of 1934 and lasted with varying intensity until late 1936 or early 1937.

Returning to Vienna: The Political Scene in Germany and Austria

We saw in the previous chapter that Kurt Gödel left New York on May 26th, on the Italian liner *S.S. Rex*, arriving in Gibraltar on the 31st, then continuing on to Naples (June 2nd) and finally docking at Genoa on June 3rd. Gödel traveled from Genoa to Milan, where he spent one night, and then on to Venice, where he remained for 3 days, evidently as a short vacation. He then proceeded to Vienna, arriving on June 7th, 1934. He returned to a city and country in turmoil.

Fascism had been on the rise in a number of European countries for a decade, and in Italy it had become firmly established under *Benito Mussolini* soon after World War I ended. Italian Fascism became a model for authoritarian governments in various places (even the Brazilian president/dictator *Getúlio Vargas*, who took power in 1930, flirted with the idea, although he finally led his country in 1942 into WWII on the antifascist side, that of the Allies). Spain and Portugal later had brutal fascist dictatorships that persisted well into the second half of the twentieth century. Austria and Germany, which had both become republics after WW I, were increasingly unstable during the 1920s and became politically polarized, with groups from the far Left (Communists) and the far Right (Fascists) carrying on street battles and competing for influence within the ruling parliamentary coalitions. Each developed a particular form of fascism: in Germany, the *National Socialist* ('Nazi') party, and in Austria, a curious alliance of right-wing populists with the Catholic Church (whose political influence had gradually declined since the days of the Holy Roman Empire, and was minimal during the early years of the *First Austrian Republic*).

In Austria, the First Republic was proclaimed on November 12th, 1918, one day after the last *k. & k.* Hapsburg Emperor, *Karl I* (Charles I) had abdicated. Its founding was influenced by the proclamation of a similar republic in Germany 3 days earlier (later known as the *Weimarer-Republik* after its capital city). The first republican Austrian government was a coalition of the Socialists and the Social Democrats. It initially claimed all the former Hapsburg lands in which the majority of the population were ethnic Germans, as well as the German-speaking regions in the (majority Slavic) former kingdom of *Bohemia* and the archduchy of *Moravia* (but as we saw in Chap. 2, unsuccessfully).

The early period was chaotic, with various 'revolutionary' forces trying to claim power in the new Republic. *Hungary* meanwhile had established a Soviet Republic, a Communist government modeled on the Russian example. To appease the various Austrian provinces, which were striving for independence, a Federal Republic was adopted, with Vienna having the status of a state. Soon after, the conservatives became the majority power in parliament, while Vienna remained the only socialist-dominated state. A new constitution was adopted on October 1st, 1920, providing for a parliamentary form of government (a *President* elected by the upper house, the *Bundesrat*, and the chief executive, the *Kanzler* (Chancellor) elected by the lower house, the *Nationalrat*). Initially, the Austrian government supported a union with Germany ('the *Anschluss*', finally carried out by the Nazis from Germany in 1938).

Similar developments took place in Germany about the same time. After the immediate post-war chaos had subsided, the two republics enjoyed about

10 years of relative stability and increasing prosperity; but the Great Depression, after 1929, accelerated the political polarization that had been increasing in the later 1920s, and by 1933, both countries were ripe for a takeover by right-wing populist forces, the *Nazi* Party in Germany and the *Vaterländische Front* ('Fatherland Front') in Austria. Authoritarian leaders had obtained a plurality of votes in elections in 1932 in both countries: *Adolf Hitler* in Germany (himself an Austrian, born in *Braunau am Inn* and schooled in Linz), and *Engelbert Dollfuss* in Austria (originally a Christian Socialist). Hitler succeeded with his *Machtergreifung* (power grab) by March of 1933 in Germany, while Dollfuss dissolved the Austrian parliament about the same time, and thereafter governed by decree, forming the Fatherland Front to unite all the conservative groupings and declaring all other parties illegal. A new, authoritarian constitution was adopted and the government became essentially a fascist dictatorship, with greater privileges for the Catholic Church.

The Austrian Nazis attempted a coup on July 25, 1934, and assassinated Dollfuss, but their takeover attempt was suppressed, and the Nazi leaders were executed. Dollfuss's successor was *Kurt (von) Schuschnigg*, a moderate who governed in a less authoritarian manner (see Fig. 10.7). The period 1933–34, when the brutally authoritarian Dollfuss regime was in power, was later referred to as *Austro-fascism*.

Fig. 10.7 *Left*: **Engelbert Dollfuss**, 1930. Photo by Max Fenichel, public domain, from Wiki Media.[17] *Right*: **Kurt von Schuschnigg** addressing the League of Nations, Geneva, September 1934. *Photo Agence de presse Meurisse*, public domain, from Wiki Media[18]

After 1934, Austria's independence was initially guaranteed by Italy, but following the Italian occupation of Ethiopia and the Italian dictator Mussolini's alliance with Hitler (1936), Austria was politically isolated within Europe. A compromise agreement with Germany that same year 'guaranteed' Austria's independence, but it had to declare itself a 'German State' in return. It was rapidly infiltrated by Nazis from Germany, and in early 1938, a new Nazi conspiracy was discovered. Schuschnigg attempted to negotiate with Hitler, and he planned a plebiscite to decide the issue of Austrian independence, but Hitler, after ascertaining that Italy, France and Britain would not intervene, demanded postponement of the plebiscite and the resignation of Schuschnigg, who capitulated. The German army entered Austria on March 12th, and a puppet president (*Arthur Seyss-Inquart*) was installed. He proclaimed the *Anschluss*, together with Hitler, on March 13th. Austria became the *Ostmark*, the "eastern province" of Germany, with Seyss-Inquart as its governor. [Seyss-Inquart was later Deputy Governor-General in occupied Poland, and after 1940, he was *Reichskommissar* of the occupied Netherlands, holding the rank of *SS-Obergruppenführer*. He was ironically called *Zes-en-een-kwart*, Dutch for 'six and a quarter', by the populace. He instituted a reign of terror in the Netherlands, in particular for the Jewish population. Near the end of the war, he was briefly Foreign Minister of the *Reich*. He was executed after the Nuremberg Trials for Crimes against Humanity.].

Gödel seems to have noted all these developments during the 1930s but not to have expressed any strong feelings about them. His mother was a vocal anti-Nazi and was probably in some danger for that; but she left Austria and returned to Czechoslovakia before the *Anschluss*, and there she no doubt found more sympathy for her views. Kurt was appalled only by the loss of his *Lehrbefugnis* in April, 1938, and the initial refusal to reinstate him as *Dozent neuer Ordnung* (which was apparently due to suspicions that he might be Jewish, or if not, a 'white Jew', a sympathizer with supposed Jewish aims). No evidence was found for that, and he was in fact formally installed in the *Dozent* position in the summer of 1940, when he was however already in the USA. He remained on the faculty rolls of the University of Vienna for some years until after the war, but never returned to Vienna and later refused various honors from the University and the Austrian government.

During the remainder of 1934, Gödel was occupied with his health problems, and apart from a comment at the Menger Colloquium (in reply to a talk by his economist friend Abraham Wald, suggesting that income as a factor in an analysis of demand be taken into account in Wald's equations), which was reported in the *Berichte* the following year, he apparently did no more significant mathematical research. In the next chapter, we will consider the following five years, 1935–39, a time of transition but also of scholarly productivity for Kurt Gödel.

Notes

1. Portrait photo of Abraham Flexner by W. M. Hollinger, 1910. From Wiki Commons, public domain. Reused from: https://commons.wikimedia.org/wiki/File:Picture_of_Abraham_Flexner.jpg.
2. Flexner (1960). See p. 356.
3. Pais (1982), pp. 450 ff.
4. Photo of Oswald Veblen, 1915, photographer unknown; source: http://legacy rlmoore.org/photos/veblen_o.html. Wiki Commons, public domain. Reused from: https://commons.wikimedia.org/wiki/File:OswaldVeblen1915.jpg.
5. Quoted from his semi-autobiographical book, '*Surely You're Joking, Mr. Feynman*', by R. Leighton and R.P. Feynman, W.W. Norton, New York (1985), p. 149.
6. Quoted by Wang (1987), p. 118.
7. Kreisel (1980), pp. 154/55.
8. Map of Princeton, 1938. Courtesy of the Princeton University Library, Cities and towns and Princeton (N.J.: Borough)—Maps; public domain, no known copyright. Reused from: https://maps.princeton.edu/catalog/princeton-2z10ws088.
9. Kleene & Rosser (1984) (their oral history interview about their Princeton days).
10. Photo courtesy of Princeton University Archives, Mudd Library, *Princetoniana* Museum; Grounds & Buildings, MP42; property of the trustees of Princeton University. Reused from: https://www.princetonianamuseum.org/artifact/db53d203-77e9-4df2-875e-59bb180c11bd.
11. Photo: Wiki Commons, 2005 by Eecc–Own work, released to the public domain. Reused from: https://commons.wikimedia.org/w/index.php?curid=161134.
12. Dawson (1997), 2005 edition, p. 98.
13. *ibid*.
14. Photo by Harold N. Hone, Madison, WI. University of Wisconsin, Madison Archives. Used with permission. Permalink: https://search.library.wisc.edu/digital/AIWRGVKTELBFFN9E Image ID S07194. Photo ca. 1950, cf. digitalcontent@library.wisc.edu.
15. Photo taken at the AMS Meeting in Denver, Winter 1965, in: J. Barkley Rosser papers, 4RM11, Photographs, Archives of the *American Mathematical Society*. Reused from: https://digitalcollections.briscoecenter.org/islandora/search/J.%20Barkley%20Rosser?type=dismax. Used with permission.

16. Davis (1965), pp. 39/40.
17. Photo: by Max Fenichel, 1930, public domain. Wiki Media: reused from: https://commons.wikimedia.org/wiki/File:Engelbert_Dollfuss.jpg.
18. Photo: *Agence de presse Meurisse*, original now at the *Bibliothèque nationale de France*, public domain. From Wiki Media, reused from: https://commons.wikimedia.org/wiki/File:Kurt_Schuschnigg_1934.jpg.

11

Transition Years, 1935–39—*A Time of Uncertainty*

The year 1935 was an active period for Kurt Gödel. He had by early spring essentially recovered from his breakdown during the previous summer, and he was able to hold his next lecture course at the University. Its topic was '*Ausgewählte Kapitel der mathematischen Logic*' ('Selected Topics in Mathematical Logic'), and his lectures began on May 4th. This course was more specialized than his first one had been, two years earlier, and was consequently attended by fewer students. Wang (1987) reports that 9 registration slips were found in Gödel's papers.

As we saw in Chap. 9, he was also active in the Menger Colloquium; there, his most important contribution in 1935 was his 'length of proofs' talk given on June 19th. It was published in the *Ergebnisse* of the Colloquium (1936), and is listed as [*Gödel 1936a*] in the *Collected Works*, Vol. I. The English translation of the report is also included in Davis (1965). Gödel discusses proofs in formal systems of varying order, which he denotes by S_i, where i gives the order of the system, i.e., $i = 1$ corresponds to a first-order system, which acts on individuals (e.g. natural numbers), $i = 2$ is a second-order system, acting on classes (or sets) of individuals, etc. His important conclusion[1] is stated very clearly:

> There are infinitely many formulas whose shortest proof in S_i is more than 10^6 times as long as in S_{i+1}. The transition to the logic of the next higher type not only results in certain previously unprovable propositions becoming provable, but also in its becoming possible to shorten extraordinarily many of the proofs already available.

© The Author(s), under exclusive license to Springer Nature Switzerland AG 2022, corrected publication 2023
W. D. Brewer, *Kurt Gödel*, Springer Biographies,
https://doi.org/10.1007/978-3-031-11309-3_11

He then gives as an example a sentence stating that a Diophantine equation $Q(x_1, x_2, \ldots x_n) = 0$ (i.e., a polynomial equation with integer coefficients) has integer solutions. (This refers to Hilbert's 10th problem, which is to find an algorithm to determine whether a given Diophantine equation has an integer solution). A remark in proof was added to the original paper by Gödel:

> It may also be shown that a function which is computable in one of the systems S_i, or even in a system of transfinite type, is already computable in S_1. Thus, the concept of 'computability' is in a certain definite sense 'absolute', while practically all other familiar metamathematical concepts (e.g., provable, definable, etc.) depend quite essentially on the system with respect to which they are defined.

This article is of (at least) historical importance, as an early contribution to computational complexity theory, the first 'speed-up theorem'. It became a classic paper, cited frequently long after its publication (see e.g., Buss (1994, 1995)).

In addition to the 'length of proofs' paper, Gödel also published at least five reviews of work in mathematical logic by others (*Skolem 1934, Huntington 1934, Carnap 1934, Kalmár 1934*, and *Church 1935*) in the years 1935/36. Thus, it seems that he was relatively active in spite of his illness the previous year, and he appeared to be on the road to a complete recovery by midsummer 1935.

Set Theory

Spring 1935 also marks the beginning of Gödel's work on set theory[2] (although he may have been thinking about it for some time already; see below). He would spend the greater part of the next ten years working on the consistency and the independence (from the other axioms) of the *Continuum Hypothesis* (CH) and the *Axiom of Choice* (AC) in Zermelo/Fraenkel (ZF) set theory.

This subject dates back to the work of Cantor, which we summarized briefly in Chap. 6. Cantor had found that there are different *orders* of infinities, described mathematically by their *cardinalities*. The set of natural numbers $\mathbb{N}$, for example, forms an infinite series, whose cardinality, the lowest known, is denoted by $\aleph_0$ (*card* $\mathbb{N} = \aleph_0$ —'aleph-sub-zero' or 'aleph-null'). But if we imagine the natural numbers to be set out along a line of infinite length, then each one occupies a discrete point, separated by an interval (of length '1'), which itself contains an infinity of points—the continuum. If we consider other classes of numbers, e.g., the *rational* numbers (given by ratios of natural numbers, e.g., $0.8 = 4/5$, lying in the first interval between 0 and 1), we see that they occupy points *between* the natural numbers on the continuum line. But nevertheless, they can be shown to represent an infinity of the *same* order as the natural numbers. This can be seen for example by coding them (take the sum of the two numbers in the ratio, which will again be an integer—and these can thus be put in a one-to-one correspondence with the natural numbers, i.e., they are a *countable infinity*).

Including the *irrational* numbers (with the *algebraic* numbers (solutions of general polynomial equations) and the *transcendental* numbers (not algebraic or rational numbers)), we arrive at the full set of *real* numbers, $\mathbb{R}$. They evidently represent an infinity of higher cardinality, since there are (infinitely many) more of them than there are of natural numbers. Cantor proved this[3] with his diagonal argument in 1891. According to Cantor, they have the cardinality $\aleph_1$: *card* $\mathbb{R} = \aleph_1$. Cantor's question concerned whether or not there is any cardinality on the continuum which lies *between* $\aleph_0$ and $\aleph_1$. The 'Continuum Hypothesis' (CH) postulates that there is *not*. We can express this by an equation: $\aleph_1 = 2^{\aleph_0}$. [This is based on the fact that the number of subsets—the elements of the *power set*—of a set with n elements is 2^n. Take for example $n = 3$: the subsets include the Null set $\{\}$, the set with only its 1st element, $\{1\}$; likewise $\{2\}$ and $\{3\}$; then $\{1,2\}$, $\{1,3\}$, and $\{2,3\}$, and finally the original set itself, $\{1,2,3\}$, for a total of eight elements, i.e. 2^3. Generalizing to $n \to \infty$, we get the above expression]. A related theorem (*Cantor's theorem*) states that the cardinality of a set is strictly less than the cardinality of its *power set* (the set of all its subsets, including the Null set and the set itself, as above). The CH claims that there is no order of infinity, no *cardinality*, between $\aleph_0$ and $\aleph_1$. Hilbert's 1st problem (1900) was to prove the Continuum Hypothesis.

Generalizing still further, we can write $\aleph_{\alpha+1} = 2^{\aleph_\alpha}$ for every ordinal number α. This is the *Generalized Continuum Hypothesis* [GCH—due to Jourdain (1905)].

The first complete axiomatization of Cantor's (naïve) set theory was given by Zermelo (1908). It was extended by Fraenkel (1922) and by von Neumann (1925). This formal system is denoted as ZF, or, if it includes the Axiom of Choice (AC), as ZFC. A simplified summary is given by Cook (2010); see also Bell (2015) on the Axiom of Choice. The detailed history of these developments is treated in the book by G. H. Moore (1982). Dawson (1997) gives a summary of work immediately before Gödel on pp. 116–120. Another good summary is given by Floyd & Kanamori (2006), in their contribution to the memorial issue of the *Notices of the AMS* in Gödel's centennial year; the sketch below is based on their article.

J. L. Bell cites Fraenkel et al. (1973), who state: "[the *Axiom of Choice* has been hailed as] *probably the most interesting and, in spite of its late appearance, the most discussed axiom of mathematics, second only to Euclid's axiom of parallels which was introduced more than two thousand years ago*". Bell goes on to state that, in spite of this 'fulsome' quote, the "[Axiom of Choice] *amounts to nothing more than the claim that, given any collection of mutually disjoint nonempty sets, it is possible to assemble a new set—a transversal or choice set—containing exactly one element from each member of the given collection*". Nevertheless, the AC has proved to be an important principle for the foundations of mathematics, and it has been widely employed in works in that field as well as in standard mathematics. It was first formulated by Zermelo in 1904. A modern formulation[4] is that (for a collection **H** of nonempty sets), a 'choice function' is a map f (on the domain **H**) such that

$$f(X) \in X \text{ for every } X \in H. \tag{11.1}$$

As an example, we might take **H** to include all those sets that each contain a pair of natural numbers. Then a choice function might pick out the lesser of the two numbers for each set. Another choice would be the greater of the two numbers. Thus, a formulation of the AC could be:

$$Any \ collection \ of \ nonempty \ sets \ has \ a \ choice \ function. \tag{11.2}$$

There are several more specific formulations for the AC. Zermelo also used the quasi-geometric concept of a *transversal*: this is a 'slice' through all of the sets in the collection **H**; for example, if **H** contains all the lines in the Euclidean plane parallel to the x axis, then a line parallel to the y axis (which cuts through all the elements of **H**, selecting one value from each) would be a transversal of **H**. In his (1908), Zermelo formulated the AC as:

$$Any \ collection \ of \ mutually \ disjoint \ nonempty \ sets \ has \ a \ transversal. \tag{11.3}$$

For *finite* collections of sets, both (11.2) and (11.3) can be proved by induction. For infinite collections of finite members, the situation is more complicated.

This is where Gödel stepped in, with his work beginning in 1935. He wanted to establish the *consistency* of the AC and of the CH (*relative* to the other axioms of ZF), and also their *independence* of those axioms, i.e., whether they might be implied by the other ('standard') axioms, or instead are completely autonomous, so that e.g., *both* AC and ¬AC would be compatible with the other axioms. He was of course not the only mathematician working on those questions. Gödel published nothing in 1935 and 1936 on his set-theoretical work, but he mentioned it to von Neumann and to Menger.

In the summer of 1935, Gödel continued working on the consistency and independence of the AC and the CH, and on a proof of the GCH. He discovered the usefulness of *constructible* sets for the latter.

Gödel's *constructible universe* builds upon von Neumann, who considered 'well-founded' sets, which obey a hierarchy, a *recursion* on the ordinals. It begins with the Null set $\emptyset$:

$$V_0 = \emptyset;\ V_{\alpha+1} = P(V_\alpha);\ \text{and } V_\lambda = \cup\{V_\alpha | \alpha < \lambda\} \qquad (11.4)$$

where P is the *power set* operator: P (X) gives the set of all subsets of X. α and λ are ordinals, and λ is called the *limit ordinal*. The universe of all the sets V is denoted as the *cumulative hierarchy*:

$$V = \cup \{V_\alpha \mid \alpha \text{ is an ordinal}\}$$

in which every set is well-founded. $\cup \{\ \}$ is the *union* of all the sets within $\{\ \}$.

We see here a certain analogy with the successor function s in number theory: repeated application of s leads from the first integer, 0, to all the integers in $\mathbb{N}$. Here, repeated application of P leads from the Null set to the universe of well-founded sets. Gödel's *constructible universe* is built in an analogous manner; he defines:

$$L_0 = \emptyset;\ L_{\alpha+1} = def(L_\alpha);\ \text{and } L_\lambda = \cup\{L_\alpha \mid \alpha < \lambda\} \qquad (11.5)$$

where def(X) is an operator which assigns to the set X all its subsets which are *definable* in terms of the first-order language of set theory. His constructible universe is then the universe of all the sets L:

$$L = \cup \{L_\alpha \mid \alpha \text{ is an ordinal}\}.$$

According to Gödel, "L *can be defined and its theory developed in the formal systems of set theory themselves*". This was confirmed by Tarski in his treatment of the satisfaction relation in set theoretical terms [see Tarski (1956)].

L is *absolute*, in the sense that for any inner model M, the construction of L within M leads to the same class L. Just as the process of defining new numbers successively in terms of those already defined can be continued to the transfinite, so can the above procedure be applied to sets of real numbers continuing to the transfinite. As Floyd and Kanamori (2006) point out, through Gödel's constructible sets, "*The jumble of the* Principia Mathematica *had been transfigured into the constructible universe* L". Furthermore, "*Gödel's work with* L *with its incisive analysis of first-order definability was readily recognized as a signal advance… As the construction of* L *was gradually digested* [by the mathematical-logical community], *the sense that it promoted of a cumulative hierarchy reverberated to become the basic picture of the universe of sets*".[5] To be continued…

Gödel in 1935/36

As we saw in Chap. 9, Gödel spent several stays in sanitoria and rest homes in the spring and summer of 1935, apparently in an attempt to gain physical and mental strength and consolidate his recovery from his breakdown in 1934. He wrote optimistically to Flexner in Princeton, saying that he felt much better, and was evidently preparing himself mentally for his next trip to America.

During his passage on the *MS Georgic* from Le Havre to New York in September, Gödel conferred with both Paul Berneys and with Wolfgang Pauli, as we noted in Chap. 9. When he arrived in Princeton, Flexner remarked that he looked much better than he had two years before.[6] His stay began well; he occupied an apartment at 23 Madison St., quite near to his dwelling two years earlier (cf. the map in Fig. 10.3). Dawson (1997) found in Gödel's papers two notebooks containing his work from autumn 1935. He mentions that the first one contains some differential geometry (was Gödel already thinking about general relativity theory?) and a systematic development of set theory, a precursor to his summary article of 1940. The second contains a (*Rein(-schrift)*, i.e., a 'finished' manuscript) on the Axiom of Choice, and a section marked *halbfertig* ('half-finished') on the Continuum Hypothesis. He was evidently already preparing a manuscript on the AC for publication.

Just why that Princeton stay, which had begun so auspiciously, was suddenly canceled on November 17th is not clear, except for the bare fact that Gödel's depression and anxiety had returned. He visited a physician with Veblen, and was advised to go back to Vienna, no doubt wisely. We heard in Chap. 9 about his return journey and his disturbed state when he finally arrived in Vienna. The administration and faculty at the IAS were sympathetic and assured him that his return to Princeton was welcome as soon as his health would permit. Gödel traveled from New York to Le Havre on the *Champlain*, and he arrived in Vienna on December 11th. His condition deteriorated still further in the following weeks. The new year, 1936, was in his opinion, as he expressed it much later in his life,[7] one of the three worst that he had experienced.

A Lost Year

As we saw in Chap. 9, 1936 was very difficult for Kurt Gödel. When he returned from his (brief) stay in Princeton in late 1935, he was already in a nearly desperate condition, depressed and suffering from anxiety attacks. This condition worsened in the first months of 1936, and he tried to regain his mental equilibrium by spending several stays in sanatoria, at first in *Purkersdorf*, where he had already spent a few weeks in October 1934. He was now being treated by the psychiatrist Dr. Otto Pötzl, the successor to Dr. Wagner-Juaregg, who had treated him in 1934. He later spent stays of varying lengths at *Rekawinkel*, in *Golling*, and in *Aflenz*, as we have seen. In December, he was at a hotel in Graz, south of Vienna, and the records found in his papers show that his reading at this point was concentrated on toxicology and pharmacology, as well as psychiatry and the laws governing the treatment and the rights of the mentally ill in Austria and Germany.

He even read a book about carbon monoxide poisoning at that time, which might suggest that he was contemplating suicide, as his brother had feared. However, it may also simply reflect his fear of and obsession with poisoning by heating gases, which had apparently already begun to bother him by late 1936. As Dawson[8] points out, this was not entirely unreasonable, since space heating in Vienna at the time was provided to a large extent using stoves that burned coal briquets or coke, and improper operation of those stoves could in fact lead to carbon monoxide poisoning. The 'coal gas' used for cooking and to some extent for heating and lighting also contained carbon monoxide. [On the other hand, rational explanations of irrational fears are not very helpful.].

The Death of Moritz Schlick

A tragic event in June of 1936 very probably had a serious and deleterious effect on Kurt Gödel's mental state. His professor and mentor, who had shown such concern about Kurt's condition earlier in the year and intervened in his favor with Dr. Pötzl,[9] was killed in cold blood while on his way to his lecture at the University on a June morning. Schlick was just 54, the same age as Hans Hahn had been when he died unexpectedly two years earlier, and the same age as Kurt Gödel's father Rudolf August, who died suddenly somewhat more than 7 years before. Gödel's fatherly mentors all seemed to die suddenly and unexpectedly, and inexplicably at the same age. (It is not certain whether Gödel noticed this detail, but with his tendency toward superstition and the esoteric, he might well have done so).

The murder—it might better be termed an assassination—of Prof. Moritz Schlick in June 1936 was an 'announced death'. Its perpetrator, *Hans (Johann) Nelböck* (1903–1954) had already threatened Prof. Schlick with death several times before. Nelböck was a former student of Schlick's, who had begun his studies of philosophy in Vienna in 1925 and finished his doctorate under Schlick's mentorship.

In fact, Nelböck barely managed to complete his doctorate in 1931 (receiving the lowest passing grade). After he finished his studies, he was diagnosed twice as a 'schizoid psychopath' by Dr. Otto Pötzl (the same psychiatrist who treated Gödel in 1936), and committed to mental institutions, on both occasions after Schlick had filed charges against him because Nelböck had threatened him with death. Both times he was later released, supposedly harmless, and each time he returned to stalking Schick.

Finally, on June 22nd, 1936, he ambushed Schlick, who was on his way to deliver his last lecture of the semester in a lecture hall in the main building (*Hauptgebäude*) of the Vienna University. Nelböck confronted Schlick on the stairway leading to the lecture room (the *Philosophenstiege*; see Fig. 11.1) and shot him four times at point-blank range. Schlick was dead before medical help could arrive.

Nelböck came from a provincial, conservative Catholic background and was evidently sympathetic to the clerical-fascist movement (organized as the *Fatherland Front* by Engelbert Dollfuss). However, his aggressive obsession with Schlick may have had more personal causes. It is not unusual for former students who are dissatisfied with their own abilities and achievements to place the blame on their professors, and cases in which they become obsessed with obtaining revenge or punishing those professors are more common than one might think. Moritz Schlick was not the first, and certainly not the last professor to be stalked and injured or even murdered by a disaffected former student.

In the Schlick/Nelböck case, there were two further complications: first, a fellow student of Nelböck's, *Silvia Borowicka*, had also done her graduate

Fig. 11.1 *Left*: the main entrance to the *Hauptgebäude* (Main Building) of the Vienna University. The 'Philosophers' Stairway' leads up to the right from the entrance foyer inside. Photo: AMZ, 2021. *Right*: the *'Philosophenstiege'*, the stairway leading to the former offices of the Philosophical Faculty, where Prof. Schlick was shot on a landing. Photo by Franz Pfluegl, *Universität Wien, Öffentlichkeitsarbeit*, CC Attribution license[10]

work in Schlick's group. Around 1928, Nelböck met Borowicka and became infatuated with her. His feelings were not reciprocated, and furthermore, she confessed to having 'a certain interest' in their mentor, Prof. Schlick. She reportedly gave Nelböck a pistol, for 'self defense', and it was the gun that he used to murder Schlick. Silvia Borowicka was also a suspect as an accessory to the murder, and she also was evaluated by Dr. Pötzl, who found her to be "*a nervous girl of slightly eccentric character*".[11]

The murder and the surrounding circumstances are described in detail in a recent book by David Edmonds (2020), '*The Murder of Professor Schlick. The Rise and Fall of the Vienna Circle*'; it also deals with the history of the Vienna Circle. The putative romances between Silvia Borowicka and Johann Nelböck (and/or Moritz Schlick) are also the subject of a chapter in Greiser (1989), '*Eine Liebe in Wien*' ('A Love in Vienna'), a book describing iconic love affairs in Vienna.

Thus, Nelböck claimed jealousy as a motive for his murder of Schlick, which he did not (and could not) attempt to deny; and the psychiatric examination had found him to be mentally accountable for his crime. He also claimed that Schlick had intervened to prevent his obtaining a position at various Adult Education Centers. Both Schlick's purported romantic involvement with Silvia Borowicka and his alleged intervention to frustrate Nelböck's efforts to find employment were most likely obsessive delusions (or defensive lies).

The second complicating factor in the case was its philosophical/theological aspect—Nelböck, supposedly originally a devout Catholic, claimed that Schlick's physicalist, anti-metaphysical philosophy had destroyed his worldview and left him in a state of desperation, in which he committed the murder. Both of these factors were cited in Nelböck's trial, perhaps to distract from the ideological background of his deed—the liberal, cosmopolitan Schlick was perceived as a threat to the fascist, conservative-Catholic ideology espoused by Nelböck.

Nelböck was found guilty of murder and sentenced to 10 years in prison, but he served less than two. After the *Anschluss*, he applied for a pardon, and the Nazi authorities exonerated him (he was released on parole) and made him into a sort of Nazi hero, under the mistaken impression that Schlick had been Jewish (in fact, he was the descendent of an old Prussian noble family), or a Jewish sympathizer and anti-Nazi (probably true). Nelböck's deed thus became a political assassination, an almost heroic act for the Nazi movement. Nelböck himself worked after his release in the petroleum industry and later in a Land Surveying Office. He sued Viktor Kraft after WW II, claiming defamation by the latter's book, '*Der Wiener Kreis*', in which he was called 'a paranoidal psychopath'. (Viktor Kraft had been a student of Heinrich Gomperz's in the early 1930s and carried on Gomperz's colloquium after the latter's politically-motivated dismissal from the University in 1934; see Chap. 9). Nelböck died in Vienna in February 1954.

Throughout the rest of 1936, Gödel was in and out of rest homes, spa hotels and sanatoria, mostly alone, sometimes accompanied by Adele. He postponed all his duties at the University and was evidently not able to accomplish much work. Late in the year, his condition improved, and by the beginning of 1937, he was able to return gradually to his normal life.

Recovering—1937. Set Theory, Continued

In the summer of 1937, after two postponements, Kurt Gödel gave his third course of lectures at the University. His topic was *Axiomatic systems for set theory*, based on the work he had begun in 1935 and his ideas about constructible sets. As we saw in Chap. 9, he had relatively few students in this rather specialized course. He had by now also obtained some results on the Continuum Hypothesis, and he reported on those in the Menger Colloquium (without Menger, who had moved to the USA, to the University of

Notre Dame, in January of 1937). Later in the summer, he corresponded with von Neumann, who was visiting in Europe, and subsequently met with him in Vienna. Von Neumann encouraged Gödel again to publish his results on the relative consistency of the AC, and later his consistency results for the GCH, in the journal published jointly by the IAS and the Princeton University mathematics department (*Annals of Mathematics Studies*).

As we know from Gödel's second incompleteness result, a formal system cannot prove its own consistency if it is indeed consistent. But the *relative* consistency of one or more axioms with respect to the others can be proved. This was what Gödel was attempting to do for the AC and the CH. Starting with the 'standard' axioms for set theory, as set out by Zermelo, Fraenkel and von Neumann (the ZF system), and *assuming* their consistency, he was able to show that adding the Axiom of Choice and/or the Continuum Hypothesis produced no additional inconsistencies.

These proofs are considered in some detail by Dawson (1997), who points out some common features with Gödel's earlier incompleteness proofs. Indeed, Dawson suggests that, on the basis of what he had been reading in the years 1928–31, Gödel had begun thinking—already before 1930—about solving *both* of the first two problems proposed by Hilbert in 1900. The second problem became his Incompleteness Theorems in 1930, and then he concentrated on the first, the proof of the Continuum Hypothesis. Dawson[12] points out the similarity in Gödel's procedures in his (1931) incompleteness paper and in his summary paper on set theory (1940). In each case, he begins with a fundamental property of predicates: in the first case, that they may be *primitive recursive*; in the second, that they are '*absolute for the constructible submodel*'. (In the latter expression, 'absolute' means that the formula in question has the same truth value in each model. Gödel's *constructible universe* is a class of sets that can be described in terms of simpler sets, as we saw above).

Both of these proofs begin with a list of lemmas, in which a number of predicates are shown to have the relevant property, but *one does not* in each case—in his (1931), it is No. 46, the predicate which asserts its provability within the formal system; and in (1940), it is the predicate which refers to being a cardinal number within a model of set theory. In each case, the whole proof depends on that predicate. The critical point is Gödel's distinction between *internal* and *external* points of view: in the case of incompleteness, between the mathematical and the metamathematical levels; and in set theory, between the functions existing within a particular submodel, and those which are outside it. This is a significant point for understanding the proofs, and it has often been misunderstood.

Gödel arrived at his concept of constructible sets in the summer of 1935; but his Axiom of Constructibility for the constructible universe, leading to his consistency proof for CH, was apparently conceived only two years later [during the night of June 14th to 15th, 1937, according to a note in one of his 'working notebooks' (*Arbeitshefte*)].[13]

How, then, did Gödel prove the consistency of the AC and CH using constructible sets? He showed that, given the axioms of ZF, and *assuming that* they themselves form a consistent system, the structure $(L, \in)$ is a model of ZF, and also of the AC; and he later showed that it is also a model of the GCH. This is sufficient to prove the *relative* consistency of the AC with ZF (leading to the consistent system ZFC), and furthermore that ZFC + GCH is also a consistent system. The first result was obtained already in mid-1935, the second in mid-1937; but both were published only in 1938 and later. [See Gödel (1938), (1939), (1939a), and (1940). We might notice that his important results were almost always obtained during the summer months. Wang[14] speculates that this is no coincidence, but rather the result of a metabolic quirk which made Gödel unusually sensitive to cold. He was later often seen in Princeton on a warm spring or autumn afternoon in his overcoat, woolen scarf and hat.].

Private Life

Kurt and Rudolf's mother Marianne, as we have heard, decided in 1937 to return to her villa (and her life) in Brünn, for whatever reasons. After she left Vienna in the autumn of 1937, Kurt had to reorganize his own life. His financial situation was not rosy, as he had hardly earned any money since becoming a *Privatdozent*; the previous year had been very costly, with his frequent stays in sanatoria and resort hotels; and his inheritance was gradually dwindling away. He had managed to save some of his honoraria and expense money from Princeton in 1933/34, but probably not much from his interrupted stay there in 1935, after paying his travel costs. It was clear that he and Rudolf could and would not keep the large apartment in the *Josefstädter Str.*, and Adele was pressing him to finally move together into an apartment of their own.

It was at this point that he chose their doubtless much less costly apartment at the outer edge of the city, in the 19th District, *Döbling*. The place he chose, as we saw in Chap. 9, was located near the beginning of the *Himmelstraße*, which starts at the center of the village of *Grinzing*, one of the subdistricts of *Döbling*. The village has a long history, dating at least back to the eleventh century, when a noble family, the *Grunzinger*, obtained rights to it and built a manor there, reportedly on the site of a Roman ruin (so that it became known as the *Trummelhof*, the 'Court of Ruins'). The region was known since antiquity for its wine, and it remains today the site of a number of *Heurigen*— wine taverns that serve the local vintages. It has long been a favored goal of the Viennese as well as tourists on outings from the city, and it is the last stop on the No. 34 streetcar line (in Gödel's time, the No. 38), still within the Vienna city limits.

The seemingly romantic—or pious—name of the *Himmelstraße* is in fact derived from an inn, called '*Am Himmel*', located on a hill, the *Pfaffenberg*, near the western end of the street in the village of *Ober-Sievering*. It probably referred to the higher elevation and the view from that inn, making it seem 'in the heavens'. The building where Kurt and Adele lived from November 1937 to November 1939 is located at the eastern end of the street, near the center of Grinzing, on the south side of the *Himmelstr.* at No. 41–45; see Fig. 11.2.

Sometime in 1937, Kurt and Adele apparently announced their engagement. Both Karl Menger and Rudolf Gödel recalled that announcement, but neither had met Adele personally up to that time. Menger described a vague memory that 'someone named Adele had visited Kurt three years ago when he was ill' (in 1934). Rudolf certainly knew who she was, but he had—surprisingly—never seen her (at least consciously; he very likely had indeed seen her

Fig. 11.2 The house at *Himmelstr.* 41–45, where Kurt and Adele lived from November 1937 to November 1939. The entrance to the right is Himmelstr. 43, and next to it are plaques listing famous former occupants, including Kurt Gödel. Photo by Clemens Mosch, February 2019, used under a CC-A-SA 4.0 license; cropped. Reused from https://commons.wikimedia.org/wiki/File:Himmelstr41.jpg

on the street at some point during the 16 months when they were neighbors in the *Lange Gasse*).

However, engaged or not, they could not marry in 1937. Since the Catholic Church had been put in charge of marriages in Austria during the *Austrofascism* period in 1934, divorce was no longer allowed or recognized in that country. Adele had divorced her first husband, Xy Nimbursky, in 1933 while it was still possible; but her divorce was no longer valid after 1934. Only after the *Anschluss* in March 1938 was divorce again legally recognized—the new Nazi masters had no interest in delegating an important aspect of civil law to the Catholic Church. This was the first, but not the last time that the presence of a Nazi government in Austria had somehow worked to the advantage of Kurt and Adele. So, after November 1937, they lived in bucolic isolation in Grinzing, no doubt presumed by their neighbors to be a married couple.

The memorial plaque beside the door of No. 43 is shown in Fig. 11.3. It mentions Gödel's Completeness results (1930) and his set-theoretical results on the relative consistency of the AC and the CH (in 1938), and two of his later honors, the Einstein Award (in 1951) and an honorary doctorate from Harvard (1952), however with some incorrect dates. Interestingly, his perhaps most important achievement, the proof of incompleteness of the logic of PM (1931), is *not* mentioned.

The pleasant environment of their dwelling in *Grinzing* is shown in Fig. 11.4—this photo was taken near the village center, at the eastern end—the beginning—of the *Himmelstraße*.

The Fateful Year—1938

Kurt Gödel's health had normalized by late 1937, and he was back at work as soon as he and Adele had settled in their new surroundings. Whether either or both of them missed the bustle of the city, where they had spent the past decade living separately but not far apart, we have no way of knowing. But the relative calm of Grinzing, especially in the off-season in the winter, gave them the opportunity to become accustomed to living together. Their move was also not inopportune, since most of Kurt's closer friends and colleagues had left Vienna by late 1937, in one manner or another. He seems to have continued working without taking too much notice of the changed environment, both politically and in terms of the people around him, with the single-mindedness that was characteristic of his personality. Adele might have suffered more from their relative isolation, but her family and friends were

Fig. 11.3 The plaque to the left of the door at *Himmelstraße* 43. It reads: '*This house was honored by the presence from 11.11.1937 to 9.11.1939, before his emigration, of one of the most significant mathematicians and logicians of the twentieth century, Prof. Dr. Kurt Gödel (1906–1978)*. [A list of Gödel's achievements and honors follows]: '*1930 Completeness of the logical calculus; 1938 Relative consistency of the Axiom of Choice and the Continuum Hypothesis of set theory; 1945 Recipient of the Einstein Award* [in fact, this was in 1951]*; 1967 Honorary doctorate from Harvard University*' [again, the correct date is 1952]. Photo: Wiki Commons, CC Attribution license[15]

just a streetcar ride away. Early in the new year, Kurt gave a talk at the Zilsel Colloquium, which he had helped to organize the previous year. His talk was for a general audience and dealt with consistency problems in set theory, his main topic of interest at the time. Budiansky (2021, p. 192) gives some details of Zilsel's colloquium, which was unfortunately short-lived and could not compare with the former Vienna Circle nor the Menger Colloquium. Dawson (1997, p. 125) gives a summary of Gödel's colloquium talk (reconstructed in the *Collected Works*, Vol. III, as [Gödel (*1938a)]). He says of this manuscript, "*The paper is typically Gödelian: incisive, thoroughgoing, and forward-looking. One can only wonder what impression it made on the mathematically unsophisticated audience to which it was delivered*".[17]

Gödel had probably planned to offer another lecture course at the University later in the spring or the summer, but as we know, one of the fateful events of the year 1938—the annexation of Austria by the *Deutsches Reich*—cut that plan short. He also conferred briefly with von Neumann, who passed through Vienna early in 1938, and who urged Gödel again to publish his set theory results and to come to Princeton, perhaps on a long-term basis.

The new year 1938 brought still more drastic changes: as we saw in Chap. 9, the pressure from Nazi Germany for Austria to unite with Germany or be occupied came to a head early in that year. There was certainly

Fig. 11.4 Near the center of the village of *Grinzing*: the eastern end of the *Himmelstr.*, with the Grinzing Church on the left. The house at No. 41–45 is a few blocks further west, on the same side of the street as the church. Photo by Martin Furtschegger, 2012. Wiki Commons, CC Attribution license[16]

a majority in Austria immediately after WWI in favor of uniting with Germany; however, that was not permitted by the victorious Allies at the time. The popular sentiment may have changed by 1938, given the political developments in the meantime. The Austrian chancellor Schuschnigg decided to hold a referendum in March 1938 to find out whether a majority were still in favor of such a union, and carried on an intensive advertising campaign in favor of Austrian independence. Hitler however put pressure on Schuschnigg to postpone the referendum and to resign, and threatened to use military force to occupy Austria. His original intention was apparently to make Austria a 'Protectorate', along the lines of what he did with Bohemia and Moravia a year later. But when German troops marched into Austria on March 12, 1938, the reception by the populace was so positive that he decided to make the country a province of the *Reich* (the '*Ostmark*'), on a par with Prussia, Saxony or Bavaria. Schuschnigg resigned and Hitler entered Vienna triumphantly on March 15th, setting up a Nazi-controlled provincial government headed by Arthur Seyss-Inquart, a Nazi loyalist who had previously been installed as Austrian Minister of Public Security under pressure from

Hitler. We have heard about the frenzy of Nazi supporters in the days that followed, a preview of what would happen the following November all over the *Reich*, during the so-called November pogrom or '*Reichs-Kristallnacht*' which is still remembered as a day of infamy on November 9th/10th.

All of this evidently did not interest Kurt Gödel very seriously. But in April, he was informed that his *Lehrbefugnis* (his position as *Privatdozent*) had been annulled, and he would have to apply for a new position, the *Dozentur neuer Ordnung*. He was incensed by this 'violation of his rights' but initially let the matter slide. From then on, he no longer had a formal position at the University of Vienna, although he still applied for a leave of absence in the coming academic year 1938/39, in order to again visit Princeton (and Notre Dame, as we have seen). During the rest of the spring and summer, he continued working on set theory, privately, and preparing his next stay in the USA.

In the meantime, another result of the *Anschluss* had been the recognition—in July 1938—of divorce, prohibited in Austria from 1934–38, so that nothing now stood in the way of the marriage of Kurt and Adele (Fig. 11.5). Budiansky (2021) describes their small, private ceremony, and mentions that there were only 9 wedding guests (as can be seen from the receipt for the reception, held in the *Wiener Ratskeller*)—all of them close friends and family. Witnesses were Kurt's cousin *Karl Gödel*, a painter, and a certain *Hermann Lortzing*, a book appraiser, both living in Vienna.

Fig. 11.5   Adele Porkert Gödel and Kurt Gödel at their wedding, on September 20th, 1938 in Vienna. This photo has often been reproduced; it seems to be the only available image of Kurt and Adele's wedding. *Photo:* September 1938, photographer unknown, public domain. Reused from [Schindler (2006)]

Just how Kurt Gödel felt about his marriage is not well documented. Wang, who knew both Kurt and Adele Gödel personally over a period of some years, saw it in a positive light[18]: "*Based on the little evidence I have, I believe that they were much devoted to each other and completely at ease together*". Dawson gives no opinion about their presumed relationship, but he does speculate that Gödel might have wanted to take Adele along to America on his subsequent trip. Budiansky, presumably on the basis of Kurt's '*Protokolle*', takes a more cynical view and supposes that Kurt had agreed to marry in September only on the condition that he be allowed to make his U.S. trip alone, as originally planned.

Both Dawson (1997) and Budiansky (2021) mention a power-of-attorney granted by Gödel to Adele; Dawson says it was to '*publish an announcement of their marriage and to obtain an official marriage certificate*', which she would have needed in his absence. Budiansky suggests that it was '*authorizing her to make all the arrangements for their wedding*'. Perhaps there was more than one, and it would not be surprising if Kurt didn't want to be tied up with organizational details so shortly before his planned departure.

There is also some disagreement over who knew about their marriage (and when). One story has it that late in Gödel's visit to Notre Dame, when Karl Menger was trying to convince him to stay in America and take a position there, he suddenly discovered that Kurt was already married, which rendered his plan unworkable. Another version, strongly supported by letters from Menger to Veblen and to Gödel, both in the later months of 1938, suggests that Menger already knew about the marriage before Gödel's arrival in the U.S. in October (but perhaps he was referring to the *engagement*). He told Veblen that he thought it would be good for Gödel (to be married), and congratulated Gödel on his (impending?) marriage, and asked why he had not brought Adele along to America.

A third version, to be found in Menger's posthumously published autobiographical book (1994), has it that in a telephone conversation in 1985 with Eckehart Köhler, Menger, shortly before his death, stated that, "…*he had only recently learned of the fact that Gödel had married Adele in September 1938*". However this may be, it is certain that events in Gödel's life in the autumn of 1938 succeeded each other in rapid order—his marriage on September 20th and his departure barely a week later for an 8-month absence (not the usual idea of a honeymoon). Adele remained in Grinzing, a '*grüne Witwe*' (roughly: a 'lonely suburban housewife').

Princeton And Notre Dame, 1938/39

Gödel had after some negotiations agreed to go to the IAS in Princeton for the last three months of the year 1938, and to deliver a series of lectures there in November and December, on the topic of his—as yet unpublished—work on set theory. He had also accepted the invitation to go to Notre Dame University in early 1939 and give at least one course there in the spring semester. The latter had been extended by his former colleague Karl Menger, and formalized by an invitation from the university's president, Father John A. O'Hara, CSC. Gödel had promised to arrive in Princeton in early October in order to have time to prepare his lectures. Somewhat surprisingly, he had no trouble obtaining an exit visa from Austria (now part of the German *Reich*), and the IAS had arranged a visitor's visa for entry into the USA.

Nevertheless, there remained some logistical problems: Kurt had limited funds, and a direct transfer of money from Princeton was not possible because of the currency restrictions that were in place. The IAS however could pay for his passage, as Flexner promised—but the second problem was that most of the ships were booked up and he had difficulty getting a berth. He finally reserved a passage on the *S.S. Hamburg*, belonging to the Hamburg-America Line and sailing from Hamburg at the end of September, scheduled to arrive in New York on October 7th.

When he arrived in Hamburg on September 29th, just over a week after his wedding, there was however some problem, so that he cabled Flexner in Princeton, asking for help. Just what the problem was, we do not know. Dawson (1997, pp. 128/129) speculates that Kurt may have brought his bride along to Hamburg, hoping that she could accompany him to New York. Of course, no-one on the other side of the Atlantic knew of his marriage at that point, as we saw above. And taking her along would require re-booking a private stateroom for two; he might have hoped to get a visitor's visa for her after arriving in New York.

The *Hamburg* in any case sailed without Gödel, but after some delay, he was able to book passage (for himself alone) on the *S.S. New York*, the sister ship of the *S.S. Hamburg* (see Fig. 11.6), which sailed from Cuxhaven to New York, arriving on October 15th.

Whether a last-minute attempt to take his new wife along to America was indeed the cause of the problems in Hamburg, as suggested by Dawson (1997), we will never know. But it seems that Kurt Gödel's departures were in any case always complicated, in one way or another. He arrived at Princeton without further incident, and—as we saw in Chap. 9—was soon established

Fig. 11.6 The *S.S. New York*, Hamburg-America Line, 1934. Kurt Gödel sailed on this ship from Cuxhaven to New York in October 1938 to begin his third stay at the IAS/Princeton, followed by a visit to Notre Dame University. Photo: photographer unknown, 1934, Wiki Commons[19]

at the *Peacock Inn* and working hard on his lectures and his first publication on the consistency of the Axiom of Choice.

Gödel's third stay in Princeton and his subsequent 5 months at Notre Dame were described briefly in Chap. 9. As we saw there, he made good use of his 2-½ months at the IAS, preparing and submitting his first set-theory paper (on the relative consistency of the AC with ZF set theory: [*Gödel (1938)*]), and delivering a set of lectures on the development of set theory including his new work—which became the kernel of his summary article [*Gödel (1940)*]. The lecture notes were taken by George W. Brown, a student of mathematical statistics, who was, according to Dawson, probably recruited for that task by Alonzo Church. Brown took it on happily, "*he 'enjoyed the experience thoroughly' and found 'the lectures … to be models of precision and organization'.*"[20] He denied ever having seen Gödel's own notes, which were found in Gödel's papers after his death. Gödel used those notes, probably his own as well as Brown's, to prepare his (1940). Dawson says of the proof of consistency given there:

After defining the constructible hierarchy of sets, three facts must be established: first, that the universe of such sets satisfies all the axioms of ZF set theory; second, that it also satisfies the Axiom of Constructibility; and third, that the latter implies both the Axiom of Choice [as indicated earlier in Gödel's (1938)] and the Generalized Continuum Hypothesis.[21]

More details of Gödel's proof are given by Dawson (1997); or, for a modern version, see e.g., Hindlycke (2017).

In Princeton, Gödel 'missed' the *Pogromnacht* in Germany (now including Austria) on November 9th—it might have raised his consciousness of the gravity of the situation, had he experienced it. Later, in March 1939, while he was at Notre Dame, he also 'missed' the German occupation of Bohemia and Moravia (this is often referred to as the 'occupation' or 'invasion of *Czechoslovakia*', but in fact *Slovakia* had been ceded to Hungary by the *First Vienna Award* (November 2nd, 1938), a diplomatic agreement pursuant to the *Munich Agreement* (September 1938), which had permitted Germany to occupy *only* the *Sudetenland* and Moravian *Schlesien*, with large ethnic German populations. On the following March 14th, the Slovak State declared its independence, and the German army occupied the rest of Czechoslovakia, that is Bohemia and Moravia. Hitler proclaimed the '*Protectorate of Bohemia and Moravia*' from Prague Castle on March 16th, 1939, violating the Munich Agreement. This meant that Brno was now again Brünn, and the Gödel family's remaining possessions there were now within the German *Reich*. Presumably, Kurt and Rudolf's mother Marianne, who still had Czechoslovakian citizenship, then also became a citizen of the *Reich*. (This was a problem for her after the War, when she was again living in Austria, since she—in contrast to her sons—had never been an Austrian citizen [see Budiansky (2021), p. 229]).

In early January, Kurt Gödel traveled to South Bend, Indiana, about 120 km (75 mi.) east of Chicago as the crow flies—and around 700 mi. from New York City, a journey by train of around 20 h. There, at the nearby *University of Notre Dame* (Fig. 11.7), he lived in a dormitory (or in a hotel room; see Chap. 9, Section on 'Princeton, 1938—Notre Dame, 1939', and backnote [31] there).

He was active in teaching there, perhaps more so than at any other time in his life. His host Karl Menger was an enthusiastic teacher, who enjoyed giving introductory courses for undergraduates as well as more specialized graduate courses. He carried Kurt Gödel along in his enthusiasm. In the event, Gödel gave a series of lectures which were evidently quite successful. Menger explains that their original intention of including the lecture notes from Gödel's own course in a to-be-published volume in honor of the Centenary of Notre Dame University in 1942—was however derailed by the advent of World War II—the celebration and the commemorative volume never came to pass. Gödel repeated that lecture series, in his own words, "*with the same* [content], *in somewhat better form*" in 1940 at Princeton.

In addition to the course offered by Gödel himself at Notre Dame on '*The Theory of Sets*', he shared the teaching of a graduate course on '*Introduction to Logic*' with Menger (his notes for both survived in his papers after his death),

Fig. 11.7 The Main Building (the 'Golden Dome') at the University of Notre Dame. Photo by Matthew Rice, 2015; Creative Commons license[22]

and also took part in Menger's new colloquium (in some sense an attempt to continue or revive the *Mathematisches Kolloquium* from Vienna days). Gödel was indeed expected to lead the colloquium; but it is reported that Menger, with his much more outgoing personality, in fact dominated the sessions. From his course notes, it seems that Gödel's course on set theory followed essentially the same pattern as the lectures that he had given in Princeton in the preceding months. Of the other course, '*Introduction to Logic*', Dawson says, "... *the course provided a remarkably thorough introduction to modern logic. Gödel's notes are eminently readable, both in terms of clarity and style, and could be used today with little alteration*".[23]

Gödel also continued his work on the consistency of the Axiom of Choice (AC) and the GCH [published as Gödel (*1938*), (*1939*), and (*1939a*)] while at Notre Dame, and wrote a manuscript for a summary article on consistency proofs in set theory [Gödel (*1940*)], based on George W. Brown's notes from Gödel's 1938 lectures at the IAS. He used his 'free time' for continuing his readings of Leibniz's philosophy, according to Dawson (1997).

After leaving Notre Dame in late May, Gödel stopped off at Princeton to confer with Veblen and von Neumann about publishing his summary paper [the monograph Gödel (*1940*)]. Von Neumann, who had read the manuscript, had written[24] to Gödel in April, 1939,

Ich möchte Ihnen vor Allem meine Bewunderung ausdrücken: Sie haben dieses enorme Problem mit einer wirklich meisterhaften Einfachheit erledigt. Und die unvermeidlichen technischen Komplikationen der Beweisdetails haben Sie … aufs Minimum reduziert. Die Lektüre Ihrer Untersuchung war wirklich ein ästhetischer Genuss erster Klasse.

[English, translation by the present author]: Above all, I would like to express my admiration: You have solved this formidable problem with a truly masterful simplicity. And you have reduced to a minimum … the inevitable technical complexities in the details of the proof. Reading your research was really a first-class aesthetic pleasure.

Back in Princeton, Gödel also conferred with Flexner about getting a long-term entry visa for the USA, since his original (immigration) visa had expired and he had entered the country in October 1938 on a visitor's visa, valid only until the end of June 1939. This turned out to be a difficult problem, and Flexner returned his passport with regrets on June 7th, a week before Gödel embarked on his return voyage to Europe.

Leaving Grinzing—And Vienna

As we have seen in Chap. 9, Kurt Gödel arrived in Germany (probably Bremerhaven) on June 21st of 1939. He would have been back in Vienna a day later (June 22nd 1939 was a Thursday)—back to *Grinzing* and Adele, but also back to seemingly endless bureaucratic problems. It is remarkable that he was able to get through the next six months without suffering another breakdown, since they must have been very stressful. His efforts in dealing with the Nazi bureaucracy were demoralizing at best, and he was indecisive as to what his future course should be during most of that time.

The visa problems with the USA continued for several months—the IAS administration, first Flexner, then his successor Aydelotte, persisted in corresponding with the US State Department and with the German consulate in Washington, trying to find a solution. The most promising path seemed to be to obtain a 'non-quota' visa for Gödel as an eminent teacher who would enrich the American educational system.

This was of course a delicate problem, since he was not technically a 'professor' in Vienna, and indeed since the annulment of his *Lehrbefugnis*, he had no teaching position there at all. And the IAS was not a traditional institution of higher education, although it did have the right to confer doctoral degrees (a right that it practically never exercised, apart from honorary doctorates). Both Flexner and Aydelotte must have engaged in some verbal acrobatics to convince the civil servants in the State Department. The German side was less complicated, and it offered no obstacles to issuing an exit visa for the Gödels. The situation was however particularly difficult after the declaration of war on September 1st.

On August 23rd, 1939, a momentous event had taken place in Moscow, hardly noticed by the rest of the world at the time (because most of it was purposely kept secret): the Hitler-Stalin (or better, *Ribbentrop-Molotov*) pact was signed. It was a non-aggression treaty between the German *Reich* and the Soviet Union, permitting each country to occupy previously independent regions in Eastern Europe, and promising their mutual support in the event of war. Strange bedfellows, and of course just another trick by Hitler, who had no intention of keeping his part of the bargain in the long run (as evidenced by the invasion of the Soviet Union by German troops on June 22nd, 1941, a sneak attack termed '*Operation Barbarossa*' by the Germans). But in the meantime, the *Reich* and the USSR were allies, which turned out to be very important for Kurt and Adele Gödel at the end of 1939. A week after the signing of the pact, on September 1st, 1939, the *Reich* invaded Poland from the west, and the Soviet Union, somewhat later, invaded from the east, Germany claiming a need to protect those of its citizens who lived there and responding to purported 'aggressive acts' by Poland, and thus beginning World War II. The USA initially remained neutral, so that diplomatic relations continued until the declaration of war against the USA by the Hitler regime in December 1941. By then, of course, the Gödels were safely ensconced in Princeton.

Kurt was however also keeping open the option of remaining in Vienna. The first problem to be solved in that direction was being re-instated as a *Dozent* at the University. A second problem developed: he was called up for military service and had to report for a medical examination to determine his fitness to serve. This was delayed over the whole summer, leaving him in suspense. In the event, he was declared fit for garrison service, to his own surprise. This caused him some worry, since on the one hand, he was clearly *not* fit for service, from a psychological point of view in any case. But that was also a delicate subject, as mentioned in Chap. 9, because the 'mentally unfit' were at mortal risk in Nazi Germany. And he was also worried that his new status as a potential inductee into the German military would make it more difficult to actually obtain an exit visa, particularly after the declaration of war in September.

Also in September, he finally applied formally for the *Dozentur* at the University. That was apparently problematic for the authorities in the Ministry of Education, now bound by Nazi rules, as well as for the academic administrators at the University. Once it was clarified that he was not Jewish, which would have been an automatic exclusion criterion, the question arose as to his previous associations with Jewish mathematicians and philosophers (Hahn, Menger, Morgenstern, and Gomperz, for example, were all 'officially' Jewish, and by 1939 were either deceased or had emigrated). His own academic record was considered to be outstanding, but there were many questions. The result was that letters went back and forth, and the decision was taken only in the summer of 1940—a positive decision, entitling him to the position of *Dozent neuer Ordnung*; but by then, he was in Princeton and not about to return, although the University officials in Vienna did not know that latter fact. He was even granted leave, and his leave was renewed several times in the subsequent years.

Furthermore, Kurt had financial problems. His inheritance was mostly used up, or inaccessible. Even the money he had saved from his Princeton stays, in a bank account in New Jersey, was difficult to transfer to Vienna, now part of the *Deutsches Reich*, due to currency restrictions. Somehow, he must have freed up some of his inheritance, or otherwise obtained some money, since in early November (while he was still indecisive as to whether to go or stay), he and Adele purchased an apartment in the Central District of Vienna, at *Hegelgasse* 5 in Wien-1. The building (Fig. 11.8) was constructed in 1877 in the *Neu-Wiener-Renaissance* style and is now a protected historical monument. Whether the apartment in question had previously belonged to a Jewish owner, who had fled after the *Anschluss*, selling the property under pressure for an unfairly low price, is not known. Kurt and Adele, in any case, do not appear to have been involved in any such transaction, but properties sold or confiscated under such conditions often came under the administration of the Nazi city authorities and were later sold at moderate prices to German citizens. The Gödels left their rented *Grinzing* apartment on November 9th, 1939 and moved to the *Hegelgasse*—Kurt Gödel's only residence in the inner city of Vienna, the 1st District. But, as it turned out, their residence there was short-lived. Figure 11.9 shows the memorial plaque recalling his brief stay at that address.

The question remains as to where the money to purchase the apartment originated—even if its price were low. Adele and her family had little money (although they must have had some influence with the regime; her father had been a member of the Nazi Party since the early 1930s, and her mother also joined later). After the *Anschluss* in 1938, the rush to join the Party was so great that no more applications for membership were accepted. Adele herself filled out an application for membership, but it was never acted upon, a result about which she was probably happy during their later life in the USA.

Fig. 11.8 The building at *Hegelgasse* 5, Wien-1, where Kurt and Adele Gödel owned an apartment after November 1939, and where they lived from November 9th, 1939 until their departure on January 10th, 1940 (they actually left Vienna on January 16th of that year). *Photo* AMZ, 2021

Fig. 11.9 *Left*: the memorial plaque for Kurt Gödel at *Hegelgasse* 5. The text reads: 'The great mathematician and logician *Kurt Gödel* (1906–1978) had his last place of residence here, from 09.11.1939 until 10.01.1940, before he left Vienna forever. *Right*: the entrance door to *Hegelgasse* 5. The plaque can be seen at the right of the door. *Photos* AMZ, 2021

It is possible that Kurt borrowed the money to buy the apartment from his brother or his mother, with the argument that it would be a good investment in any case. There seems to be no evidence of that, however, and Kurt and Adele continued as owners until after the end of WWII.

By late November of 1939, shortly after their move to the *Hegelgasse*, two events occurred which evidently ended Kurt Gödel's indecision about their future and convinced him to return with Adele to Princeton and to stay there indefinitely. The first was that their visa problem appeared to be solvable, so that they had some hope of entering the USA on long-term visas.

The second was an unprovoked attack on Kurt on the street in Vienna, on the *Strudelhofstiege*, a public stairway in the *Alsergrund* district, near the buildings where the Vienna Circle had held its meetings and where mathematics and philosophy lectures were given. That event was described in Chap. 9, in the Section on '*A Brief and Hectic Interlude in Vienna*', and Fig. 9.5. As we saw there, he was attacked by a gang of young Nazi supporters, and no doubt shaken up, although not injured. It was a foreboding of things to come should he stay in Vienna. If he had not been accompanied by Adele, who drove off the attackers with her umbrella (no doubt more due to surprise than to any real threat from her blows), the outcome might have been much more serious for Kurt, and he probably realized that.

In any case, in December, Kurt traveled to Berlin, apparently to apply for exit visas from Germany. This was a sign that he had finally made a serious decision to leave Vienna. He had originally planned to make the trip in early December, but he corresponded with the head of the Mathematics Institute in Göttingen (Helmut Hasse, Weyl's successor; Weyl had in turn been Hilbert's successor when the latter retired in 1930) about making a stop there and giving a talk. In the end, he spent December 14th–17th in Göttingen, where he gave a lecture[25] [later found in his archives and reconstructed as (*1939b*) in the *Collected Works*, Vol. III], and went to Berlin afterwards. His lecture in Göttingen was about his GCH results. Dawson[26] examined his notes for the lecture, written in the *Gabelsberger* shorthand. He reports that Gödel took care to make the connection to Hilbert's work, in particular the latter's lecture '*On the Infinite*' [Hilbert (1926)], and to point out the analogies with his own work. There is no record of who attended Gödel's talk; Hilbert was by then nearly 77 and had been retired since 1930, and he was not well. Hermann Weyl had been forced to leave Germany and was working at the IAS, and Paul Bernays had also left. *Gerhard Gentzen*, who met a tragic end shortly after WWII, was an assistant in Göttingen in 1939, but was temporarily away for military exercises. Whether he heard Gödel's lecture is not known; he would have understood it, in any case. Gödel spoke in particular about the properties of the *constructible sets* that he had introduced.

Gödel's subsequent visit to Berlin on December 17th–19th was successful: He and Adele were granted exit visas to travel to Princeton, valid until April 30th, 1940. He attributed that success to a letter written by Aydelotte from the IAS to the German Embassy in Washington, emphasizing Gödel's standing as a mathematician and the importance of his work. The American entry visas were issued on January 8th, 1940, and Kurt and Adele wasted no time in leaving Vienna after that. The only complication was that they would have to take the eastern route, through Siberia to Japan and thence by ship to the Pacific Coast of the USA, and finally across the North American continent to the East Coast. This was stipulated in their exit visas, and due to the war on the North Atlantic (German submarine activity and the danger of German citizens being arrested through British naval operations), it was the wisest course.

The Gödels still had to obtain transit visas for the Baltic countries and the USSR, but the latter was no problem due to the non-aggression pact between the *Reich* and the Soviet Union. The Russian visas arrived by January 12th, but those from Lithuania and Latvia were finally received three days later. In the end, the Gödels left Vienna by train on January 16th, and they were to arrive in Princeton nearly two months later, on March 10th. Chap. 13 will deal with their adventurous but grueling journey around nearly 75% of the earth's circumference (about 19,000 km at 50° north latitude). But before that, we will have a look at an important effect of Gödel's work in the 1930s on an entirely different field, *computability* (now a part of *computer science* or *information technology*).

Notes

1. Quoted from Davis (1965), pp. 82–83.
2. Dawson (1997, 2005 ed., p. 108) tells us that Gödel reportedly cried out, "*Jetzt, Mengenlehre!*" ("Now, set theory!", in the manner of "Eureka!") when he obtained his first consistency proof of the AC relative to the standard axioms of ZF, during the time when he was delivering his lectures in spring/summer of 1935.
3. See e.g. Gray (1994) for a discussion.
4. Bell (2015).
5. Floyd & Kanamori (2006), p. 426.
6. Dawson (1997, ed. 2005), p. 109.

7. As quoted by Wang (1987), p. 98, from a statement made by Gödel in 1976. His other two 'worst years' were 1961 and 1970.

8. Dawson, *op. cit.*, p. 112.

9. See Chap. 9, section on '*The disastrous Year 1936*'.

10. Photo of the *Philosophenstiege*, University of Vienna Main Building, by Franz Pfluegl, Univ. Wien, *Öffentlichkeitsarbeit*. Creative Commons Attribution-Non-Commercial 2.0 Austria License. Cropped, color reduced. Reused from: https://phaidra.univie.ac.at/view/o:12072, courtesy of the University of Vienna.

11. Quoted from Edmonds (2020), p. 171.

12. Dawson, *op. cit.*, p. 121.

13. *ibid.*, Endnote 293.

14. Wang (1987) , p. 99.

15. Photo by GuentherZ, 29.03.2017; Wiki Commons, Creative Commons Attribution 3.0 license. Reused from: https://commons.wikimedia.org/wiki/File:Wien19_Himmelstrasse043_2017-03-9_GuentherZ_GD_G%C3%B6del_0638.jpg.

16. Photo by Martin Furtschegger, 2012, Wiki Commons. Licensed under Creative Commons Unported 3.0 license. Reused from: https://commons.wikimedia.org/wiki/File:Die_Grinzinger_Kirche_-_panoramio.jpg.

17. Dawson, *op. cit.*, p. 126.

18. Wang, *op. cit.*, p. 100.

19. Photo: Wiki Commons, photographer unknown, 1934; public domain, fair use. Reused from: https://commons.wikimedia.org/wiki/File:New_York_(Schiff,_1927)_.jpg.

20. Quoted from Dawson, *op. cit.*, p. 132.

21. *ibid.*, p. 133.

22. Photo by Matthew Rice, 2015; Wiki Commons, Creative Commons License 4.0. The original is at https://de.wikipedia.org/wiki/Datei:Main_Building_at_the_University_of_Notre_Dame.jpg.

23. Dawson, *op. cit.*, p. 136.

24. *ibid.*, endnote [337] on p. 137: von Neumann's letter to K. Gödel, dated April 2nd, 1939; Folder 01/198 in the Gödel archive.

25. See Dawson (1997), pp. 147 ff. for details about Gödel's visa problems and his trip to Göttingen.

26. *ibid.*, p. 148.

12

Computability: *Post, Gödel, Church, Turing (and Many Others)*

Kurt Gödel's work had far-reaching effects, beyond the rather special-ized fields of mathematical logic and fundamentals of mathematics. It had long been a dream of scientists, mathematicians and philosophers to build 'computing machines' which, in the simplest case, would automate mathematical calculations, initially the simple operations of addition and multiplication. After the development of precision clockworks in the later seventeenth century, that dream seemed to be within reach. Gödel's sometime hero *Leibniz* actually built such a calculating machine (around 1670–1700) using gears and ratchets, as well as an encoding machine. His calculating (multiplying) machine extended ideas suggested by Blaise Pascal, and was called the '*Stepped Reckoner*' (see Fig. 12.1). Leibniz himself is quoted with the rather elitist statement, "… *it is beneath the dignity of excellent men to waste their time in calculation when any peasant could do the work just as accurately with the aid of a machine*".[1]

Nearly 150 years later, the English mathematician *Charles Babbage* built more sophisticated machines, his 'Difference Engines' and the 'Analytical Engine'. The latter was apparently the first programmable computer, and Babbage collaborated with *Ada Lovelace* (the daughter of Lord Byron), who wrote algorithms to be programmed into the machine. She is thus known as the first computer programmer. Around the same time, *George Boole* was working out his Boolean Algebra, based on the binary system, which had already been suggested by Leibniz as being suitable for mechanical computers. This can be seen as the beginning of computer theory.

© The Author(s), under exclusive license to Springer Nature
Switzerland AG 2022, corrected publication 2023
W. D. Brewer, *Kurt Gödel*, Springer Biographies,
https://doi.org/10.1007/978-3-031-11309-3_12

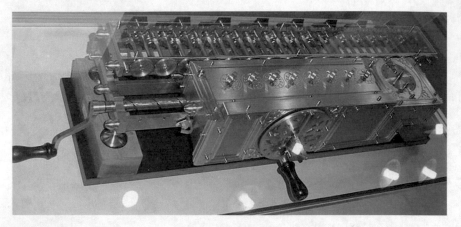

Fig. 12.1 A replica of Leibniz's 'Stepped Reckoner', a machine for multiplication by mechanical means. This replica is in the *Technische Sammlungen* in Dresden. Photo from WikiMedia, CC Documentation license[2]

But that topic really came into its own only a century later, in the 1930's, when a group of younger logicians and mathematicians began considering seriously what the foundations of a theory of computability might be like. Their immediate predecessors, logicists and formalists such as Frege, Russell and Hilbert, had long since entertained the possibility of representing logical sentences and proofs in a purely formal way, so that they could be manipulated by 'mechanical' means. But they had no possibilities of testing their ideas with real machines; while the next generation—which included Emil Post, Kurt Gödel, Alan Turing, Alonzo Church and Jacques Herbrand, and, somewhat later, the younger logicians Stephen Kleene and J. Barkley Rosser, both students of Church—for the most part lived to see that possibility become a reality. We met up with all of them in earlier chapters in connection with formal logic. In this chapter, we want to consider their contributions to the topic of 'computability', or, more generally, to theoretical computer science. Readers who wish to delve deeper into the subject of computability and machine intelligence should consult the book '*Computability*' (2013),[3] edited by B.J. Copeland, C.J. Posy, and O. Shagrir (CPS). As an introduction, we take a brief look at the biographies of Turing and Church, key figures in the development of computability theory.

Alan Mathison Turing (1912–1954) was born in London. His father was a civil servant in the British colonial service in India, and his parents gave Alan and his older brother John to foster parents in Hastings to be cared for while they were in India. After 1916, their mother returned to England and took the boys back into her home. Alan showed considerable talent as a schoolboy, in particular for mathematics, and he was sent at 14 to a private school; however, it was more inclined to a humanistic education than a scientific one, so that Turing's marks were not the highest. Nevertheless, he was able to study at King's College, Cambridge under the mathematician Geoffrey Harold Hardy.

In 1936, at age 23, Turing (cf. Fig. 12.2) published a seminal work, '*On Computable Numbers, with Application to the "Entscheidungsproblem"*' [Turing (1936)]. The latter term refers to Hilbert's 'Decision Problem', formulated in Hilbert and Ackermann (1928), which proposes the existence of an algorithm that can decide whether a given formal-logical statement is valid in every system satisfying a certain set of axioms.[5] The algorithm would thus 'mechanically' make a *decision* as to the validity of the input statement. Gödel's (1930) demonstration of the completeness of first-order predicate logic suggests that every statement derived from a given formal-logical system is provable or not provable within that system, and Hilbert's problem is to find a formal, i.e., automatic or mechanical method of showing this for any particular statement, given the axioms and the rules of inference of the system. This is based on the intuitive notion of 'effectively calculable' (the term is due to Church), and the question is whether this notion is contained in those functions that are computable by an idealized machine (e.g., a 'Turing machine'). At the same time as Turing's work, Alonzo Church independently published an article on the same topic [Church (1936)], based on determining whether 'effectively calculable' is captured by the functions expressible in Church's 'λ-calculus'. Both came to the conclusion that a general solution to the *Entscheidungsproblem* is not possible. Their assumptions about 'effective calculability' (in more modern terms, the possibility of designing an algorithm to perform the intended calculation) are now called the *Church-Turing thesis*. A similar conclusion may be reached from Gödel's incompleteness theorems, combined with a definition of *general recursive functions* due originally to Jacques Herbrand and introduced by Gödel in his 1934 lectures at the Princeton IAS. Subsequently, all three approaches were shown to be equivalent by Church's student Stephen Kleene, with contributions by Church and Turing themselves.

Turing's (1936) thus offers an alternative formulation of Gödel's (1931), replacing Gödel's arithmetized formal language by a hypothetical machine (the '(*Universal*) *Turing Machine*') which sequentially reads symbols from an (endless) paper tape, and changes its state accordingly, printing out the result on the tape before going on to the next symbol. The machine halts when it has arrived at the correct value of the function that it is computing. A Turing 'proof machine' could also be envisioned, which evaluates strings of symbols intended to prove a particular sentence in an axiomatic system. The

Fig. 12.2 Alan Turing at age 16, a passport photo taken in 1928. Photo courtesy of Archive Centre, King's College, Cambridge[4]

Turing Machine became the starting point for computability theory, the basis of modern theoretical computer science.

In September 1936, after submitting his paper, Turing went to Princeton on a fellowship and continued his graduate work under Alonzo Church at Princeton University. He was granted the Ph.D. there in 1938 for his thesis,[6] '*Systems of Logic based on Ordinals*'. He was then offered a postdoctoral position at the IAS by John von Neumann, but decided instead to return to England, where he was a Fellow at King's College, Cambridge. In 1939, after his return to Cambridge, he attended Ludwig Wittgenstein's lectures on the fundamentals of mathematics and had discussions (and disputes) with him about the nature of formalism and of mathematical truths. Turing and Gödel did not meet at Princeton, mainly due to Gödel's illness in 1935/36, since he otherwise would have remained at the IAS and/or returned there later in 1936 and would certainly have encountered Turing. In the event, Turing had left by the time Gödel arrived back in October of 1938.

After the outbreak of WW II in September 1939, Turing volunteered to help break the German codes used for transmitting secret messages to submarines and naval vessels. He was one of the leaders of the codebreaking group at *Bletchley Park*, and he was instrumental in breaking the code of the German '*Enigma*' coding machines, based on information obtained from the Polish codebreaking group in July 1939. The story of his accomplishments at Bletchley, which included designing and building an electromechanical computer called '*bombe*' (after an earlier Polish machine), is worthy of any wartime thriller. A large number of those machines were subsequently built, and they were able to decode much of the German communication with their submarine fleet, leading to a vast reduction of damage to shipping and the capture or sinking of many submarines. Turing was appointed an Officer of the British Empire (OBE) in 1946 for his contribution (although the precise reason for the award was kept secret for many years). This story has been told in several documentaries, including a 1996 BBC film called '*Breaking the Code*', and in books as well as a feature-length film, '*The Imitation Game*' (2014). Turing also worked on breaking the *Lorenz cipher*, another German code used for teleprinters; and based on his ideas, a programmable computer, called '*Colossus*', was constructed (by several of his colleagues).

In November 1942, Turing went to the USA, where he consulted with Navy cryptographers on breaking the Naval *Enigma* code, and visited the Computing Machine Laboratory; and he also consulted with Bell Labs on developing a secure speech encoding mechanism. He returned to Bletchley in March of 1943, but moved to the Radio Secret Service (at Hanslope Park, Milton Keynes) in the last years of the war, where he continued to work on voice encoding. After the end of the war, he lived in Hampton and helped to design the *Automatic Computing Engine* (ACE) at the National Physical Laboratory (the British national standards laboratory); and during this period, he published one of the first articles describing the design of a stored-program computer. Military secrecy delayed that project, however, and Turing returned to Cambridge in 1947, where he wrote a treatise on '*Intelligent Machinery*'. That same year, he also met with German computer pioneer *Konrad Zuse* in Göttingen. In 1948, he moved to Manchester where he occupied a university position in mathematics, and also became deputy director of the Computing Machine Laboratory. In 1950, he published another important article, entitled '*Computing Machinery and Intelligence*', in which he discussed the mind-machine problem. It was in this paper that he introduced the famous 'Turing Test' to determine whether a machine could be termed 'intelligent'. He also wrote one of the first chess-playing programs around 1950. In 1951, he began investigating mathematical biology and developed a description of pattern-forming biochemical reactions, having considerable success in his new field. But even as late as 1954, he still published an article on mathematical logic (with no mention of machines).[7]

The end of Turing's life was tragic. He himself recognized early on that he was homosexual. He had a deep and serious friendship with *Christopher Morcom*, whom he met at school. Morcom was one year older than Turing, and they formed a strong relationship that undoubtedly had a sexual aspect. Morcom died early of bovine tuberculosis, at that time not uncommon (it has since been essentially eliminated by pasteurization of milk products). He was only 18 years old (and Turing was only 17) at the time of Morcom's death. Turing apparently idealized this relationship and was perhaps dubious of ever finding anything similar in his later life. This is reminiscent of Ludwig Wittgenstein's early relationship with *David Pinsent*; the two of them were older than Turing and Morcom (Wittgenstein was 23/24 at the time of their relationship, Pinsent two years younger). It was likewise ended by the tragic and early death of the partner; Pinsent became a pilot in WWI and was killed in 1918 without having seen Wittgenstein again. Wittgenstein later referred to him as 'My only friend'.

During WWII, while he was working at Bletchley, Turing became engaged to a female coworker, Joan Clarke. He confessed his homosexuality to her, and she was still willing to marry him, but he himself backed out of the engagement, evidently plagued by self-doubts. After the war, in 1952 at age 39, he had a brief affair with a younger man, and later admitted this to the English police (rather naively). Homosexuality was officially a crime at that time, and both were charged with 'indecency'. Turing agreed to a hormone treatment intended to reduce his sexual drive, and completed it with apparently no ill effects. But in June 1954, he was found dead by his housekeeper; the cause of death was determined to be cyanide poisoning. Shortly before, he had visited a fortuneteller, who apparently gave him bad news, according to friends who accompanied him but did not hear what his 'fortune' entailed. His death is usually considered to be a suicide, but there are divergent opinions suggesting that it was due to an accident with an unprotected gold-plating apparatus found in his workroom.

Alonzo Church (1903–1995) was born in Washington D.C., where his father was a judge. He was sent to a private school in Ridgefield, CT with the help of his uncle, after whom he was named. He completed his high school diploma with high marks and then studied mathematics at Princeton University from 1920–24. Already as an undergraduate (cf. Fig. 12.3), he published a paper on Lorentz transformations, and graduated with honors. He then began graduate work at Princeton in the group of Oswald Veblen, completing his Ph.D. in three years with a thesis entitled '*Alternatives to Zermelo's Assumption*'. After obtaining his doctorate, he spent a brief stay at the University of Chicago, then two postdoctoral years as a National Research Fellow, the first at Harvard, the second at Göttingen and Amsterdam (visiting the two great opponents, Hilbert and Brouwer, just at the time of their serious controversy). In 1929, he became assistant professor of mathematics at Princeton, advancing to associate professor in 1939 and full professor in 1947. Among his early Ph.D. students were *Stephen C. Kleene* and *J. Barkley Rosser* (cf. Fig. 10.6). After Veblen moved to the IAS in 1932, Church was the principal logician at Princeton University.

Fig. 12.3 Alonzo Church, as an undergraduate student at Princeton, around 1923. Photo from the private archive of Alonzo Church[8]

In the first years of his tenure at Princeton, Church worked on formulating a new formal-logical system, a *language* in the sense of Chap. 7. When Gödel's (1931) burst upon the scene (at Princeton, it was introduced by von Neumann's talk in the fall of that year), Church tried to find a logical system for which the incompleteness theorem would not hold. His students Kleene and Rosser were able to show that his system was not consistent, but a part of it, termed the λ-*calculus*, remained valid, and they spent the next few years developing and applying it. Church investigated what he called 'effective calculability', the ability to evaluate a function 'automatically' by means of an algorithm; it is now often termed 'intuitive computability'. This led to *Church's thesis* in 1936, now called the *Church-Turing thesis*, later found to be equivalent to Gödel's analysis of computability in terms of general recursive functions [Gödel-Herbrand (1934); see the notes from Gödel's spring 1934 lectures, in Davis (ed., 1965)]. In his (1936), Church provided the first example of a function that is *not* intuitively computable. (This paper appeared just as Turing was preparing his own article '*On Computable Functions...*'. Turing, who had worked quite independently of Church, hurried his article to publication; it was submitted in May, read—presumably by the editorial staff—in November, and published in two parts in December 1936). After the equivalence of the two approaches with that of Gödel was shown, Gödel himself, and also Church, stated that they preferred Turing's formulation, which they found to be clearer and more convincing.

Church himself apparently had little direct contact with Gödel in Princeton, perhaps due to Gödel's reclusive nature, and to the fact that they had very different personalities. Church's students Kleene and Rosser had a much closer relation to Gödel—at least to his manner of thinking and his logical arguments, if not personally [cf. Kleene & Rosser (1984)]. This was mainly due to their note-taking at Gödel's lecture series in Princeton in spring 1934, and their interactions with him in correcting and editing the notes (finally published by Davis in 1965). Gödel famously criticized Church's hypothesis, calling it '*thoroughly unsatisfactory*'.[9] Church mentioned this in a letter to Kleene on November 29th, 1935 [quoted by Davis (1982) and again in Copeland et al., (eds., 2013) (CPS), p. 9]:

> In regard to Gödel and the notions of recursiveness and effective calcula-bility, the history is the following. In discussion with him *[on]* the notion of λ-definability, it developed that there was no good definition of effective calculability. My proposal that λ-definability be taken as a definition of it he regarded as thoroughly unsatisfactory.

Gödel however soon changed his opinion, writing in an unpublished paper [10] around 1938 that,

> When I first published my paper about undecidable propositions *[1931]* the result *[the Church-Turing thesis]* could not be pronounced in this generality, because for the notions of mechanical procedure and of formal system no mathematically satisfactory definition had been given at that time. This gap has since been filled by Herbrand, Church and Turing.

'Herbrand' is modestly cited here; in fact, it was Gödel's (posthumous: Herbrand died in an accident in 1931) development of an idea of Herbrand's from their correspondence in 1931, and further evolved by Gödel in his 1934 lectures, that is referred to. Indeed, it was Turing's formulation that convinced Gödel, and he later states, "*... that this really is the correct defini-tion of mechanical computability was established beyond any doubt by Turing*"[3]. Church himself, about the same time, in a review of Turing's (1936), wrote,[11] "*computability by a Turing machine ... has the advantage of making the iden-tification with effectiveness in the ordinary (not explicitly defined) sense evident immediately*".

In his later career, Church devoted himself to the philosophy of mathemat-ical logic, introducing what he called the *logistic method*, in which he dealt with higher-order logics and model theory, developing a new version of type theory, expressed within model theory. He remained at Princeton Univer-sity until 1967, retiring there at the age of 64, and then taught at UCLA in Los Angeles, CA for the next 23 years. He died at age 92, in 1995, and is buried at the Princeton Cemetery. Church and Turing together are considered to be the founders of modern theoretical computer science (although Gödel, Herbrand, Rosser, Kleene and Post are also often mentioned in this connection).

As we saw in Chap. 7, *Emil Post* (see Fig. 7.3) anticipated the work of Gödel (1931) and Turing (1936) to a considerable extent, although his published and unpublished work in the 1920s aroused little attention [cf. Stillwell (2004)]. Post started from the concept of *computability*, and made use of Cantor's (1891) *diagonal lemma*. (A similar lemma was also used by Gödel in his (1931) proofs, but in a more limited way and without specifically naming it). Post's program, established while he was a postdoc at Princeton in 1921/22, included four steps[12]:

1. Describe all possible formal systems.
2. Diagonalize them.
3. Show that some of them have unsolvable deducibility problems.
   – And he also saw one step further the incompleteness theorem, because:
4. No Formal System Obtains All the Answers to an Unsolvable Problem.

Stillwell (2004) continues by pointing out the complications in Post's program:

> It is true that there are several matters arising from this argument. What is the significance of consistency? Are there unprovable sentences in mainstream mathematics? But for Post, incompleteness was a simple consequence of the existence of unsolvable problems. He also saw unsolvable problems as a simple consequence of the diagonal argument...
>
> Thus Post's proof of incompleteness was delayed because he was trying to do so much: The task he set himself in 1921 was in effect to do most of what Gödel, Church, and Turing did among them in 1931-36. In 1936, Church published a definition of computability and gave the first published example of an unsolvable problem. But 'Church's thesis', that here was a precise definition of computability, was not accepted until the equivalent Turing machine concept appeared later in 1936, along with Turing's very lucid arguments for it .

As a result, Post's proof of incompleteness, which would have anticipated Gödel's proof by almost 10 years, was never published in any convincing form. As we saw in Chap. 7, he accepted his own failure to publish his results in a timely manner *vis-à-vis* Gödel, albeit not without regrets. His anticipation of Turing's results was, in contrast, made public at the last possible moment:

In 1936, Post published an important paper entitled [13] '*Finite combinatory Processes—Formulation 1*'. In it, he also addressed the problem of computability and arrived at essentially the same results as Turing. His paper was accompanied by a note from Church, verifying that Post had obtained his results independently of Turing (who was at the time a graduate student in Church's research group). Turing's article has however remained more well-known, perhaps because it was more complete and more clearly written.

The German computer scientist and artificial intelligence (AI) expert *Jürgen Schmidhuber*, in an article published in the *Frankfurter Allgemeine Zeitung* (FAZ) in June 2021, and on his own blog @SchmidhuberAI,[14] in memory of the publication of Gödel's 'most famous paper' 90 years before, compares the achievements of Turing and Post to those of Gödel and Church, who preceded them in publishing their versions of the answer to the *Entscheidungsproblem*. He poses the question as to whether Post (1936) and Turing (1936) added anything really new to what Gödel (1933/34) and Church (1935/36) had already published.

And his answer is 'yes': both Gödel and Church paid no attention to the question of computational efficiency. As we saw in Chap. 8, Gödel's methods involved the manipulation of very large numbers, and he didn't concern himself with the computational effort that would be required to carry them out in practice. This was true of Church, also. But Post and Turing used very simple binary machine models in their considerations, along the lines of the suggestion for a binary computing machine made much earlier by Leibniz (in 1679!). In the mid-1930s, nearly no-one was actually building computing machines (other than simple office machines which were not programmable); an exception was Konrad Zuse, who applied for a patent on such a machine in 1936 (and built several working models in the following years), based on similar principles to those of Turing and Post in their theoretical models.

Gödel's and Church's concepts of computability were indeed equivalent to those of Turing and Post in a formal-logical sense, and they both found Turing's explanation in terms of a simple machine most convincing. But years later, when computer theory was applied to modeling and improving real machines, the more efficient methods of Turing and Post provided the basis for computational complexity theory.

Gödel's Contribution

Just what, exactly, was Gödel's contribution to 'effective calculation' in the years 1931–1934? After the publication of his incompleteness paper in March

1931, Gödel carried on a brief correspondence with *Jacques Herbrand*, who was at the time visiting Hilbert's group in Göttingen. Herbrand mentioned his ideas on *general recursive functions* and on the 'characterization of classes of finitistically calculable functions'. Gödel was not able to continue the discussion with Herbrand, because, as we have seen, the latter was killed in a mountaineering accident soon after his return to France. See e.g., Sieg (2005, 2006) for a more complete description of their correspondence and its repercussions. Gödel developed Herbrand's ideas further, and he spoke of them in his lecture to the Mathematical Association of America in Cambridge, MA in late 1933 (See [*Gödel (*1933o)*]; the 'Cambridge lecture'). And by the time of his lecture series at the IAS in the spring of 1934 (the 'Princeton lectures'), he had conceived his own concept of 'effective calculability' and included it in the lectures, attributing the ideas on general recursive functions to Herbrand. We can quote Wilfried Sieg[15] on Gödel's interpretation of Herbrand's letter of April 7th, 1931 and their ensuing correspondence:

> Nowhere in the correspondence does the issue of general computability arise. Herbrand's discussion, in particular, is solely trying to explore the limits of consistency proofs that are imposed by the second theorem. Godel's response focuses also on that very topic. It seems that he subsequently developed a more critical perspective on the very character and generality of his theorems. This perspective allowed him to see a crucial open question and to consider Herbrand's notion of a finitist function as a first step towards an answer.

In his Princeton lectures, Gödel points out that 'the rules of inference and the notions of formula and axiom have to be given constructively', and he is at pains to stress the generality of his (1931) proofs. He then concentrates on 'the *computability* of general number-theoretic functions', mentioning that 'primitive recursive functions (PRFs) are computable by finite procedures', and in a footnote[16] to that last remark, he says,

> The converse seems to be true if, besides recursions according to the scheme (2) *[i.e., primitive recursion as given above]*, recursions of other forms (e.g., with respect to two variables simultaneously) are admitted. This cannot be proved, since the notion of finite computation is not defined, but it can serve as a heuristic principle.

Gödel, in the last section of his 'Princeton lectures', gives a more precise definition of 'general recursive functions' (GRFs), containing 'mechanical' rules for deriving them:

If φ denotes an unknown function, and ψ_1, ..., ψ_k are known functions, and if the ψ' s and φ are substituted in one another in the most general fashions and certain pairs of resulting expressions are equated, then, if the resulting set of functional equations has one and only one solution for φ, φ is a recursive function.

Gödel later wrote in a letter to van Heijenoort (August 14th, 1964) that, "*it was exactly by specifying the rules of computation that a mathematically workable and fruitful concept was obtained*". He attributed the above definition of GRFs to Herbrand and insisted that it was taken from Herbrand's letter 'exactly as written there, with no reference to computability'. After Gödel's death, when his correspondence with Herbrand came to light, this statement proved to be incorrect: it was Gödel himself who had further generalized the concept of GRFs from the intuitionist sense intended by Herbrand. As emphasized by Sieg (2005, p. 183):

At this earlier historical juncture the introduction of the equational calculus with particular computation rules was important for the mathematical development of recursion theory as well as for the underlying conceptual motivation. It brought out clearly what Herbrand (according to Gödel in his letter to van Heijenoort) had failed to see, namely 'that the computation (for all computable functions) proceeds by exactly the same rules'.

Gödel repeatedly praised Turing's approach to 'effective calculability' and called it the clearest and most understandable definition of that notion, and 'correct beyond any doubt'. However, much later (indeed, over 16 years after Turing's death), Gödel wrote a note which he sent to Wang[17] in 1972. In it, he states:

Turing ... *[in his (1936)]* gives an argument which is supposed to show that mental procedures cannot carry any farther than mechanical procedures. However, this argument is inconclusive, because it depends on the supposition that a finite mind is capable of only a finite number of distinguishable states ... although at each stage of the mind's development the number of its possible states is finite, there is no reason why this number should not converge to infinity in the course of its development ... But it must be admitted that the precise definition of a procedure of this kind would require a substantial deepening of our understanding of the basic operations of the mind.

This would seem to rest upon a misinterpretation of Turing's intentions on the part of Gödel. Indeed, Turing considers all the individual steps taken by a human 'computer' in carrying out an effective calculation, e.g., in evaluating a particular computable function. On the basis of that analysis, he concludes that each step could have been taken by an automatic process, i.e., a machine following an algorithm, and thus that the function could have in fact been evaluated 'mechanically'; this is the origin of his 'Turing machine'. He makes no statements about the mind-machine problem as such, neither that machines might be constructed which would equal the performance of a human mind, nor that the mind is limited to processes that could also be carried out by machines, as Gödel suggests. Copeland and Shagrir, in their chapter '*Turing versus Gödel on Computability and the Mind*' in CPS, give a rather complicated explanation of Gödel's accusation that Turing had committed a 'philosophical error'. It involves two distinct approaches to understanding computability, which they call the '*cognitivist*' and the '*non-cognitivist*' approaches. Whether their explanation is correct or will eventually be accepted as such remains to be seen. Gödel first wrote this note in 1970, as he told Wang, and 34 years had passed since the publication of Turing's article. He may have simply read an intention into Turing's procedures which was not meant in that sense by their author.

In the conclusion of his (2005), Wilfried Sieg ascertains,

> We see, finally, three specific and important points drawn from that [*historical*] evolution, listed in order of their increasing significance: (i) the Godel-Herbrand notion of general recursive function is really Godel's; (ii) in the early 1930s finitist mathematics was viewed as going significantly beyond primitive recursive arithmetic; (iii) at that time, finitist mathematics was viewed as coextensive with intuitionist mathematics. Each point is counter to broadly held contemporary views and, indeed, undermines deeply held convictions concerning our logical past .

In short, Gödel's early contribution was by no means negligible, and can indeed be considered, at least historically, as a decisive step towards a theory of computability.

Post's Later Work

Emil Post had originally intended to publish a series of computation models of increasing sophistication—that is the reason for the title of his (1936): '*Finite combinatory Processes—Formulation 1*'. The number 1 signifies that this was to be the first in a series of related papers. But in the event, he did not pursue writing that series immediately. He continued to have relations with Church's group in Princeton, not far from his home institution, CCNY in New York City. But they were not particularly warm or close. Turing, as we have seen, left Princeton and returned to Cambridge in mid-1938, shortly after receiving his Ph.D. After the beginning of WWII in September, 1939, he was completely occupied by his work in military cryptography, and even after the war, he was involved in 'secret' projects for some time, so that his contributions to computability theory ceased. Gödel and Church were also no longer working actively on that topic; thus the field was left open for Post.

The latter's most important paper was doubtless his (1944), the written version of an invited talk that he gave to the *Mathematical Society* in New York in February 1944, and published as [Post (1944)]. There, he developed the concept of 'computably enumerable' sets and showed how one set could be *reducible* to another. He used Turing's (1939) idea of 'oracle machines', i.e., Turing machines which could consult an 'oracle', an external data source (today, we would speak of a data bank, perhaps online) to define his concept of reducibility of a set A to a set B, denoted by B $\leq_T$ A (the subscript 'T' refers to 'Turing'). During the last decade of his life, Post continued to develop his ideas of Turing reducibility and data (information) content. He gave his notes on that work to S.C. Kleene, who published them as Kleene & Post (1954) (q.v.). These contributions were important for the later development of the field of computability theory.

Mind and Machine—*Intelligent Machines and Artificial Intelligence (AI)*

Alan Turing, in his article on '*Computing Machinery and Intelligence*' (1950), posed the question of whether an 'intelligent machine' could be constructed, and how to identify such a machine in the case that it could. The question of whether the human mind surpasses any machine which might be assembled had already been raised in connection with Gödel's incompleteness theorems, which place intrinsic limits on the decidability of all the derivable sentences in consistent formal-logical systems (suggesting to some that constructible machines would always be more limited than human thought). This discussion continues today, and the issue can hardly be considered to be definitively settled. Gödel himself tended toward the belief that no machine could ever be constructed that would be comparable to a human mind. However, as we have seen, he was very insistent on having mathematically precise definitions before reaching a judgment on any topic, and in the mind-machine debate, there are still today no such precise definitions of many important concepts (we might mention *'intuition'*, *'understanding'*, *'consciousness'*, *'self-awareness'*, to name a few). The question of *consciousness* is doubtless the most difficult; even if a machine could be built whose computational ability (and *'intuition'!*) are equal to those of human minds, would that machine also be *conscious* (self-aware)?

These questions have led to whole new areas of research; in philosophy, the 'mind-machine' problem (closely related to the much older 'mind–body' problem), and in applied mathematics/computer science, to the field of artificial intelligence (AI), which today has become practically an industry. But quite apart from potential applications and practical implementations, the debate still continues as to the theoretical possibility of machine intelligence, and the relevance of Gödel's results (in particular his first incompleteness theorem) to that possibility.

A certain group of logicians and philosophers has adopted the position that Gödel's theorem rules out true machine intelligence, in the sense that machines might someday 'think' and 'understand' in a way comparable to humans, and specifically, that they could develop and carry out mathematical proofs independently of human intervention. The related question—of whether the human mind can be considered to be a sort of computing machine, based on nerve cells and networks, nerve impulses and neurotransmitters instead of microstructures in semiconducting substrates and clocked changes in electrical voltage levels—is also still under discussion, especially as approached from the other side, i.e., from the viewpoint of neurophysiologists, cognitive theorists and psychologists.

As mentioned, Gödel himself indicated several times that he doubted the possibility of constructing intelligent machines which could reproduce the potentialities of the human mind. But he initially refrained from making concrete statements, no doubt due to the abovementioned lack of precision in the definitions. In his chapter in CPS, entitled '*Gödel's Philosophical Challenge (to Turing)*', Wilfried Sieg traces the evolution of Gödel's position on this question, from his [(1934), 'Princeton lectures'] through his [(1939b), lectures on logic at Notre Dame] and continuing on to his unpublished article of 1951 on the foundations of mathematics (his Gibbs Lecture, [*Gödel (*1951)*]). In the latter, he states very definitely that "*either ... the human mind (even within the realm of pure mathematics) infinitely surpasses the powers of any finite machine, or else there exist absolutely unsolvable Diophantine problems*". [The latter phrase refers to a class of problems in number theory involving the solution of polynomials with integer coefficients—Hilbert's 10th problem deals with their computability.]

This 'anti-computational' position was also taken—at least in part –by Nagel and Newman in their (1956)[18] [and (1958)]:

> Gödel's conclusions have a bearing on the question [of] whether a calculating machine can be constructed that would equal the human brain in mathematical reasoning. Present calculating machines have a fixed set of directives built into them, and they operate in a step-by-step manner. But in the light of Gödel's incompleteness theorem, there is an endless set of problems in elementary number theory for which such machines are inherently incapable of supplying answers, however complex their built-in mechanisms may be and however rapid their operations. The human brain may, to be sure, have built-in limitations of its own, and there may be mathematical problems which it is incapable of solving. But even so, the human brain appears to embody a structure of rules of operation which is far more powerful than the structure of currently conceived artificial machines … Nor does the fact that it is impossible to construct a calculating machine equivalent to the human brain necessarily mean that we cannot hope to explain living matter and human reason in physical and chemical terms. The possibility of such explanations is neither precluded nor affirmed by Gödel's incompleteness theorem. The theorem does indicate that the structure and power of the human mind are far more complex and subtle than any non-living machine yet envisaged.

On the other hand, researchers in theoretical computer science and AI tend, understandably, to doubt that Gödel's results conclusively rule out the eventual construction of a 'thinking machine', and many cognitive scientists and neurophysiologists are still pursuing a physical–chemical explanation of the underlying structure and functioning of 'mind'. (This view is sometimes referred to as 'computationalism', or the 'anthropic mechanism thesis'; it was previously called 'mechanism' and is an aspect of 'physicalism'). One wonders what the teenaged Douglas Hofstadter thought of the above- quoted conclusion of Nagel & Newman when he read and was inspired by their (1958), where similar statements are made—and how his many years as a cognitive scientist may have changed his opinion (for the latter question, one can turn to his later book, 'I am a Strange Loop' [Hofstadter (2007)], where he takes up some of the questions posed in his *GEB* (1979) after 25 years of intervening professional work in cognitive science). Turing himself clearly inclined to the 'computationalism' view.

Very vocal support of the idea that Gödel's results preclude the construction of a 'thinking machine' comparable to the human mind (and the concomitant idea of explaining the mind in terms of physical and electrochemical processes)—which we may call the 'anti-computationalism' (or 'anti-AI') view—was expressed by the Oxford philosopher *John R. Lucas*. In

1961, Lucas published an article[19] based on a lecture given to the *Oxford Philosophical Society* in October 1959, entitled '*Minds, Machines, and Gödel*'. It became his most famous work, and was cited in numerous obituary notices after his death in April of 2020. Early in that paper [19], he makes the amusing remark, speaking of Gödel's proof of incompleteness, "*The foregoing argument is very fiddling, and difficult to grasp fully… the argument remains persistently unconvincing: we feel that there must be a catch in it somewhere…* ".

Lucas was religious, and the concept of free will for humans was thus important to him. His 'anti AI' argument served to show that humans would have free will if the mind is more powerful than any Turing machine. He used Gödel's first incompleteness theorem to demonstrate the limitations of proofs within any (deterministic) logical system and therefore the impossibility for a Turing machine to 'go outside' the logical system for which it was programmed. But humans, with their greater thinking ability, or intuition, should be able to do so, and for example to determine the truth of a proposition which was—according to Gödel—*undecidable* within the given logical system. They can thus escape from the determinism of the system, and ergo have free will.

In fact, it has been shown[20] that 'recurrent neural networks' indeed have more computing power than Turing machines, and this may apply to human minds. (But recurrent neural networks can be simulated on digital computers, i.e., Turing machines—which then would presumably have a similar 'computing power' to a mind, although this would not *make* them 'minds'; a simulation is not the same as 'being', although it may deliver similar formal results). Hofstadter allegedly saw Lucas' arguments as a challenge, and it has been suggested that his *GEB* was intended as a reply to and a refutation of Lucas' thesis.

Another distinguished scientist who has taken the 'anti AI' stance is *Roger Penrose*, an eminent theoretical physicist who carried out early work on gravitation and GR and received the 2020 Nobel prize in physics for his result, obtained in 1965, that the gravitational collapse of stellar remnants of a certain mass leads inevitably to the formation of a gravitational singularity, a 'black hole', irrespective of their spatiotemporal symmetry.

Penrose later worked extensively in quantum field theory and quantum gravitation, and became interested in the 'mind-machine' problem and the origin of consciousness, collaborating in the 1990s with a medical doctor (*Stuart Hameroff*) on the 'microtubule' theory, which explains consciousness as a result of quantum effects in microtubules within the brain. This theory has however encountered considerable resistance within the field of human cognition.

Penrose has written several books on the subject: '*The Emperor's New Mind*' (*ENM*) (1989), in which he explores the possible origin of consciousness and proposes to search for 'new physics' which might explain it. He points out that a process might be deterministic but not algorithmic (a result of the negative answer to Hilbert's *Entscheidungsproblem*), and he posits the importance of such processes for the functioning of the human mind. In *ENM*, Penrose took up J.R. Lucas' 'anti-AI' thesis and worked it out in great detail, leading to what is now called '*the Penrose-Lucas argument*'. The latter was strongly criticized by logicians and computer scientists after the publication of *ENM*, inspiring Penrose to publish two more books in reply, '*Shadows of the Mind*' (*SM*) (1994), in which he reiterates and expands on his arguments involving the 1st incompleteness theorem, and '*The Large, the Small, and the Human Mind*' (1997), where he develops the Penrose-Hameroff microtubule theory in detail.

Logicians Martin Davis (1993) and Solomon Feferman (1995) have dealt with Penrose's arguments from the mathematical-logical point of view in their reviews of *ENM* and *SM*. They each find a number of errors of detail in Penrose's exposition of the incompleteness theorems, but both admit that these are not decisive for his argument. Davis, reviewing *ENM*,[21] writes:

> ...the technical side of what Penrose asserts is certainly correct ... Iteration of the 'Gödelization procedure' was first studied systematically in Alan Turing's doctoral dissertation (Turing 1939) and has been the subject of interesting research by Sol Feferman (1988) and others. But it adds not one iota to Penrose's erroneous claim that Gödel's theorem can be used to show that mathematical insight cannot be algorithmic ... It is certainly worth discussing its *[Gödel's theorem's]* philosophical implications, but without being carefully grounded in what the theorem actually states, it is all too easy to go astray. ... Penrose goes on to say, "the original algorithm is thus limited in what it can achieve: It is unable to incorporate the particular insight 'F sound $\Rightarrow$ G(F) true'.". Calling this fact an insight helps lead Penrose down the 'slippery' path to mystification.

Feferman, in his (1995) review of *SM*, praises Penrose's writing, especially his explanations for non-specialists, but questions the conclusiveness of his arguments against computationalism based on Gödel's theorems, as well as his call for a microscopic quantum–mechanical explanation of consciousness. He gives a detailed analysis of the first topic, discovering a number of minor but sometimes surprising errors of fact or interpretation on the part of the author. About Penrose's '*Gödelian argument*' itself, he writes,

I must say that even though I think Gödel's incompleteness theorems are among the most important of modern mathematical logic and raise fundamental questions about the nature of mathematical thought, and even though I am personally convinced of the extreme implausibility of a computational model of the mind, Penrose's Gödelian argument does nothing for me personally to bolster that point of view, and I suspect the same will be true in general of similarly inclined readers. On the other hand, I'm sure that those whose sympathies lie in the opposite direction will find reasons to dismiss the Gödelian argument quickly on one ground or another without wading through its painful elaboration. If I'm right, this is largely a wasted effort—diligent as it is.

Penrose places great emphasis on what he calls '*understanding*', an aspect of the human mind which he believes to be lacking in machines, no matter how sophisticated they may become. Feferman, after six pages of discovering various errors in Penrose's detailed treatment of Gödel's (1931) arguments, writes,

> However, the following comments by Penrose about the significance of Turing's and my work are correct: '… there is no algorithmic procedure that one can lay down beforehand which allows one to do this systematization for all recursive ordinals once and for all', and that '… repeated Gödelization … does not provide us with a mechanical procedure for establishing the truth of Π_1 sentences'.

Before the heading '*What follows from Gödel's incompleteness theorem?*', Feferman continues,

> The main question, though, is whether these errors undermine the conclusions that Penrose wishes to draw from the Gödelian argument. I don't think that they do, at least not by themselves. That is, I think that the extended case he makes from Sect. 2.6 on through the end of Chap. 3 would be unaffected if he put the logical facts right; but the merits of that case itself are another matter.

In the following section of his review, Feferman gives a very interesting and enlightening summary of how mathematicians' thought processes work: he describes the process of finding mathematical truths, for example proving theorems, as a "*marvelous combination of heuristic reasoning, insight and inspiration (building, of course, on prior knowledge and experience)*"—which we might subsume under 'mathematical intuition + ingenuity' (keeping in mind Turing's (1939) distinction between intuition and ingenuity); and this is probably what Penrose means by '*understanding*'. Feferman describes

Penrose's effort to show that '*mathematical thought cannot even be* re-represented *in mechanical terms, as a result of the Gödel theorem*' with the words, "… *this effort raises more questions than it answers and leads one off into dead-end dialectics*' (p. 29 in the review). Later in this section, in answer to Penrose's professed Platonist position, he quotes Gödel from his unpublished lecture to the AMS in December, 1933 ([*Gödel *1933o*] in his *Collected Works*, Volume III): '*The result of the preceding discussion is that our axioms* [for set theory], *if interpreted as meaningful statements, necessarily presuppose a kind of Platonism, which cannot satisfy any critical mind and which does not even produce the conviction that they are consistent*'. This is an interesting quote [22] which sheds some light on Gödel's often-claimed Platonism (in this context, see the article by Jaakko Hintikka (2005) entitled '*What Platonism? Reflections on the thought of Kurt Gödel*'). It also casts doubt on Penrose's '*Gödelian argument*', even if his anti-AI stance may prove to be correct in the long term.

It should be mentioned that, in addition to Church, Gödel, Post and Turing, also—30 years later—*Gregory J. Chaitin* [born 1947; for many years at the IBM Thomas J. Watson Research Center in Yorktown Heights, NY; now at the *Federal University* in Rio de Janeiro (UFRJ)] has developed a proof of incompleteness based on *computer science* (information theory and computational complexity). He suggested a new measure of complexity (now called *Kolmogorov-Chaitin-Solomonoff* complexity) and used it to show that incompleteness is in fact rather common. He was apparently, like Douglas Hofstadter, inspired at an early age by Nagel & Newman's (1958) book on Gödel's incompleteness proofs, and developed some of his ideas while still in high school. His arguments have been criticized, however, e.g., by Franzén (2005); see Chap. 8 there. Many of the notions in his development have no precise definitions of the kind that Gödel insisted upon. Compare his book, '*Thinking about Gödel and Turing*' [Chaitin (2007)] (Item 31 in Appendix B).

Notes

1. See Kidwell & Williams (1992).
2. WikiMedia, uploaded by user 'Kolossus', reused under a GNU Free Documentation License (Creative Commons 3.0). Reused from: https://commons.wikimedia.org/wiki/File:Leibnitzrechenmaschine.jpg
3. See Copeland et al., eds. (2013) (CPS), Item 39 in Appendix B.
4. Photo cropped and enhanced. Author unknown, copyright status unknown, fair use. Image courtesy of Archive Centre, King's College, Cambridge,

image AMT/K/7/4. Reused from: https://turingarchive.kings.cam.ac.uk/material-given-kings-college-cambridge-1960-amtk/amt-k-7-4.

5. This problem was first considered by Leibniz, who hoped to construct a machine capable of determining the truth value of any statement in symbolic logic (which he subsequently tried to develop).

6. The thesis was published in the *Proceedings of the London Mathematical Society*; see Turing (1939).

7. Turing (1954): '*Solvable and unsolvable problems*', a popular article published in *Science News* in 1954.

8. From the archive of Alonzo Church; photographer unknown, copyright status unknown, fair use. Reused from: https://alchetron.com/Alonzo-Church#alonzo-church-d4067efd-43cb-4b48-afa4-d3a5bdd903a-resize-750.jpg.

9. Cf. Davis (1982).

10. See M. Davis, introduction to [*Gödel *193?*] in the *Collected Works*, Vol. III (1995), pp. 156–163.

11. Church (1937)

12. Adapted here from Stillwell (2004), p. 9.

13. See Post (1936).

14. Schmidhuber (2021); see online at: https://people.idsia.ch/~juergen/goedel-1931-begruender-theoretische-informatik-KI.html (consulted 15.11.2021) and also the *FAZ* article from 16.06.2021, "*Als Kurt Gödel die Grenzen des Berechenbaren entdeckte* ' ('*When Kurt Gödel discovered the limits of the computable*'); In German, paywall. Translation of the quotation in the text is by the present author. [Schmidthuber (2021)].

15. Sieg (2005), p. 180.

16. Gödel (1934), in Davis (ed., 1965), p. 44. This is a famous footnote, since it was initially considered to be an anticipation of Church's thesis. Gödel later contradicted that idea; as Sieg (2005, footnote 11 on p. 181) states: "*Gödel emphatically rejected in the sixties (in a letter to Martin Davis) that this formulation anticipates a form of Church's Thesis: he was not convinced that his notion of recursion was the most general one*".

17. See Wang (1974), Item 2 in Appendix B, p. 325.

18. Nagel & Newman (1956), quoted from p. 86.

19. See Lucas (1961).

20. See e.g. H.T. Siegelmann, '*Computation Beyond the Turing Limit*', in: *Science*, Vol. 238, No. 28 (2012), pp. 632–637; or J. Cabessa and H.T. Siegelmann, '*The Computational Power of Interactive Recurrent Neural Networks*', in: *Neural Computation*, Vol. 24, No. 4 (1995), pp. 996–1019.

21. See Davis (1993).

22. We have seen this sentence already in Chap. 10 (section on '*Gödel's First Stay in Princeton*'). Dawson and other authors have noted that it is ambiguous due to the placement of the comma after 'Platonism', perhaps due to Gödel's inexperience with the English language at the time.

13

1940: *A Long Journey Eastward to the West*

The journey of Kurt and Adele Gödel from Vienna to Princeton in early 1940 is a saga in itself. Unfortunately, we have no first-hand reports. They were both no doubt too busy coping to keep records, and neither of them wrote a diary in any systematic way. Stamps in their passports give some information about where they were, and when. Years later, Adele spoke to her friend and nurse *Elizabeth Glinka* about the trip, but her only memories were that they were worried (probably at the beginning, while they were still in the *Reich*) that they might be stopped and sent back, and that they traveled mostly at night (also probably at first).

The first half of January in the new decade of the 1940s was no doubt hectic and nerve-wracking for Kurt and Adele. The decision to leave Vienna was taken, most of the arrangements had been made, but they had to wait for all the documents (rail tickets, visas) to arrive—while taking care of endless last-minute details—so that they could begin their journey. The 'official' date of their departure (shown for example on the memorial plaque at *Hegelgasse* 5) was January 10th; but that was just the date on which they registered their move out of the *Hegelgasse* apartment, for which they had most probably found renters before leaving Vienna. Registration with the official Registry of Residents (*Einwohner-Meldeamt*) was legally required, usually within two to four weeks of a move into or out of a dwelling. The dates on the memorial plaques in Vienna are those given in the Registry and are not necessarily the actual dates of a move. In fact, they departed by train on January 16th, leaving

© The Author(s), under exclusive license to Springer Nature
Switzerland AG 2022, corrected publication 2023
W. D. Brewer, *Kurt Gödel*, Springer Biographies,
https://doi.org/10.1007/978-3-031-11309-3_13

many of their belongings behind (Rudolf shipped some of them to Princeton after WWII).

Precisely which route they took on the first part of the journey is not recorded. They certainly traveled north to Brünn, and perhaps continued on northward to *Breslau* (now the Polish city of *Wrocław*, in 1940 still the capital of Upper Silesia (*Oberschlesien*) and part of the *Reich*). From there, or from Brünn, they could have taken the eastern route, through Warsaw, which would have been the shortest path to Moscow but would require a long journey through occupied Poland; or the western route, through Berlin, Stettin (now *Szczecin*) and Danzig (*Gdansk*), still German cities at that time, and then on to Königsberg, the easternmost city in the *Reich* (this latter route was the one taken by Kurt Gödel 9½ years earlier, on his way to the conferences in Königsberg in 1930; but he went by boat from Stettin to Königsberg). From Königsberg (now *Kaliningrad*, Russia), the rail line continues straight eastward to *Vilnius*, the capital of Lithuania, and then northeastward into Latvia, where it intersects the main line from *Riga* (the capital of Latvia, on the Baltic coast) to Moscow. That line crosses the border from Latvia to Belarus (in 1940, part of the USSR) and continues on to Moscow, another 700 km eastward. According to the stamp in his passport, Kurt crossed that border (at *Bigosovo*, the first town in the USSR, about 6 km from the border) on January 18th, two days after leaving Vienna, having covered around 1000 km thus far. The trip on to Moscow would have taken another 18–20 h. There, they could board the Trans-Siberian Railway, which would take them directly to *Vladivostok* (Russian for 'Ruler of the East'), on the eastern coast of the USSR (on the Pacific coast, or more precisely, the coast of the Sea of Japan, also known as the East Sea).

The Trans-Siberian Railway (or '*Trans-Sib*', as it is known familiarly) was a prestige project of the last Russian Tsars, constructed between 1891 and 1916. It is the world's longest railway line, covering almost 9,300 km (5,800 mi.), and is today in fact a network rather than a single line (see the map in Fig. 13.1). The modern trains take about 7 days direct from Moscow to Vladivostok, but the travel time would have been considerably longer in 1940, when steam engines were still used. A direct rail connection to Tokyo is currently planned, following the northern route of the Trans-Sib and crossing over two bridges from the mainland to the island of *Sakhalin*, and from Sakhalin to *Hokkaido*, the northern-most Japanese island. The Moscow terminus of the Trans-Siberian is shown in Fig. 13.2.

The Gödels' trip in fact took 15 days from Moscow to Yokohama, the harbor city of Tokyo. They traveled by ship and train from Vladivostok to Yokohama, arriving on February 2nd.

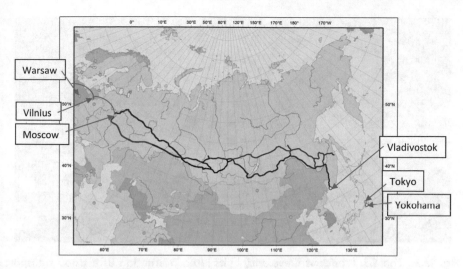

Fig. 13.1 Map of Eastern Europe, Central Asia and East Asia, showing the lines of the Trans-Siberian Railway. The arcs at the left indicate schematically the route taken by Kurt and Adele Gödel, passing through occupied Poland and Königsberg to *Vilnius* (the capital of Lithuania), then through Latvia, entering the USSR at *Bigosovo* (now in Belarus) and continuing by rail to Moscow. There, they boarded the Trans-Siberian, which covers more than 9,000 km enroute to Vladivostok. Map from Wiki Commons, public domain[1]

Fig. 13.2 The Ярославский Вокзал, (*Yaroslavskii Voksal*), or Yaroslavsky Station: the beginning of the Trans-Siberian Railway line in Moscow. The station was built in 1862, and later became the Moscow terminal of the Trans-Sib. Photo by Sergey Ashmarin, Wikimedia Commons[2]

Fig. 13.3 The *S.S. President Cleveland*, in its later incarnation as a troop transport (1942: '*USS Tasker H. Bliss*'). Photo from U.S. Navy, public domain[3]

The Gödels had just missed their planned trans-Pacific ship connection, on the *S.S. President Taft*, a ship that was operated at that time by the American President Lines. It had been launched (under another name) in 1920 and operated by several different steamship lines. After 1941, it was converted into a troop transport and was in service for the U.S. Army until 1947.

Having missed their connection, the Gödels waited 18 days in Yokohama, a time that they used well for rest and recuperation. Kurt apparently enjoyed the hotel where they stayed, and Adele "*took a 'boxful' of purchases*" with her when they left.[4] On February 20th, they boarded the *S.S. President Cleveland*, another President Lines ship (see Fig. 13.3). It had a similar history to that of the *President Taft*, and operated as a passenger liner until 1942, when it likewise became a troop transport. It was torpedoed and sunk on November 12th, 1942, and a new *President Cleveland* was launched in 1947.

After leaving Yokohama, the *President Cleveland* sailed for Honolulu, arriving on February 28th (1940 was a leap year), and then continuing on to San Francisco, where it docked on March 4th. Dawson (1997) quotes the captain's roster of disembarking passengers, which gives the physical descriptions of Kurt and Adele and the captain's opinion that both were in good (mental and physical) health. They stated truthfully that they did not plan to return to their country of origin, but they made the mistake of saying that they were also not planning to become U.S. citizens, and of declaring themselves German citizens (not realizing that the USA still recognized Austria as a separate country), which caused them some trouble later. By March 5th, they were ready to depart on the final leg of their journey. Kurt and Adele, having traveled for 50 days, still had to cross another continent to reach their goal.

After what was apparently a relaxing transcontinental train ride, in comparison to their previous travels, the Gödels finally arrived in Princeton on March 10th, 1940, a Sunday. This was the beginning of a completely new chapter in both of their lives; and Kurt, in particular, never again traveled very far from their new home. Adele did travel back to Vienna and other destinations in Europe to see her family and friends a number of times in the 1950s and 1960s, a fact which annoyed Kurt's mother Marianne, in particular since Kurt failed to visit *her*—but later, she and Rudolf also traveled several times to Princeton on visits.

The Trans-Siberian Railway as an Escape Route from Fascist Europe

The Gödels were by far not the only Europeans who escaped from the horrors of the war and the fascist regimes in Europe on the 'Trans-Sib'. Dawson (2002) has written an account of the Gödels' journey and compared it to that of *Max Dehn* (1878–1952), a German mathematician a generation older than Kurt Gödel (see Fig. 13.4). He had Jewish forebears but was secularized and assimilated. He obtained his doctorate in Göttingen under Hilbert in 1900 and then solved Hilbert's Third Problem, for which he was granted the *Habilitation* in Münster. He was a *Privatdozent* there until 1911, then associate professor in Kiel, and from 1913–1921 full professor in Breslau, moving to Frankfurt in 1921. He published significant works in topology and on group representations.

Dehn was forcibly retired in Frankfurt in 1935 at age 56 by the Nazis, but he remained there until 1938, although he sent his children abroad. He spent some time in England with his daughters, but he returned to Frankfurt in the spring of 1938. There, he was arrested after the November Pogrom, but was released and escaped with his wife to the home of friends in nearby Bad Homburg, where they sheltered for some time until the situation had calmed down. They then traveled to Hamburg, hid out there for several weeks, and finally escaped via Denmark to Norway, both countries not yet occupied. Dehn had a temporary academic position in Trondheim in 1939/40.

In March 1940, Norway was occupied by Nazi forces, and the Dehns were once again in mortal danger. They planned their journey across Europe, Asia and the Pacific, leaving in late October of 1940, but first spent 3 weeks in Stockholm (delayed for 'obscure political reasons'). They then were able to

Fig. 13.4 Max Dehn, photographed around 1945. Photo: *Oberwolfach Photo Collection*. Creative Commons[6]

fly to Moscow, where, after three more days, they boarded the *Trans-Sib*, and after another delay of 6 days in Vladivostok, they crossed to Japan where they spent some more time in Kobe before boarding a ship for San Francisco. They finally arrived on January 1st, 1941. Thus, their total travel time was around 70 days, somewhat longer than that of the Gödels for their even longer passage to Princeton.

Max Dehn's account of their arduous journey was recorded in a lecture that he gave at Idaho Southern University (ISU) (its text is in the archive of his papers at the University of Texas[5]). Dehn had a temporary appointment at ISU, in Pocatello, ID in 1941/42 (where the present author spent part of his youth, 10 years later). He then taught for a year at the Illinois Institute of Technology (IIT) in Chicago, then for another year at St. John's College in Annapolis, MD, finally settling in 1944 at Black Mountain College, near Asheville, NC. After a 'trial period' there, he was hired to the tenured faculty in 1945.

The latter two schools were both in some sense 'experimental': St. John's is an older institution, founded as King William's School in 1696 and chartered as St. John's College in 1784. It offers a BA in Philosophy and History of Science, with a double minor in Classical Studies and Comparative Literature, using the study of 100 'great books' as a guideline for learning; and it also grants MA degrees in Liberal arts and Eastern Classics.

Black Mountain College was founded in 1933. Like St. John's, it offered a liberal arts education, but based on the principles of John Dewey. It was highly successful for a time, and many of its graduates and former teachers were later influential in the arts, literature and education. It provided the model for many other innovative educational institutions. It was forced to close in 1957 due to insolvency, but it left behind a considerable intellectual legacy. Max Dehn taught there from 1945 until he retired in 1952, and he died on June 27th, 1952, soon after his retirement. He was content with being a teacher there, but he was also able to spend two sabbaticals (a semester in 1946/47 and the academic year 1948/49) at the University of Wisconsin, Madison (where he had a last graduate student, *Joseph Engel*). Dawson (2002) quotes Engel [7] on Dehn: "*Working under* [Dehn's] *kind and understanding guidance was a joy and a privilege... Looking back at that wondrous time, I still love him, and am in awe of his wisdom and humanity and humor and compassion*". More information on Dehn's life can be found in the references given below.[8]

As a contrasting example, we mention another scholar who escaped Nazi persecution—as a young child—via the *Trans-Sib*. *Jorge André Swieca* was born *Jerzy Andrzei Swieca* on December 16th, 1936 in Warsaw, Poland, and he was thus a generation younger than Kurt Gödel (Fig. 13.5). His parents were Michał Swieca and Renata Theophila (Szporn) Swieca. They were of Jewish origin and therefore in great danger after the invasion of Poland by Nazi armies on September 1st, 1939. However, they were also unusually foresighted, and departed immediately, accompanied by three of Renata's siblings (Bronja, Janusz and Ignacy Szporn), making their way to Moscow and from there via the Trans-Siberian Railway to Japan. There, they spent some time before continuing on by ship to Buenos Aires, Argentina, where they had relatives. They had to hide the not-quite-three-year-old Jerzy Andrzei when their passports were examined, since they had no papers for him. They finally settled in Rio de Janeiro, Brazil, in July of 1942 (just before Brazil entered WWII on the side of the Allies). Renata's siblings all lived in São Paulo and had no descendants.

Young Jerzy Andrzei became *Jorge André*, and grew up in Rio, where his parents had a beautiful apartment on a side street in Copacabana, in view of the south Atlantic Ocean. He became a Brazilian citizen when he turned 18, studying physics and mathematics at the *Faculdade nacional de Filosofia* (FNFi) of the *Universidade do Brasil* (now *Universidade Federal do Rio de Janeiro*, UFRJ) from 1954–58. The following year, he went at the invitation of *Mario Schenberg*, a well-known theoretician, to the *Universidade de São Paulo* (USP), the foremost Brazilian institution for theoretical physics.

At USP, Swieca began work on his doctorate under *Werner Güttinger*, who recognized his talent for both pure and applied mathematics and sent him to work at the *Max-Planck-Institut für Physik und Astrophysik* in Munich, at that time still directed by Werner Heisenberg. Swieca returned to USP

Fig. 13.5 *Jorge André Swieca*, during the 1970s. Photo: Courtesy of the private collection of J.A. Swieca

to defend his doctoral thesis in 1963, and he was subsequently invited by *Rudolf Haag* to go with a postdoctoral fellowship to the University of Illinois (Champaign/Urbana).

He spent three years there, working with Haag on degrees of freedom in phase space, and returned to Brazil in 1967, where he obtained his '*Livre Docente*' (comparable to the *Habilitation*) with a thesis on '*Spontaneous symmetry-breaking in Quantum Theories*'. For the thesis, he received the prestigious '*Prémio Moinho Santista*' prize in 1968, as the second physicist to do so. He joined the newly-founded group of *Jayme Tiomno* [9] at USP (Tiomno was the *first* physicist to receive the prize; see Chap. 19), but after Tiomno's blacklisting by the Brazilian military dictatorship in May, 1969 (along with hundreds of other scholars, journalists, military officers and politicians who were judged to be 'subversives'), Swieca moved in 1970 to the *Catholic University* in Rio de Janeiro (PUC/RJ), where political pressure was not so omnipresent. He was considered to be one of the most promising of the younger theoreticians in Brazil, and was revered as a teacher.[10] His scientific work is summarized in two review articles.[11] Swieca later moved to the Federal University at *São Carlos* (UFSCar, in the State of São Paulo). He died there in late 1980, at only 44 years of age. His early death was felt to be a serious loss to Brazilian physics, and his memory is kept alive as the name of a series of summer schools sponsored by the *Brazilian Physical Society* (SBF).

These three examples, of people of very different ages and backgrounds: Gödel, Dehn, and Swieca, each of them scientists devoted to mathematics and physics, stand for the many thousands of people who were able to use the

Trans-Siberian Railway as an escape route from the madness which prevailed in Europe in the 1930s and 40s. Dawson (2002) also mentions the well-known example of the Japanese diplomat *Chiune Sugihara* (1900–1986), who was a Vice-Consul for the Japanese Empire in *Kaunas*, Lithuania (the second-largest city in Lithuania, in the south-central region of the country, around 80 km northwest of Vilnius), beginning in 1939. After the Soviet occupation of Lithuania on June 16th, 1940, many Jewish refugees from occupied Poland and further west, as well as from Lithuania itself, were looking for an escape route. The majority accepted Soviet citizenship, but a significant minority needed to escape to other countries. Sugihara issued thousands of transit visas to such people, who—by prior agreement and by paying several times the usual ticket price—were allowed to travel on the *Trans-Sib* to Vladivostok and thence to Japan, where they could then go on to other destinations where they would be safe. Sugihara supplied an incredible number of people with visas from July 18th through August 28th, 1940, until he was required to leave Lithuania, as the consulate was being closed. He was honored by Israel as a '*Righteous among the Nations*' in 1985, the only Japanese person to be granted that honor. The year 2020 was declared '*The Year of Chiune Sugihara*' in Lithuania in his honor.[12] It has been estimated that up to 100,000 people now living are descendants of persons saved by Sugihara's efforts. None of them would exist today if he had not saved their parents/grandparents in 1940.

The escape route via the *Trans-Siberian* was open for only a relatively short time. Before the *Ribbentrop-Molotov* pact in August of 1939, most countries in Western and Central Europe were not on especially good terms with the Soviet Union, and obtaining a transit visa through Poland or the Baltic countries to the USSR and a ticket on the *Trans-Sib* was not easy for refugees who had to remain very discreet and keep a low profile in order to avoid attracting the attention of the wrong people. After the German invasion of the USSR in June of 1941, the western USSR was a war zone, and the *Trans-Sib* was suspended as a civilian passenger service for the duration of the war. So, it was a matter of luck to be able to use that 'time-window', as the Swiecas did right at its beginning, the Gödels in the middle and the Dehns near the end of the brief period while it was open. Once again, Nazi politics somehow favored the Gödels, while at the same time threatening them.

Notes

1. Map: from Wiki Commons, reused from: https://commons.wikimedia. org/wiki/File:Map_Trans-Siberian_railway.png. Author: Stefan Kühn; public domain.
2. Photo: by Sergey Ashmarin, May 19th, 2013. Wikimedia Commons, used under Creative Commons Attribution-Share alike 3.0 Unported License. Reused from: https://commons.wikimedia.org/wiki/File:Yaroslavsky_Train_ Terminal,_Moscow,_Russia_-_2013-05-19_-_90650652.jpg.
3. Photo: U.S. Navy, public domain. Reused from: http://www.navsource.org/ archives/09/22/22042.htm.
4. From the correspondence of Kurt Gödel with his mother, November/December 1964; quoted by Dawson (1997), 2005 edition, p. 150 and endnote [367].
5. Quoted by Dawson (2002); the text (1941) is in the Dehn Papers, *Archive of American Mathematics*, U. Texas Austin.
6. Photo: *Oberwolfach* Photo Collection. Freely usable (Wiki Media; Creative Commons License Attribution-Share Alike 2.0 Germany). See: https://opc. mfo.de/detail?photo_id=13906. Reused from: https://commons.wikimedia. org/wiki/File:Max_Dehn.jpg.
7. Dawson, *op. cit.*, p. 1074, quoting an unpublished manuscript, '*Professor Dehn*', by Joe Engel (1997).
8. See e.g. J.J. O'Connor and E.F. Robertson, 'Max Wilhelm Dehn', in the MacTutor biographies (2014), online at: https://mathshistory.st-andrews.ac. uk/Biographies/Dehn/. For Dehn's time at Black Mountain, see R.B. Sher, '*Max Dehn and Black Mountain College*', in *The Mathematical Intelligencer*, Vol. 16 (1994), pp. 54–55.
9. See Brewer & Tolmasquim (2020).
10. See '*Jorge André Swieca*', by A. Luciano L. Videira, online at http://www.sbf isica.org.br/v1/portalpion/index.php/fisicos-do-brasil/15-jorge-andre-swieca; and also https://jaswieca.if.uff.br/ (both in Portuguese).
11. Schroer (2010), in English, and Marino (2015), in Portuguese.
12. See Levine (1996) and Sugihara (1993) for biographies of Chiune Sugihara.

14

Princeton. *Settling in at the IAS. Gödel in America, 1940s*

Kurt Gödel had gradually lost most of his friends and colleagues in Vienna during the 1930s, and he was—by 1940—one of the last of the original group around the Vienna Circle and the *Mathematisches Kolloquium* to leave Austria and seek a safer place to live and work. He was not under mortal threat as were his Jewish colleagues, but he was severely disappointed by the treatment that he had received from the University (their refusal to automatically convert his *Privatdozentur* into a *Dozentur neuer Ordnung* in 1938, unlike many other colleagues) and what he considered to be an incorrect judgment of his fitness for military service. This disappointment was gradually converted into an active dislike and even fear of all things Austrian during the first 5 or 6 years of his exile in Princeton, so that by the time that it was again possible to travel there, he was not at all inclined to do so. But in fact, by 1945, the vast majority of his former colleagues and friends were also in the USA. It was only in the summer of 1951 that he *almost* went to Vienna to visit his mother and brother, but he let the opportunity pass, along with a few others in the following years.

The Viennese Diaspora

The fates of Gödel's former associates were very diverse: Apart from those who had died (both prematurely, and tragically: *Hans Hahn* and *Moritz Schlick*)— the members of the Vienna (Schlick) Circle had mostly left Austria by 1939.

W. D. Brewer, *Kurt Gödel*, Springer Biographies, https://doi.org/10.1007/978-3-031-11309-3_14

The participants in the old, or *first* Vienna Circle included *Philipp Frank* and *Otto Neurath*, as well as Hans Hahn. We met up with them, and many other members of the later Schlick Circle, in previous chapters. But what happened to them as a result of the political upheavals in Austria during the 1930s?

Philipp Frank (1884–1966), as we saw earlier, was offered a position in 1912 as Einstein's successor at the *Karlsuniversität* in Prague, where he remained until 1938. He helped Rudolf Carnap to obtain a position there in 1931. He maintained his interest in logical positivism/empiricism and the Unity of Science movement for the rest of his life, and he, like Carnap, participated in the Vienna Circle 'from afar' for some years. Forced to emigrate by Hitler's occupation of Bohemia and Moravia in early 1939, he went to the USA and became a lecturer in physics and philosophy at Harvard University, in Cambridge, MA, where he remained for the rest of his life, revered by many of his students.

Otto Neurath (1882–1945), who had closer ties to Kurt Gödel, although he was also a generation older, led a restless life. He was another member of the 'old' Vienna Circle, and after the death of his first wife before WWI, he married *Olga Hahn*, the sister of Hans Hahn, who was his age, a mathematician and philosopher, and was blind. We have seen that Neurath spent the last years of WWI in Germany, where he finished his *Habilitation* (in Heidelberg) and directed a museum (in Leipzig) before doing some time in prison in Munich during the revolutionary days following the War. Back in Vienna, he was among the most active of the Schlick Circle members, and his wife was also a participant. Neurath was the chief author of the 'manifesto' that alienated Gödel and Menger (and some others) from the Circle; in it, he defined the views of the Vienna Circle for the rest of the world. He was working in Russia during the '*Austrofascist* revolution' in 1934, and—being warned by the code sentence, "*Carnap is waiting for you*"—he went directly to the Netherlands to avoid being arrested in Vienna. His wife joined him there, but only for a few years; she died in 1937. Neurath was surprised in Rotterdam by the German '*Blitzkrieg*' in the spring of 1940, but escaped by small boat to England, together with his assistant and later third wife, *Marie Reidemeister*. After being interned on the Isle of Man, he was released with the help of international appeals, and spent his last years in Oxford, where he set up an institute for isotype charts, a visual presentation method that he had developed. He died unexpectedly in late 1945, only 63 years old.

Herbert Feigl (1902–1988), a still closer friend of Gödel's during his Vienna years, was an early member of the Schlick Circle and one of its few members to have direct contacts to Ludwig Wittgenstein. He went on a fellowship to Harvard in 1930, and after his marriage the following year to

Maria Kasper, another member of the Circle, he emigrated to the USA, first taking a faculty position at the University of Iowa, then at the University of Minnesota, where he continued in the philosophy department for 31 years, remaining true to logical empiricism.

Marcel Natkin (1904–1963), also a close friend of Gödel's during their student years, completed his doctoral thesis under Schlick in 1928 and moved soon after to Paris, where he became a well-known photographer. He survived the German occupation and remained for another 20 years in Paris, until his death in 1963. He and Feigl famously held a reunion with Kurt Gödel in 1957, in New York City.

Rudolf Carnap (1891–1970), one of the most active members of the Vienna Circle, and its 'front man' for years after the actual Circle had ceased to exist, had been invited to Vienna by Moritz Schlick, as we saw in Chap. 4. He played an important role for Gödel as mentor and teacher in the years 1927–29; but their roles were thereafter reversed, Gödel becoming Carnap's 'advisor' on matters concerning mathematical logic and metamathematics.[1] Carnap was granted the *Habilitation* in 1928 in Vienna for his book '*The Logical structure of the World*', and he went to Prague with the aid of Philipp Frank in 1931. After the *Austrofascist* episode and with the rise of Nazism, he emigrated to the USA in 1935, working initially in Chicago and at Harvard, and in the postwar era, he spent almost two years as Gödel's colleague at the IAS (in 1952–54), apparently meeting occasionally with Gödel during that time, but without the close contacts they had during the Vienna Circle days. He finally settled in California, at UCLA, where he spent the last years of his career on the philosophy faculty as the successor to Hans Reichenbach. He died there in September 1970. His diaries have been an important source of information, but unfortunately[2] the volumes after 1935, which might contain comments about his later relations with Gödel, in particular during his time at the IAS, have yet to be transcribed.

About **Karl Menger** (1902–1985)—perhaps the most long-term of Gödel's close associates, along with Oskar Morgenstern—we have already heard a great deal in previous chapters. Following his move to the University of Notre Dame in early 1937 and his collaboration with Kurt Gödel there in 1939, their relations cooled somewhat because Menger could not accept Gödel's apparent lack of empathy for the victims of fascism in Europe. However, they maintained contact in later years, and Menger also collaborated with Gödel's friend Oskar Morgenstern on game theory during the 1940s. Menger moved from Notre Dame to the Illinois Institute of Technology (IIT) in Chicago in 1946, and spent the rest of his career there, retiring in 1971. He was awarded an honorary doctorate by the IIT in 1983, and a scholarship as well as an annual lecture there are named in his honor.

In the end, nearly all the participants in the Circle and the Colloquium managed to escape from Austria prior to or just after the *Anschluss*; we mentioned some of the others, including *Heinrich Gomperz* (1873–1942), *Edgar Zilsel* (1891–1944), *Viktor Kraft* (1880–1975), *Friedrich Waismann* (1896–1959), *Abraham Wald* (1902–1950), *Franz Alt* (1910–2011), and *Olga Taussky-Todd* (1906–1995) in Chap. 9 and in some previous chapters. All of them except Viktor Kraft and Friedrich Waismann settled in the USA after WWII. Kraft stayed in Austria, ultimately losing all his academic qualifications during the Nazi era; but after the War, they were restored, and he became a professor in Vienna until his retirement in 1952. He died in Austria in 1975, at age 94. Waismann emigrated to England in 1937, following Ludwig Wittgenstein, by whom Waismann was fascinated for many years; but after a few years there, he moved from Cambridge (where Wittgenstein spent the last 22 years of his life) to Oxford, and died there at 63 years of age. Edgar Zilsel moved to England with his family in 1938, and later went to the USA on a Rockefeller Fellowship; there he had a promising beginning, accepting a position to teach at Mills College in Oakland, CA; but he committed suicide in 1944, at age 52. Abraham Wald became a professor at Columbia University, but he died in a plane crash in India in 1950, only 48 years old. At the other end of the age spectrum, Olga Taussky-Todd lived to the age of 89, and Franz Alt, true to his name, to 100, both of them having had very successful scientific careers. Taussky-Todd taught in her later years at Caltech, in Pasadena, CA, where she remained until her death in 1995; and Alt worked at the U.S. National Bureau of Standards/NIST on computer development, and later at the *American Institute of Physics*; he died in New York City in 2011.

However, most of Kurt Gödel's former colleagues and associates had few or no interactions with Gödel himself after leaving Vienna, even though they may have had opportunities to do so.

In Princeton, Gödel initially made two new friends: *Oskar Morgenstern* and *Albert Einstein*. Morgenstern, who had been a successful mathematician/economist in Vienna, was in the USA on a lecture tour at the time of the *Anschluss*, and he remained there, soon obtaining a position as a faculty member at Princeton University. He reconnected with Gödel shortly after the latter's arrival at the IAS in early 1940, as we saw in Chap. 9, and became one of Gödel's closest long-term friends.

Oskar Morgenstern (1902–1977) (cf. Fig. 14.1) was an economist, born in Görlitz, Germany (in Silesia; see Fig. 14.4). He moved with his family at age 12 to Vienna and later studied there, obtaining his doctorate in political science in 1925. He spent three years (1925–28) on a Rockefeller Fellowship in England, Canada and the USA, working at Columbia and Harvard Universities, among other places. Back in Vienna, Morgenstern's *Habilitationsverfahren* became a comic opera[3] due to the stubborn opposition of a single faculty member (doubtless for political or racial-bigotry reasons). He was finally named *Privatdozent* in 1929, later associate professor, and from 1931 he was head of the *Österreichisches Institut für Konjunkturforschung* (Austrian Institute for Economic Research).

Morgenstern had attended the meetings of the Vienna Circle and sometimes the *Mathematisches Kolloquium,* and had doubtless met Gödel there, at least informally.

In 1938, Morgenstern was caught in the USA by the *Anschluss* (in a situation similar to that of *Alfred Tarski,* 1-½ years later, who was also on a lecture tour in the US when his country, Poland, was occupied by the Nazis). Morgenstern was subsequently stripped of all academic offices and honors in Austria. He was however soon able to join the faculty of Princeton University, and there he met John von Neumann, a founding professor at the IAS. They later collaborated on a significant and influential book, '*Theory of Games and Economic Behavior*' (1944), and they are considered to be the founders of Game Theory. Morgenstern was in demand as a consultant during and after WWII. His diaries are also an important source of information about

Fig. 14.1 Oskar Morgenstern (in Vienna, about 1935). *Photo Universität Wien,* picture archive[4]

Kurt Gödel's later life at the IAS. Budiansky (2021, pp. 219–220) gives a touching account of their friendship, which was important to both of them and continued until Morgenstern's death, just 5-½ months before Gödel's. It may have contributed to Gödel's final depression.

Kurt Gödel continued his work on mathematical logic in his first years as a regular member of the IAS (but with only annual contracts). In the summer of 1940, he gave several lectures in Princeton which mirrored his course on set theory given at Notre Dame in early 1939. He also summarized them in a lecture to the Mathematics Colloquium at Brown University on November 15th, 1940 (the manuscript of that talk was reconstructed as [*1940a] in the Collected Works, Vol. III). Wang (1987) reports that Gödel read Brouwer's work in 1940, and that he had an idea for his independence proof of the Axiom of Choice while doing so. He also found a 'general consistency proof for the Axiom of Choice based on syntactical considerations' in that year. In early 1941, he obtained an interpretation of intuitionistic number theory using primitive recursive functionals, which he however failed to publish (it was mentioned in a much later paper in honor of Paul Bernays in 1958). But he lectured on it in Princeton, and also gave a single lecture at Yale University (April 15th) on the topic: 'In What Sense is Intuitionistic Logic Constructive?' ([*1941] in Volume III of the Collected Works). In the years between 1941 and 1944, Gödel gave no more public lectures, nor did he publish any papers. He seems to have been at an impasse in trying to prove the independence of the Continuum Hypothesis, and he was also gradually shifting the focus of his interests from formal mathematical logic to more philosophical topics.

Gödel's Chapter on Bertrand Russell's Logic, 1944. Philosophical Works

Gödel published an important article in 1944, a contribution to a volume in honor of Bertrand Russell, edited by Paul Arthur Schilpp. Schilpp had visited Gödel at the IAS in November 1942, at a time when the latter was frustrated with his lack of success with the GCH problem, and had obtained his agreement to write the chapter. Gödel's contribution was entitled "Russell's Mathematical Logic", and it contains a critique of Russell's 'no-class' theory. He submitted a draft chapter in May of 1943, and promised the final version before the Gödels' summer vacation on the New Jersey shore that year; but in the event, he delivered it only in September. That was very late in the process of preparing the book, so that Russell, in his own opinion, had too little time to reply within the book itself. He did make a brief reply in which he stated that he had left the field (of mathematical logic) 18 years before (in fact he mentioned in other places that he had effectively left the field after the publication of the third volume of the *Principia mathematica* in 1913), and thus he was no longer able to '*form a critical estimate of Dr. Gödel's opinions*'. [See Gödel (1944) and Schilpp (ed.) (1944); compare also Fig. 14.2]. This essay was very forthright, by Gödel's standards.

Indeed, Russell had spent some months at the IAS in the spring of 1943, during which he met with Einstein, Pauli and Gödel for discussions at Einstein's house. Recollections of the frequency of those meetings later varied; Russell recalled that they took place every week while he was in Princeton, but Gödel could remember only one such meeting. By the time this discussion arose (after the publication of the second volume of Russell's autobiography in 1956), Einstein had passed away and could not contribute to the discussion, and Pauli apparently was not interested.

Paul Bernays reviewed Gödel's chapter on Russell [cf. Bernays (1946)], and he gives a clear exposition of what Gödel's intentions were:

> Gödel, discussing, with some criticism, the leading ideas of *Principia Mathematica*, treats mainly of the devices which were chosen by Russell as a means of overcoming the difficulties connected with the logical paradoxes… Gödel

Fig. 14.2   A modern edition (1989) of Schilpp (1944) which contains Russell's autobiography in addition to the original content of the 1944 first edition. Other editions were issued in 1971, 1963, 1951 and 1946. *Photo* private archive WDB

points especially to the circumstance that it is just Russell's refraining from a more decided realism towards the logical and mathematical objects, to which are due the known difficulties in *Principia Mathematica*. All these difficulties, he argues, can be avoided by admitting that classes and concepts may be conceived as real objects, classes as structures consisting of a plurality of things, and concepts as the properties and relations of things existing independently of our definitions and constructions.

A more modern discussion of the reception of Gödel's chapter on Russell, his first significant paper on a philosophical topic, was given for example by Crocco (2013), who emphasizes the role of Leibniz in Gödel's work (this essay on Russell's work was the only one of Gödel's publications in which he explicitly mentions Leibniz, although he did so in many unpublished sources, e.g., in the *Grandjean questionnaire*). Another modern interpretation is given by Urquhart (2016). He points out that Gödel was initially impressed by Russell's realism:

…he *[Gödel]* had in mind the sentence '*Logic is concerned with the real world just as truly as zoology, though with its more abstract and general features*' *[Russell (1920), p. 169]*, a passage that he quoted in his Schilpp volume *[Gödel 1944, p. 127]*; Hao Wang later observed that '*the sentence quoted by Gödel is his favorite and corresponds pretty closely to his own view*' *[Wang (1987), p. 112]*.

But in his essay on Russell's logic, Gödel accuses Russell of '*refraining from a more decided realism…*' (see above quote from Bernays' review). Bernays also points out that Gödel here makes the distinction between ideas which he holds to exist in the universe independently of human thought—called '*concepts*'; and constructed ideas, deriving from human thought, which he calls '*notions*'. Crocco (2013) mentions in this connection,[5]

… a very basic question: in what way are Gödelian *concepts* tied to Leibnizian ideas and what does Gödel actually mean by *concepts*? What does Gödel have in mind when, in 1972 during his conversation with Wang, he speaks of concepts in *intension* ['Sinn'], as opposed to concepts in *extension* ['Bedeutung']? … The painstaking, almost maniacal attention that Gödel always paid to the editing and publication of his writing commands our respect. We need to reconstruct Gödel's thought-process, taking as a starting point his published work, which is very often misunderstood because of Gödel's style (concise and sometimes enigmatic) but also because of the content of his writings, which are generally contrary to what the *Zeitgeist* considers as standard. We need to go from the published works to the unpublished materials and then back again. [emphasis and comments added].

The essay Gödel (1944) remains today an important source of information on Gödel's philosophy of mathematics.

Somewhat later, Kurt Gödel gave a lecture at the Princeton Bicentennial in 1946 [see the *Appreciation* by Juliette Kennedy, in Kennedy (ed., 2014)]. It concerned the topic of computability [cf. Chap. 12], and in it, he compared Turing's notion of *computability* to his own concepts of *definability* and *provability*.

Furthermore, Gödel contributed to mathematical logic and philosophy in 1947 with his article "*What is Cantor's Continuum Problem?*", published in the *American Mathematical Monthly*. It was a presentation for a general (mathematical) audience, dealing with philosophical aspects as well as mathematical-logical details; see Gödel (1947). In that paper, for example, he states,

> ... This negative attitude towards Cantor's set theory *[held by the intuitionists]*, however, is by no means a necessary outcome of a closer examination of its foundations, but only the result of certain philosophical conceptions of the nature of mathematics, which admit mathematical objects only to the extent in which they are (or are believed to be) interpretable as acts and constructions of our own mind, or at least completely penetrable by our intuition. For someone who does not share these views there exists a satisfactory foundation of Cantor's set theory in its whole original extent, namely, axiomatics of set theory, under which the logical system of *Principia Mathematica* (in a suitable interpretation) may be subsumed.

After the mid-1940s, however, Gödel's attention shifted away from mathematical logic, and his later writings dealt more and more with philosophy, apart from his brief excursion into theoretical physics and cosmology (Chap. 15). Much of his philosophical work in fact remained unpublished, and it was only later found in his copious notes and 'first drafts', after his death. In his conversations with Hao Wang (1975/76), he summarized his work on set theory[6] in retrospect:

> In 1942 I already had the independence of the Axiom of Choice. Some passage in Brouwer's work, I don't remember which, was the initial stimulus ... Details of the proof are very different from the proof that uses forcing *[Cohen (1963)]* ... The independence of the Axiom of Constructibility is easier to prove. If it were provable in ZFC, it would also be provable in ZF... Cohen's models are related to intuitionistic logic and double negation. I tried to use my method to prove the independence of CH *[in 1942 and 1943]* but I could not do it. The method looked promising ... At the time I developed a distaste for the whole thing ... I was then more interested in philosophy, more interested in the relation of Kant's philosophy to relativity theory and the universal characteristic of

Leibniz … I am sorry now. If I had persisted, the independence of CH would have been proved by 1950 and that would have speeded up the development of set theory by many years.

Friendship with Albert Einstein

Gödel's second close friendship formed in Princeton, one of the relatively few in his life, began in 1940–42.[7,8] It was with *Albert Einstein*, perhaps the most famous scientist of the twentieth century. Kurt Gödel was included among 'the 100 most influential people' of the twentieth century by *Time* magazine—but Einstein was *Time*'s '*Person* of the 20th Century', representing the highest achievements of human civilization in that century.

There has been much speculation[9] about the nature of their friendship— the two scholars, a generation apart, could be seen walking to the Institute together, lost in conversation, and then back to their respective homes for a late lunch every day (cf. Fig. 14.3). Their personalities were very different: Einstein was outgoing, humorous, musical, and friendly, although he clearly was often preoccupied with his 'inner world'. He at various times suggested that he 'had become a scientist to escape the trivial banality of everyday life'.[10]

Kurt Gödel was born just after Einstein's '*annus mirabilis*' of 1905, when the latter published four articles that guaranteed his place in the history of science, still a young patent officer 3rd class who had not yet completed his doctoral thesis. (Two of them were on his Special Relativity Theory (SRT), which revolutionized the way that physicists look at space and time, and the relation of mass to energy, abolishing the concept of *absolute time* and *absolute space* forever; a third was the quantum explanation of the photoelectric effect, for which he was later awarded the Nobel prize; and the fourth dealt with the microscopic structure of matter, the motions of atoms, and diffusion and transport; this became his doctoral dissertation). By the time that Kurt Gödel matriculated at the University of Vienna in 1924, Einstein was world famous, having completed his General Relativity Theory (GRT) in 1915, today still the best theory of gravitation, and even more revolutionary than SRT.

And yet, the works of the young Kurt Gödel in the next 7 years after his matriculation were just as revolutionary with respect to the fundamentals of mathematics as those of Einstein had been to fundamental physics in 1905– 1915. Einstein was 26 in 1905, when he burst upon the world of theoretical physics, while Gödel had not yet turned 25 when his 'most famous paper' was published in March of 1931. But it was certainly not their respective scientific/mathematical achievements that formed the basis of their unusual

Fig. 14.3 Albert Einstein and Kurt Gödel, 1954 in Princeton. Photo by Richard Arens, courtesy of AIP *Emilio Segrè Visual Archives*[11]

friendship. That might have been the case for example between Gödel and von Neumann; *their* friendship was based on mutual respect and admiration.

With Einstein, there was evidently something deeper, a mutual affinity that transcended their scientific work, as important as that was to each of them. Gödel himself said that he differed from Einstein in many of his opinions, and he was not afraid or hesitant to say so and to defend his own positions; and that was pleasing to Einstein, who was accustomed to overblown respect and even awe from other people, who would hypocritically refuse to express their true opinions or challenge his views. And that was no doubt true, but there were certainly other factors that contributed to their spontaneous affinity. It has been suggested[8] that their origins in the southern part of the German-speaking world, or also their very early scientific successes, were important factors—a common basis for their conversations.

An aspect of this which deserves emphasis is their *language*, specifically the dialect with which they grew up. Although Einstein was born in Ulm, in the region of Alemannic-Swabian dialects, his family moved to Munich when he was three, and it was there that he spent most of his childhood and his school years. He left at the age of 16, in 1895. Gödel was born 11 years later in

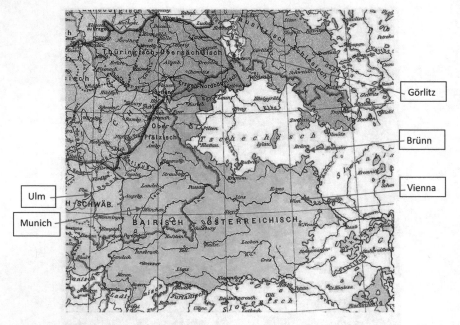

Fig. 14.4 Map of southeastern Germany and Austria, showing the distribution of various dialects. *Karte der Deutschen Mundarten*, from *Brockhaus Konversation-slexikon (1894)*, 14th edition. Map used under a CC A-SA license[12]

Brünn, and spent his childhood and school years there, as we know. Munich and Brünn belonged to the same dialect region, and although there were of course local variations, both Einstein and Gödel grew up in an environment where a similar dialect was spoken, *Mittelbairisch* (Middle Bavarian). Einstein was fond of his dialect and occasionally used it in correspondence with people who would understand it.

The map in Fig. 14.4 shows the geographical distribution of various German dialect families in Southeastern Germany and Austria in 1894. Einstein's birthplace, *Ulm*, belongs to the Alemannic-Swabian region, while Munich, Brünn and Vienna are all within the Bavarian-Austrian region—specifically, the *Mittelbairisch* region. Note that the *Sudetenland* (labeled '*Erzgeb(irge)-Nordböhmisch*' on the map) is a distinct dialect region, belonging to the *Ostfränkisch* (East Franconian) dialect family, while Moravian Silesia is in the region labeled '*Lausitzisch-Schlesisch*' ('Lausitzian-Silesian') and thus again had a distinct dialect, belonging to the Silesian family.

While this connection should not be overly belabored, it certainly contributed to their affinity. Many of the other European refugees in Princeton, even those whose mother tongue was German—e.g., von

Neumann or Weyl—came from different linguistic regions and their accents/dialects would not have evoked childhood recollections in Einstein or Gödel.

And others, such as Morgenstern, were too worldly and too involved with their ongoing projects, very much anchored in the 'external' world, to have the time and patience for long talks with either Einstein or Gödel. Morgenstern was of course also a close friend of Gödel's, but on a different level, a more social and less introspective/intellectual one. Morgenstern in any case also grew up, at least until the age of 12, in a different dialect region: his hometown, *Görlitz*, is in the 'Lausitzian-Silesian' region (cf. Fig. 14.4). Both Einstein and Morgenstern evidently felt the need to 'take care' of Gödel, to help him deal with his new environment and not to become involved in endless bureaucratic snarls.

An aspect which is emphasized by Goldstein,[9] and which may have fascinated Einstein, was Gödel's insistence—referred to as his 'governing axiom' by E.G. Straus, one of Einstein's later assistants—on the existence of a *logical explanation for every fact in the universe*. In other words, he believed, in an even more rigorous way than Einstein, that the world is fundamentally *understandable* using logic, mathematics and reason. Einstein, some years before, had remarked that the most surprising aspect of the world is that it is comprehensible.[13] In fact, it would appear that Gödel had a deep-seated *need* for the world to be comprehensible in those terms [cf. Chaps. 9 and 15, final section].

Freeman Dyson, who was a Fellow of the IAS in 1948/49 and later a permanent member (from 1953 onward), observed that "*Gödel was ... the only one of our colleagues who walked and talked on equal terms with Einstein*". And Einstein himself famously remarked (he was quoted by Oskar Morgenstern), "that he only went to the Institute '*um das Privileg zu haben, mit Gödel zu Fuss nach Hause gehen zu dürfen*' (only—'to have the privilege of being allowed to walk home with Gödel')". At the same time, he said that his own work was no longer of great importance.

Einstein had been attempting to find a *unified theory* that would combine gravitation (i.e., GRT) with electromagnetism. By the late 1940s, when he made these remarks, that would not have been a truly fundamental theory, even if he had been successful. This was also another common aspect of both Gödel's and Einstein's work at the time their closer friendship began, and perhaps a factor that united them—each was struggling with a seemingly impenetrable wall in his chosen problem. Gödel could not find the final proof of the independence of the General Continuum Hypothesis, although he felt himself to be very near to it; and Einstein had been working for nearly 20 years, unsuccessfully, on finding a 'unified theory' by 1942, when his deeper friendship with Gödel began, and he may have begun to doubt that he would ever be successful in finding it.

[When Einstein started his unification program in the early 1920s, gravitation and electromagnetism were the only known 'fundamental interactions' which govern all of physics. But by the second half of the 1940s, it was becoming clear that there were (at least) two other fundamental interactions, the 'nuclear' interactions, now called the *strong force*, responsible for binding the subnuclear particles together, and the component particles within atomic nuclei; and the *weak interaction*, which gives rise to a whole series of decays and transformations of nuclei and elementary particles. Ironically, a unified theory of those latter three interactions, electromagnetism and the weak and strong forces, was achieved by the 1970s (now called the *Standard Model of Particle Physics*); in it, the electromagnetic and weak (unified) interactions are described by a quantum field theory, the *electroweak theory*, and so is the strong interaction (by quantum *chromodynamics*). But *gravitation*, Einstein's original contribution, has still not been combined with the others—the search for 'quantum gravity' continues to this day].

A longer quote from Albert Einstein may shed some more light on his friendship with Gödel; it is from his address to the *Physical Society in Berlin* (PGzB) on the occasion of Max Planck's 60th birthday on April 26th, 1918 (we already quoted it briefly above). He began his speech with the following words[14]:

> In the temple of science are many mansions, and various indeed are they that dwell therein and the motives that have led them thither. Many take to science out of a joyful sense of superior intellectual power; science is their own special sport to which they look for vivid experience and the satisfaction of ambition; many others are to be found in the temple who have offered the products of their brains on this altar for purely utilitarian purposes. Were an angel of the Lord to come and drive all the people belonging to these two categories out of the temple, the assemblage would be seriously depleted, but there would still be some men, of both present and past times, left inside. *Our Planck is one of them, and that is why we love him.*
>
> I am quite aware that we have just now lightheartedly expelled in imagination many excellent men who are largely, perhaps chiefly, responsible for the buildings of the temple of science; and in many cases our angel would find it a pretty ticklish job to decide. But of one thing I feel sure: if the types we have just expelled were the only types there were, the temple would never have come to be, any more than a forest can grow which consists of nothing but creepers. For these people any sphere of human activity will do, if it comes

to a point; whether they become engineers, officers, tradesmen, or scientists depends on circumstances. Now let us have another look at those who have found favor with the angel. Most of them are *somewhat odd, uncommunicative, solitary fellows, really less like each other, in spite of these common characteristics, than the hosts of the rejected*. What has brought them to the temple? That is a difficult question and no single answer will cover it. To begin with, I believe with Schopenhauer that *one of the strongest motives that leads men to art and science is escape from everyday life with its painful crudity and hopeless dreariness*, from the fetters of one's own ever-shifting desires. A finely tempered nature longs to escape from personal life into the world of objective perception and thought; this desire may be compared with the townsman's irresistible longing to escape from his noisy, cramped surroundings into the silence of high mountains, where the eye ranges freely through the still, pure air and fondly traces out the restful contours apparently built for eternity. [emphasis added].

At the time of this speech for Planck's birthday, Einstein was 39 and had accomplished essentially all of his revolutionary theoretical achievements. While he still published a number of important works in later years, his fame would not be significantly diminished if those had never appeared. He also belonged to that minority of scientists who would have been 'left inside the temple'. And so did Gödel. Indeed, the adjectives '*somewhat odd, uncommunicative, solitary*' apply more precisely to Gödel than to either Einstein himself or to Planck; '*… and that is why we love him…*'.

The Later 1940s

Whether Kurt Gödel felt at home in his new location, where in the end he spent his last 38 years—more than half of his life—whether he ever really 'arrived' in America, is a matter of controversy. *Stephen Budiansky* and *Rebecca Goldstein*,[15] Americans by birth, judge this question rather differently. Budiansky maintains that he did adapt well and found his new home to be a desirable place to live and work, a place where he *belonged* (in contrast to his wife Adele, who never really felt at home in Princeton and apparently was not accepted there by many of the Princetonians). Goldstein, in contrast, finds that he was in exile in Princeton, not only the external exile from his country and language, but a local exile within the Institute, where mistrust of his 'eccentricities' by his colleagues left him isolated (and perhaps for that reason scientifically unproductive, especially after about 1950). In a similar vein, *Verena Huber-Dyson*, a European by birth and education, although a longtime resident of Canada and the USA, who spent her latter years in

Berkeley/CA, believed[16] that he '*probably never found a home in Amerika.*'
(Her use of the German spelling is ambiguous; it of course refers to Gödel's
native language and implies that his origins remained more important to
him than his adopted home. But it also refers to a common ironic usage
(especially in Berkeley!) of that spelling to suggest that the USA was showing
autocratic and despotic, even fascist tendencies; that usage originated during
the Vietnam War in the 1960s). *Georg Kreisel*[17] takes a different approach:
he reserves judgment about whether Gödel was actually 'at home' in America,
but he quotes Gödel's defense of his adopted country in practically every letter
that he wrote to his mother in Vienna over 20 years from 1946 to 1966. (She
of course was somewhat perplexed by Kurt's passive refusal to come to Austria
even for a visit, and his rejections of all the honors and recognition which
various Viennese and Austrian organizations tried to give him after the War).

The Gödels applied for and received U.S. citizenship in 1947/48. Kurt's
citizenship hearing was scheduled for December 5th, 1947, and in typical
fashion, he prepared himself months in advance and in great detail, learning
obscure aspects of local government and studying the U.S. constitution. In
the latter, he found a legal/logical loophole which would permit the country
to become a dictatorship *legally*, and mentioned it to Morgenstern and
Einstein, who were to accompany him to the examination. They both advised
him not to bring it up during the questioning, and Einstein tried valiantly
to distract him while they were driving to the hearing. Dawson recounts
the occasion, based on Morgenstern's diary entries. Einstein jokingly said to
Gödel that he was taking his next-to-last test[18] (the German word '*Prüfung*',
which he undoubtedly used, means not only 'test' or 'examination', but also
'trial' in at least two of the many senses of that word. 'Test' misses some
of that). Gödel asked, "*What do you mean, next-to-last?*"—to which Einstein
replied, "*Very simple. The last will be when you step into your grave*". The story
of how Gödel nevertheless started out to explain to Judge Philip Forman,
who administered the hearing, how the U.S. constitution might permit the
country's becoming a dictatorship, is legendary and need not be repeated
here. Just what Gödel had in mind has never been discovered; there has been a
suggestion[19] that he might have meant that while amendments to the constitution must pass a high barrier, amendments *to the amendments* are, in theory,
much easier to pass (specifically, to Amendment V). His suggestion that the
USA might become a dictatorship must have seemed weirdly unrealistic and
unthinkable in 1947, but as of this writing (2022), it seems less improbable.[20]

Adele had her hearing a week later, since she had not yet returned from
Austria on December 5th. Both were formally granted US citizenship the

following April 2nd. It was Kurt Gödel's fifth citizenship, after the Austro-Hungarian Empire, the Czechoslovakian Republic, the Austrian Republic, and the *Deutsches Reich* (the first, second and fourth by birth and the accidents of war and politics; Austria and the USA by his own choice). In this, his history was similar to that of Einstein, who had become an American (U.S.) citizen eight years earlier, officiated by the same Judge Forman.

Private Life in Princeton

The most detailed published account of how the Gödels lived and what their preoccupations were in the later 1940s in Princeton is probably that given by Budiansky (2021), who relies in particular on the diaries of Oskar Morgenstern, a keen observer who was close to Gödel and sympathetic to him, even if at times skeptical of some of his stranger ideas. The immediate postwar period was of course difficult on both sides of the Atlantic, and it is remarkable how little Kurt understood of what went on in Europe in the last 5 years of the 'Third Reich', and equally remarkable how little his mother and brother understood of Kurt's renown as a mathematician, scientist and philosopher. This became apparent during Morgenstern's journey to Vienna in the autumn of 1947, when he also visited Kurt Gödel's mother and brother there.

An important event for the Gödels during the later 1940s was the purchase of their own house in Princeton. Adele discovered a property which she desperately wanted to buy, and Kurt supported her wishes, perhaps because he realized that life was not easy for her in provincial Princeton. Morgenstern, an economist, was skeptical of the correctness of its price ($12,500), which he felt to be too high; and Gödel had to take out a mortgage and also to borrow an advance on his salary from the Institute (which was approved by its new director, J. Robert Oppenheimer).

In the event, the house proved to be very satisfactory for both of them, and Adele in particular spent many (evidently happy) hours improving it and decorating the house and garden to her own taste, which ran to *Kitsch*—a fact that was jeered at by many more 'sophisticated' observers,[22] but was indeed shared by Kurt himself. The purchase of their house, in the northeast corner of Princeton on Linden Lane (Fig. 14.5), in a 'less fashionable' neighborhood (cf. the map in Fig. 10.3) was, like his marriage to Adele, a decision which may have seemed dubious to many people at the time, but which was in the long run an excellent move by Kurt Gödel.

Fig. 14.5 The Gödels' house at 145 Linden Lane (originally 129; the number was later changed). *Photo* 1950, private archive[21]

An Excursion into Physics

In the years 1946–1950, Gödel occupied himself with theoretical physics, the field that he had originally intended to study. Specifically, he searched for, and found, a new—and novel—solution to Einstein's field equations, themselves first published in 1915. (Gödel's solution—for a *rotating* universe— was however not entirely novel—cf. Rindler (2009), who points out that Cornelius Lanczos *almost* anticipated Gödel's Universe in 1924). It may seem surprising that new solutions to the Einstein equations were still being found more than 30 years after their initial publication; but indeed, even today, new solutions are still being found. *Roger Penrose* received the Nobel prize in physics in 2020 for a solution that he obtained and published in 1965, which demonstrates that not only *can* a collapsing massive star create a 'black hole' (this possibility was suggested already in 1916 by Karl Schwarzschild's early solution of the Einstein equations, showing a *gravitational singularity* at a point mass in spherical symmetry [see Schwarzschild (1916)], and underscored in 1939 by the paper of Oppenheimer and Snyder [q.v.] on 'gravitational collapse'); but that it *must* do so under the proper circumstances, and not only in idealized geometries [as in Schwarzschild (1916)].

Just what motivated Gödel to take up this challenge, rather far removed from his work in mathematical logic and in philosophy, is not certain; for him, it undoubtedly had a strong philosophical aspect, namely the conception of *time* in General Relativity, and its comparison to Kant's ideas—but it is also telling that he published it in the *Festschrift* for Einstein's 70th birthday in 1949 [Schilpp, ed. (1949]. We will take a closer look at this unique and significant 'sidestep' of Gödel's in the following chapter.

Notes

1. See E.-M. Engelen on '*Rudolf Carnap and Kurt Gödel...*', Engelen (2021).
2. *ibid.*, p. 224.
3. Cf. the informational page of the University of Vienna on Morgenstern, at: https://geschichte.univie.ac.at/en/persons/oskar morgenstern prof-dr-rcr-pol-dr-jur-hc. Consulted on Nov. 25th, 2021.
4. Photo: *Universität Wien*, picture archive; by *Photo Dietrich, Wien*; public domain. Courtesy of the Uni Wien Picture Archive; reused from: https://geschichte.univie.ac at/en/node/28208
5. See Crocco (2013). On the points mentioned in this quote, see also the *Collected Works of Kurt Gödel*, Volume I, Introductory note to Gödel's [1931]; and also Huber-Dyson (2005).
6. Wang (1997), p. 87.
7. See Graham (2018). The story of how Paul Oppenheim introduced Gödel to Einstein is told there, without giving a date or naming a place. Dawson (1997) and Yourgrau (2005) also mention this event, evidently from a different source. They both suggest that it took place in 1933, when Gödel was first in Princeton. That, however, is contradicted by the life history of Oppenheim: He indeed left Germany in 1933, but he went to Belgium and lived in Brussels for more than 5 years, until he emigrated with his family to the USA in 1939, settling in Princeton, where he was Einstein's neighbor and became his friend. Thus, he could not have introduced Einstein and Gödel before the latter arrived in March of 1940. And the event would have taken place in Fuld Hall, occupied by the IAS after its completion in the autumn of 1939, and not in Fine Hall. Gödel and Einstein did meet during Gödel's first Princeton visit in 1933/34, but were then probably introduced by von Neumann, who knew both personally; or by Abraham Flexner or Oswald Veblen.
8. Budiansky (2021, p. 216/17) cites a remark made by *Paul Oppenheim* (1885–1977), a chemist-philosopher in Princeton who had emigrated from

Germany and was a neighbor and friend of Albert Einstein's (compare the previous backnote). He introduced Gödel and Einstein to each other at Fuld Hall in 1940, and he held that act to be 'his only scientific contribution'. But their close friendship apparently began later, around 1942. They had in fact met for the first time in 1933, when Gödel was first in Princeton.

9. There are varying opinions about the reasons for the unquestionably close friendship between Gödel and Einstein; see e.g. Budiansky 2021, pp. 217 ff.; Goldstein 2005, pp. 33 ff.; Huber-Dyson 2005.

10. For example, in his address to the *Physical Society* in Berlin for Max Planck's 60th birthday in 1918. He was quoting Schopenhauer on '*everyday life with its painful crudity and hopeless dreariness*'. Cf. backnote [14].

11. Photograph by Richard Arens, courtesy of AIP *Emilio Segrè Visual Archives*, cropped. Reused from: https://repository.aip.org/islandora/object/nbla%3A2 90524.

12. Map used under the Creative Commons Attribution-Share Alike 3.0 Unported license. Reused from: https://commons.wikimedia.org/wiki/File: Brockhaus_1894_Deutsche_Mundarten.jpg.

13. The correct Einstein quote is, "*The eternal mystery of the world is its comprehensibility…The fact that it is comprehensible is a miracle.*" It appears in an essay by Einstein on '*Physik und Realität*' ('*Physics and Reality* '), published in the *Journal of the Franklin Institute.*, Vol. 221, No. 3 (1936), pp. 313–447.

14. The English translation of Einstein's 1918 address can be found in the *Collected Papers of Albert Einstein* (CPAE), Volume 7, English translation supplement, '*The Berlin Years: Writings 1918–1921*', p. 42 (Princeton University Press, 2002), translated by Alfred Engel. Online at: http://press.prince ton.edu/titles/7185.html.

15. See Budiansky (2021) and Goldstein (2005).

16. Huber-Dyson (2005). See also Huber-Dyson (1991).

17. Kreisel (1980).

18. Dawson (1997, 2005 edition, p. 180). See also Pais (1982), p. 453, and Budiansky (2021), p. 232.

19. See e.g. Guerra-Pujol (2013).

20. In this connection, a recent *New Yorker* comment is pertinent: See the article by Adam Gopnik, Nov. 26th, 2021 issue.

21. Photo: private archive (photo dated 1950). Copyright status unknown, fair use. Reused from: https://is.muni.cz/www/jannovotny/o_case/aaGodel-Brno-Cas-Vesmir2018.pdf.

22. See e.g., Goldstein 2005, pp. 207/208.

15

A Bizarre Birthday Present: *Gödel's Universe*

Einstein's Relativity Theory

The roots of relativity theory go far back—perhaps to Galileo Galilei in the seventeenth century, perhaps to James Clerk Maxwell in the mid-nineteenth century, but at least to Ernst Mach in his book '*Die Mechanik in ihrer Entwicklung*' (1883, translated as '*The Science of Mechanics*'), whose ideas were to some extent anticipated by Bishop Berkeley (early eighteenth century). At the very latest, they date from 1895, the year that Albert Einstein turned 16, and when he decided to leave Munich, where he had been left to finish the *Gymnasium* after his family had moved to Milan. Einstein later said that this was the age at which he started trying to imagine what it would be like to chase a light beam at nearly the speed of light, or to hold a lantern while moving at that speed. (In Fig. 15.1, we see Einstein at about 40, 25 years after his insight about light propagation, with his sometime mentor Hendrik A. Lorentz).

The *Principle of Relativity* certainly goes back to Galilei, and no doubt had even earlier precedents. It is a simple idea, almost obvious: every observer should see the same laws of physics, regardless of his or her state of motion; or in more modern language, the laws of physics should be invariant with respect to the frame of reference from which they are observed.

The *constancy of the speed of light* is a newer concept, which became apparent from James Clerk Maxwell's theory of electromagnetism, around 1865. It predicted the existence of electromagnetic waves, traveling at the constant speed c. In Maxwell's theory, c is determined by two other constants

© The Author(s), under exclusive license to Springer Nature Switzerland AG 2022, corrected publication 2023
W. D. Brewer, *Kurt Gödel*, Springer Biographies,
https://doi.org/10.1007/978-3-031-11309-3_15

Fig. 15.1 Albert Einstein and Hendrik Antoon Lorentz (for whom the *Lorentz transformations* are named), at the entrance to Ehrenfest's house in Leiden/NL, in 1921. Photo by Paul Ehrenfest (1921); from Wiki Media[1]

of nature, the 'electric field constant' ε_0 (originally known as the *permittivity of vacuum*) and the 'magnetic field constant' μ_0 (*the permeability of vacuum*). Since these two constants are simply properties of space itself ('the vacuum'), they should be universal throughout the cosmos, and thus so should the speed of light c (in vacuum). However, this particular aspect of c as the propagation velocity of electromagnetic radiation is not the reason for its Einsteinian property as a universal constant in all frames of reference—that derives from a more general aspect of c, as we shall see.

Einstein, in 1905, combined those two principles—relativity and the universal constancy of c—and arrived at his *Special Relativity* Theory (SRT).

'Special', because the theory describes how measured properties can vary for different observers traveling at different *uniform velocities*, i.e., at constant speeds along straight-line paths, with no accelerations, neither along their directions of travel nor perpendicular to them. The frames of reference of such observers are called 'inertial systems' or *inertial frames*. A typical example, used by Einstein himself, is that of one observer standing on the platform of a railroad station (the 'rest frame') and another in a passing train, moving at a precisely constant speed along a completely flat and perfectly straight track (the 'moving frame'). Those terms were put in single quotes because they are in fact *relative*; whether it is the train that is moving and the platform which is at rest is a matter of definition (although it is an automatic convention in everyday speech).

When the two observers compare their measurements of time intervals and the lengths of objects along the direction of relative motion, they will find differences, in such a way as to precisely conserve the constancy of the speed of light in both frames. This is described mathematically by the *Lorentz transformations*, which define the changes in space and time coordinates on going from one inertial frame to another, moving at a different speed (called a 'boost' in relativity theory). In addition to lengths and time intervals, a third property thought to be unvarying in classical physics, the *mass* of an object, is also found in SRT to vary, depending on its motion. The 'rest mass', i.e., the mass as measured in a reference frame where the object is at rest, is indeed an unvarying property of that object in SRT; but in a frame in which it is *moving*, its observed mass increases, depending on its speed. As the speed approaches c, the observed mass tends toward infinity (and thus also the energy required to accelerate it still further, so that it can never be accelerated to a velocity equal to—or greater than—c). Mass and energy are related by Einstein's famous equation $E = mc^2$, given in his second publication on relativity in 1905. This results from the essential connection between space and time in SRT.

Thus, SRT leads to a new view of space and time—the old, Newtonian view of *absolute* space and *absolute* time, fixed (and *independent*) axes which provide a background for processes occurring in the universe, became obsolete. After Albert Einstein's two publications on SRT, his former mathematics professor in Zurich, *Hermann Minkowski*, formulated the new interrelation between space and time as a four-dimensional *spacetime*, a mathematical construct called 'Minkowski space'. (Einstein reputedly said that he no longer understood his own theory after Minkowski had finished with it; and Minkowski, for his part, supposedly said of Einstein, "*So eine schöne Sach' hätte ich dem Kerl nicht zugetraut.*" ("*I wouldn't have thought that guy could ever be capable of producing such a beautiful thing*"!).

While he was writing a review article on SRT in late 1907, Einstein came to the realization that he had to continue working on relativity theory; it would be completed only when he had *generalized* it to include frames of reference moving with *arbitrary velocities*, i.e., including *accelerations*. (These can be accelerations in the direction of travel, causing an increase or decrease in speed, or also perpendicular to that direction, giving rise to a curved path for the motion).

Einstein soon realized that an essential basis for this extended theory (the *General* Relativity Theory, GRT) was the *equivalence principle*, which says, most simply stated, that *inertial* mass and *gravitational* mass are equivalent, and—for all practical purposes (FAPP)—indistinguishable. In other words, a person in a closed compartment cannot distinguish through any physical measurements or observations whether the force pressing them down toward the floor is due to a gravitational field (the compartment is sitting at rest on the Earth's surface, for example), or due to inertia in an accelerated frame (the compartment is far out in space, being pushed by a rocket engine, i.e. accelerated in the 'upward' direction, with an acceleration just equal to the acceleration of gravity on the Earth). This of course meant that he had to include *gravitation* in his more general theory. SRT deals directly only with the motions of inertial frames and with electromagnetic phenomena. Abraham Pais, in his (1982), Chap. 9, gives a detailed account of how Einstein arrived at these conclusions (the chapter is called '*The happiest Thought of my Life*', which is how Einstein characterized his realization that the equivalence principle would be the basis of *General Relativity*, as the constancy of *c* had been for SRT). The motto of that chapter is '*Muss es sein? Es muss sein.*' ('Must it be? It must be.'—also the motto of Beethoven's Opus 135, his String Quartet No. 16 in F major). So, Einstein set out with hope and perhaps a little trepidation on an intellectual journey that would take 8 years and considerable 'blood, sweat and tears' before reaching its conclusion.

Based on his early ideas, Einstein was able to deduce some physical (and, he hoped, observable) consequences of General Relativity: (i) Light rays will be 'bent' in a strong gravitational field (near a large mass, for example the Sun). In fact, the light rays themselves are not *bent*; rather, they follow *geodesics*, the shortest paths between two points. That had long been known in classical physics (Fermat's principle). In General Relativity (which we abbreviate as 'GR'), the geodesics—even in vacuum—are in general not straight lines, as in Euclidean space; spacetime is curved, and the degree of its curvature is determined by the mass distribution in a given region. Where there is a large mass (e.g., near a star), the curvature is greater, and this is the origin of the 'bending' of light passing near the Sun (and of the 'gravitational lensing' effect of distant galaxies, now well established observationally). (ii) The precession of the orbit of the planet Mercury can be precisely explained. Mercury's elliptical orbit changes its orientation with respect to the Sun in a regular manner; its long axis rotates gradually around the Sun in a way which cannot be explained by classical (Newtonian) physics. GR solves this problem (which Einstein realized in a qualitative way in 1908; but it took several more years to achieve quantitative precision). (iii) There is a *gravitational red shift*: Clocks tick more slowly in a strong gravitational field, and spectral lines emitted near the surface of the Sun are 'red-shifted' relative to lines from the same atoms in a terrestrial laboratory or in a diffuse gas in interstellar space. This is because the non-simultaneity of events occurring at distant points in moving frames, known from SRT and due to 'time dilation', has a gravitational analogue in GR (equivalence principle!).

In early 1908 Einstein, who was still working at the patent office in Bern, submitted his *Habilitationsschrift* to the University of Bern (he had received his doctorate from the University of Zurich in 1905 for his work on molecular motion and diffusion). By this time, he had published 17 articles, several of which were later among the most cited papers in physics history. His *Habilitationsschrift* was accepted in record time, and on February 28, he was granted the *venia docendi* by the University and became a *Privatdozent* there. He gave only two courses in Bern, in the summer of 1908 and the winter semester 1908/09. In 1909, he published two important papers on the quantum theory of light, showing that the fluctuations of 'black-body' radiation contain both a quantum part (as if the radiation consisted of discrete particles) and a classical part (the fluctuations due to wave interference).

By now, he had acquired a reputation as a 'rising star' in theoretical physics, and he was hired to a newly-created associate professorship at the University of Zurich, beginning in the fall of 1909. He also received (his first) honorary doctorate, from the University of Geneva, in the summer of 1909. But he found his teaching load in Zurich onerous, and early in 1911, he accepted the offer of a full professorship in Prague, at the *Karlsuniversität* (officially the *'Karl-Ferdinand-Universität Prag'*), where he would have a lighter teaching load. During the years 1908–1912, he published actively on various topics, in particular the 'old' quantum theory, to which he made important contributions. But in the background, he was still thinking about general relativity, although he initially published nothing on the subject, and the quantum theory was perhaps uppermost in his mind. He famously took a visiting colleague in Prague to the window of his office, which looked out on the garden of an adjoining psychiatric clinic, and remarked, "*There you see that portion of the lunatics in Prague who are* not *working on the quantum theory*".

About Einstein's time in Prague,[2] Pais [(1982), p. 193] says the following: "*It is mildly puzzling to me why Einstein made this move. He liked Zurich, Mileva liked Zurich. He had colleagues to talk to and friends to play music with. He had been given a raise...* ". Pais quotes Otto Stern, who was Einstein's assistant from 1912–1914, in Prague and later back in Zurich: "*At none* [of the four universities in Prague] *was there anyone with whom Einstein could talk about the matters which really interested him ... he was completely isolated in Prague*". Nevertheless, he published his first paper specifically on general relativity and gravitation in June of 1911, shortly after arriving in Prague. It was called '*On the Influence of Gravitation on the Propagation of Light*'. Pais again (pp. 194/195): "*... he used once again the approximate methods of 1907. Thus in 1911, the three coordinate systems S, Σ, and S'* ... *reappear* [S is the 'rest' system, where a gravitational field is present; S' is an inertial system moving at constant velocity relative to S, so that they are related by a Lorentz transformation; Σ is an accelerated system which coincides at a particular moment with S'. The system S' is used to estimate clock readings in Σ in a very short interval after they coincide, an approximate procedure]. ... *In 1911, the four main issues were the same as in 1907: the equivalence principle, the gravity of energy, the red shift, and the bending of light. However, Einstein now had new thoughts about each one of these four questions*".

Pais then gives a brief summary of Einstein's approximate results for the four issues mentioned. He (following Einstein) uses semiclassical arguments based on the known effects in SRT and on the equivalence principle, combined with Newtonian gravitational theory. Gravitation can be described classically in terms of a *vector field* $\mathbf{g}$ $(\mathbf{r})$ which has a direction and a magnitude at every point $\mathbf{r}$ in space. Multiplying the field by the mass m of a test object gives the *force* of gravity (direction and strength) acting on the object at that point. One can also define a gravitational *potential* $\varphi(\mathbf{r})$, which is a scalar quantity that gives the (gravitational, potential) energy of the test object at the point $\mathbf{r}$ when multiplied by m. The *gradient* of the potential at $\mathbf{r}$ (spatial derivative in three dimensions, pointing in the direction of the greatest rate of change) yields the *field* at that point. These quantities are familiar from electrostatics, where the description is analogous. As an example, we consider the *gravitational red shift*.

In the accelerated frame Σ, consider a light source at a point $\mathbf{r}_1$ and a detector at point $\mathbf{r}_2$, separated (along the direction of motion of the frame) by a distance d. The source emits a light beam of frequency ν_1. Its frequency (as observed in S′) will be Doppler shifted due to the relative motion of the source. By the time the light is detected, the relative velocity will have changed, due to the acceleration of frame Σ, so the detector (again observed from S′) will register a different frequency, v_2. The difference in *velocities* v = $v_2 - v_1$ is given to a good approximation by at, where a is the acceleration of the frame Σ and t is the (very brief) elapsed time between emission and detection of the light, i.e., $t = d/c$. Applying the (linear) Doppler formula for the frequencies, we find:

$$v_1 = v_2(1 + v/c) = v_2\left(1 + ad/c^2\right). \tag{15.1}$$

This frequency shift is due to the (accelerated) motion of Σ and is thus an *inertial* effect. From the equivalence principle, in the rest frame S where a gravitational field is present, there must be an analogous effect due to *gravity* (principle of relativity + equivalence). It will then be given by an analogous formula:

$$v_1 = v_2\left(1 + \varphi/c^2\right), \tag{15.2}$$

where φ is the difference in gravitational potentials between the points r_1 and r_2: $\varphi = \varphi(r_2) - \varphi(r_1)$. Einstein used formula (15.2) to estimate the gravitational red shift on the Sun as compared to the Earth (e.g., the shift Δv of the frequency v of a spectral line as observed in the solar spectrum and in a terrestrial laboratory) to be about $\Delta v/v \approx 10^{-6}$.

[After the publication of the complete GRT and the end of WWI, an astrophysical observatory with a solar telescope and spectrograph was built with Einstein's support at the observatory on the *Telegraphenberg* near Potsdam (now called the *Albert Einstein Science Park*) to verify this effect by measuring the shift $\Delta v/v$ for lines in the solar spectrum. It was initially operated by the astronomer *Erwin Findlay-Freundlich*, a friend of Einstein's. The tower built to house the telescope was designed by the architect *Erich Mendelsohn*, and it is now called the *Einsteinturm* (see Fig. 15.2). The original search for the gravitational red shift was unsuccessful due to turbulence in the solar atmosphere and the resulting Doppler line-broadening of the spectral lines; the effect was observed much later after the turbulence effects were well understood, and in other ways (e.g., by using nuclear radiation and resonant gamma-ray absorption, the *Mössbauer Effect*)]. The Potsdam solar telescope is still in use today, mainly for instrument testing and for training students.

After this first GR publication, Einstein worked more and more intensively on general relativity. He was not happy in Prague, nor was his family,

Fig. 15.2 The *Einsteinturm* (Einstein Tower) in Potsdam, housing a solar telescope and spectrograph (the latter is in the basement, protected from temperature variations by the earth walls). Photo courtesy of the *Astrophysikalisches Institut Potsdam*, September 21st, 2005[3]

and when offered a professorship at his *alma mater*, the *Federal Polytechnical Institute* (ETH) in Zurich, he accepted, leaving Prague after only 16 months, and moved back to Zurich in the autumn of 1912 (his wife Mileva and their two sons returned in July and found a suitable apartment). As we have seen in Chap. 14, Philipp Frank, one of the original members of the 'old' Vienna Circle, accepted the open position in Prague which Einstein had vacated, and remained there until 1938, when he emigrated to the USA, spending his later years at Harvard University.

Einstein had been unable to find the proper mathematical formulation for general relativity, where spacetime is in general not flat, like Minkowski space, but rather *curved* in response to the matter-energy density. The curvature in turn produces gravitation, which then affects the motion of the matter. He corresponded with his old friend *Marcel Grossmann*, by now a professor of mathematics at the ETH Zurich, and Grossmann not only supported his obtaining a professorship there, but also helped in his search for the correct mathematical description of GR. (The position that Einstein occupied at the ETH was previously vacated by his former mathematics professor, Hermann Minkowski, who 'completed' SRT. Minkowski had returned to Göttingen— one of those 'bizarre threads' which intertwine in our lives, according to

Gödel). This period in Einstein's career and his collaboration with Grossmann in 1912/13 are described from Grossmann's point of view in the scientific biography of Grossmann written by his granddaughter, Claudia Graf-Grossmann (2018); and from Einstein's point of view, they are described for example by Pais (1982) and by Fölsing (1993).

When he returned to Zurich and met Grossmann, Einstein reportedly said, "*Sie müssen mir helfen, Grossmann! Sonst werde ich verrückt!* " ("*Grossmann, you have to help me!—Otherwise, I'll go crazy!* "). This quote was saved for posterity by Louis Kollros, a colleague at the ETH, who later, in his obituary for Marcel Grossmann, wrote,[4]

In 1913, he *[Marcel Grossmann]* was pulled into a scientific spiral that gave him a great deal of work and considerable pleasure. His friend and fellow student Albert Einstein had already arrived at the theory which is today known as special relativity, but without using the tools of higher mathematics. But when he dared to proceed to general relativity and the theory of gravitation, he found himself confronted with such great mathematical difficulties that one day, he consulted his friend. Marcel Grossmann was able to show him that the mathematical tools for the new physics had already been developed in 1869 by Christoffel in Zurich, the founder and first dean of the mathematics and physics institute of the Swiss Federal Technical Institute *[ETH]*.

And Grossmann indeed helped. His own specialty was projective geometry, somewhat distant from differential geometry, which was what Einstein needed to describe the curved, four-dimensional spacetime of GR. But Grossmann knew where to look: He found what was required in the work of Gauss, Riemann and Lobachevsky, but in a difficult form; however, the founder of his institute at the ETH, *Elwin Bruno Christoffel* (1829–1900), who occupied the chair vacated by Dedekind in 1862 and published important results on differential geometry, proved to be the most fruitful source. Christoffel introduced *covariant differentiation* and the *Riemann-Christoffel tensor*, both of which were indeed relevant to Einstein's problem. Einstein and Grossmann collaborated on formulating GR for about 18 months, and published two joint papers, the first of which in fact contained the (almost) correct equations, but Einstein and Grossmann failed to recognize that at the time, and rejected them, believing that they would not reduce to the Newtonian limit. Einstein 'rediscovered' them in 1915, after moving to Berlin. Their first joint paper was entitled[5] "'Outline of a Generalised Theory of Relativity and of a Theory of Gravitation', *published in the* Zeitschrift für Mathematik und Physik *early in 1913. This paper contains a physical and a mathematical part;*

Albert was responsible for the physical part and Marcel for the mathematical part".

In the event, it took nearly another three years before Einstein finally found the solution to his mathematical and physical problems. In March of 1914, he accepted an invitation to move to Berlin, where he took a post with the *Prussian Academy of Sciences* and a joint professorship at the University of Berlin (with essentially no teaching duties). Neither his family nor his colleagues at the ETH were pleased by the move, but his colleagues in Berlin were persuasive, and he had private reasons for the move, as well. The last months of his efforts to complete GRT were hectic, a sort of cliff-hanger, in particular because he had delivered a lecture series at Göttingen in the summer of 1915, which aroused considerable interest in the problem and spurred Hilbert to look for a mathematical solution. In October and November of that year, Einstein gave almost weekly updates to the Prussian Academy, which duly noted them in its *Sitzungsberichte* (Meeting reports); but, formally at least, Hilbert essentially beat him to the conclusion—although by a different route, and undoubtedly motivated by Einstein's lectures in Göttingen. However, Hilbert quite fairly noted that, "*Every street urchin in Göttingen knows more Riemannian geometry than Einstein, but the physics that Einstein created is his own and original"*.

There has been an extended 'academic dispute' about whether Einstein or Hilbert deserves the final (or the complete) credit for GRT. Among many other sources, we can quote Ivan Todorov (2005), who in lectures in Trieste and Bremen proposed a wiser path. His written version ends with the words,

Einstein and Hilbert had the moral strength and wisdom—after a month of intense competition, from which, in a final account, everybody (including science itself) profited—to avoid a lifelong priority dispute (something in which Leibniz and Newton failed). It would be a shame to subsequent generations of scientists and historians of science to try to undo their achievement.

Theories of Gravity

Here, we summarize some important aspects of GRT without going into great detail or claiming any sort of mathematical rigor. Interested readers who wish to penetrate more deeply into its mysteries can find many sources—the classical introduction and textbook is *Gravitation*, by Misner, Thorne, and Wheeler (1973/2017).[6] The scientific biography of Einstein by Abraham Pais (1982) gives many details of his journey to GRT. A more recent textbook which is relatively accessible was written by *Wolfgang Rindler* (1924–2019), a theoretical physicist and educator originally from Vienna; see his (2001/2006).[7] We will also quote from some of his articles in the *American Journal of Physics* in the following. His (2009) gives a very clear introduction to Gödel's Universe. The story of Gödel's cosmological work is also told in some detail in the biography by *Palle Yourgrau* [Yourgrau (2005); see also his earlier publications (1991), (1999), listed under Item 24 in Appendix B].

The set of equations at which Einstein finally arrived on November 25th, 1915, relates the geometry of spacetime to the density of matter and energy. Since GRT deals with a four-dimensional, curved ('warped') Riemannian spacetime, the description of its geometry and of the matter/energy density requires *tensors*, which can be represented as 4×4 matrices $[A_{ij}]$. The *metric tensor* $g = [g_{\mu\nu}]$ describes the fundamental geometric structure of the space-time (μ and ν are indices that run over the four dimensions of spacetime and enumerate the components of g). It was introduced by Minkowski for his (flat) four-dimensional space of SRT, but it comes into its own in the curved spacetimes of GRT. There, it plays a role analogous to the gravitational potential in Newtonian gravitation (see below).

As we can see from the wall painting in Fig. 15.3, the equation(s) in their final form are given by

$$R_{\mu\nu} - 1/2 R g_{\mu\nu} + \Lambda g_{\mu\nu} = \left(8\pi G/c^4\right) T_{\mu\nu}. \qquad (15.3)$$

In the following, we shall try to see what this equation means and how Einstein arrived at it.

Fields and Potentials—Electrostatics and Newtonian Gravitation

We begin our considerations of GRT and Gödel's universe with a comparison of the mathematical description of Newtonian gravitation and of electro-statics—since the latter may be more familiar to many readers and their descriptions are analogous. The following table (Table 15.1) lists some quantities of interest and their mathematical representations in classical theory. Here, r is a position vector in three-dimensional (Euclidean) space, and q and m are (infinitesimal) electric charges/masses, respectively.

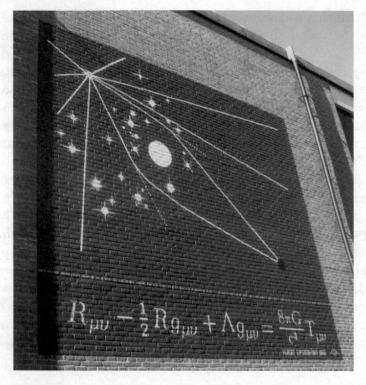

Fig. 15.3 A painting by the *Stichting Tegenbeeld* art project on a wall of the *Museum Boerhaave* in Leiden, NL (the home of *Hendrik Antoon Lorentz* during most of his life). The upper portion illustrates the gravitational lensing effect, below are the field equations of GRT (in their modern form), developed by Einstein. Photo by *Vysotsky*, July 13th, 2016, Wiki Media[8]

Table 15.1 Comparison of electrostatic and Newtonian-gravitational quantities

Quantity	Electrostatics	Gravitation
(Vector) *Field*	(Static) electric force field $\boldsymbol{E}(r)$	Gravitational field $\boldsymbol{g}(r)$
Force law	$F_E = q\boldsymbol{E}$	$F_G = m\,\boldsymbol{g}$
Potential	φ_E	φ_G
(Potential) Energy	$q\varphi_E$	$m\varphi_G$
Source equation	$\nabla.\boldsymbol{E} = \rho_E/\varepsilon_0$	$\nabla.\,\boldsymbol{g} = -4\pi G\rho_M$
Potential-field relation	$-\nabla\varphi_E = \boldsymbol{E}$	$-\nabla\varphi_G = \boldsymbol{g}$
Poisson equation	$-\nabla^2\varphi_E = \rho_E/\varepsilon_0$	$\nabla^2\varphi_G = 4\pi G\rho_M$

In the table, ρ_E and ρ_M are the electric *charge density* and the *mass density*, respectively. The gravitational force is only attractive, and there is only one kind of mass; in contrast, there are positive and negative electric charges and correspondingly attractive (for unlike charges) and repulsive (for like charges) electrostatic forces—thus the different signs in the Poisson equations. ε_0 is the electric field constant (SI units), and G is Newton's universal gravitational constant. ε_0 enters the basic force law as $1/4\pi\varepsilon_0$, while the Newtonian gravitational force is simply proportional to G.

The symbol ∇ ('nabla') stands for a three-dimensional space derivative, which is itself a vector: $\nabla \equiv \partial/\partial x\, \boldsymbol{i} + \partial/\partial y\, \boldsymbol{j} + \partial/\partial z\, \boldsymbol{k}$, where $\boldsymbol{i}, \boldsymbol{j}$, and $\boldsymbol{k}$ are unit vectors in the three space directions x, y, z. The operation $\nabla\cdot\boldsymbol{A}$ is called the *divergence* of the vector field $\boldsymbol{A}$ (abbreviated as *div A*), and it is a measure of how the lines of constant field strength behave: *div A* > 0 $\rightarrow$ divergent field lines; *div A* < 0 $\rightarrow$ convergent field lines; *div A* $= 0$ $\rightarrow$ parallel field lines (homogeneous field). Nabla operating on a scalar quantity gives the *gradient*: $\nabla f = grad\, f$ is itself a vector and points in the direction of greatest slope. $\nabla^2 \equiv$ div grad is the *Laplace operator*, often denoted simply by Δ. Electric charges are the *sources* (or sinks) of electric fields, masses are the sources of Newtonian gravitational fields. A third operation involving ∇, given as the *vector product* $\nabla \times \boldsymbol{A}$, is called the *curl* (here *curl A*) and is itself a vector that is perpendicular to both ∇ and $\boldsymbol{A}$. Incidentally, we use boldface italic ($\boldsymbol{A}$) to denote vectors, and non-italic capitals (A) to denote tensors.[9]

We begin with the notion of a *line element* (or 'interval')[10] within a given space(time). It is an infinitesimal 'distance' ds pointing in an arbitrary direction in the n-dimensional spacetime. For n $= 3$, flat Euclidean space, ds is easy to visualize. Figure 15.4 shows a rectangular prism in 3-dimensional space with axes x, y, z. Its edges are taken to be infinitesimals in the three space directions, dx, dy, dz. A face diagonal da, given by the line segment A'–C', is found from Pythagoras' law: $da^2 = dx^2 + dy^2$. The body diagonal, representing an arbitrary displacement in the space, is found using Pythagoras' law again for the triangle $A'C'C$ to be $ds^2 = da^2 + dz^2 = dx^2 + dy^2 + dz^2$.

Considering n $= 4$, the spacetime of SRT (Minkowski space), with three spacelike components and one timelike component, we need to make the 'scale' the same for all the components, and to decide on a sign convention.

Often, the timelike component is given as ct and is treated as an *imaginary* number ict, so that its square will be negative (a convention abbreviated as $(- + + \,+)$; the components are enumerated by the indices 0, 1, 2, 3).

Thus, the universal speed of light c becomes the *coefficient of time*, which relates the units of time to those of distance (e.g., seconds to meters, so that c

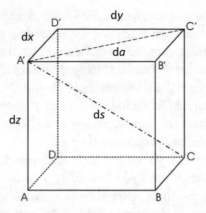

Fig. 15.4 A rectangular prism, showing a face diagonal A'–C', which we denote as da, and a body (or 'space') diagonal, A'–C, denoted by ds. *Drawing* WikiCommons, courtesy of the copyright holder[11]

is given in units of m/s. Physicists often choose a unit system in which c has the numerical value 1, so that it can be left off in the formulas). This is the real reason why c is a universal constant in SRT: it unifies the 'scale' of the four components of spacetime, and it can thus be considered to be a *property* or quality of the latter.

Summarizing: In flat Minkowski space, we can write the equation for a line element ds^2 as:

$$ds^2 = -c^2 dt^2 + dx^2 + dy^2 + dz^2 \qquad (15.4)$$

and we can make the notation more compact by introducing the *metric tensor* (in Minkowski space, it is often denoted by η, with components $\eta_{\mu\nu}$, rather than the generic notation g):

$$ds^2 = \eta_{\mu\nu}\, dx^\mu\, dx^\nu, \qquad (15.5)$$

where dx^μ etc. are generalized coordinates, and the Einstein convention is used: repeated indices (μ,ν) imply a summation. The matrix representation of the metric in this flat space is simple: it is a diagonal 4×4 matrix (all its off-diagonal elements are zero), with diagonal elements $-c^2$, 1,1,1. Corresponding coordinates can readily be given for spherical or cylindrical symmetry.

In GRT, the spacetime is *curved*, so the off-diagonal elements in the metric no longer all vanish. The curvature of spacetime is contained in the metric and is described by partial derivatives of the elements of the metric tensor with respect to the coordinates. After a somewhat involved procedure, Einstein was able to describe the geometry of curved spacetime in terms of the Ricci curvature tensor $R_{\mu\nu}$, which roughly speaking describes the local deviations of the metric from that of a flat Euclidean space. The second term on the left in the Einstein field Eq. (15.3) contains the *scalar curvature* or *Ricci scalar R*, which is derived from the Ricci tensor: $R = g^{\mu\nu}R_{\mu\nu}$—and is associated with volume changes in the curved space relative to a Euclidean space in the same locality. The first two terms in the field equations thus contain the information on the curvature of spacetime, and therefore determine the gravitational attraction acting locally; they are sometimes combined into an 'Einstein tensor' $G_{\mu\nu}$. The third term on the left as shown in Eq. (15.3), $\Lambda\, g_{\mu\nu}$, is the 'cosmological constant' term introduced by Einstein in 1917 to allow the field equations to describe a *stationary universe*, i.e., one that is neither expanding nor contracting (not 'evolving'). This term can be considered to represent the energy density of empty spacetime,[12] or 'the vacuum'. It has had a long and controversial history which is still unfolding, but it is generally considered to 'belong' to the field equations, even if the value of its coefficient Λ should turn out to be zero.

The right-hand side of the field equations, $(8\pi G/c^4)T_{\mu\nu}$, contains the *stress-energy tensor* $T_{\mu\nu}$, which is the GRT analogue of Newton's mass density term. Its contravariant form $T^{\mu\nu}$ represents the flux of the μ-th component of the relativistic energy-momentum four-vector through a surface of constant ν-th component. Roughly speaking, it contains the information on the density of mass and energy in a local region of spacetime, and it is the *source* of the curvature of spacetime in that region. This is the basic content of the field equations: an energy and/or mass density in a spacetime region produces curvature, and that in turn gives rise to gravitation in the region.

In GRT, the stress-energy tensor is *symmetric*, i.e., its off-diagonal elements above and to the right of the diagonal are identical to the corresponding elements below and to the left of the diagonal, so that it contains at most 10 independent elements (and not 16, as in the case of an arbitrary, non-symmetric tensor of second order). This means that the Einstein field Eqs. (15.3) are in fact at most 10 simultaneous differential equations; finding a solution to them is correspondingly hard.

In the following section, we will follow Gödel's path to a rotating universe as a solution of the Einstein field equations, as he described it in his lecture at the IAS in May of 1949. The Newtonian description of gravitation given above will be helpful, since it is a first approximation to what Gödel found, and it can be used to make his solution plausible—that is why he himself used it in his lecture.

Gödel's Universe—Relativity and Cosmology

Precisely how Kurt Gödel acquired his interest in a *rotating universe* as a solution to Einstein's equations, and in the corresponding cosmology, is not clear in detail.[13] He had a long-standing interest in GRT, as documented by various events: early in his studies at the University of Vienna, in 1925/26, he attended the course on relativity theory[14] given by Hans Thirring. Thirring himself was a contributor to relativity theory, having described the 'frame dragging' effect in rotating relativistic systems in 1918, together with the Viennese mathematician Josef Lense (the 'Lense-Thirring effect'; cf. Menger 1994). He probably did not go into details about that rather specialized effect in his survey course, however. Later, Gödel also attended lectures in the Mathematics Institute on differential geometry, required for the mathematical formulation of GRT. Furthermore, Gödel's sometime mentor Moritz Schlick was also an expert on relativity theory and occasionally lectured on it after arriving in Vienna. Karl Menger[15] recalled that:

> Gödel became interested *[in early 1932]* in the direct metric definition of curvature $\kappa(p)$ of an arc A at a point p in a general metric space... Mention is also made *[in Issue 5 of the Berichte]* of a discussion between Gödel, Wald, and myself on higher-dimensional coordinate-free differential geometry but no details about Gödel are given *[there]* or are now present to my mind ... I do remember that in 1932, Gödel began to study general relativity theory and to look for integrals of the field equations—a topic on which he published profound results 17 years later.

In Gödel's 'Working Notebooks' from the autumn of 1935 in Princeton, there are also pages on differential geometry, indicating that he was still interested at that time, although he was concentrating mainly on the mathematical logic of set theory (see Chap. 11). Still later, in his unpublished notes on philosophy ('Philosophical Maxims', '*Max Phil X*', 1943/44), quoted here from Audureau (2016), he wrote,[16]

> *Bem[erkung] (Phil[osophie]): Welches ist die richtige Auffassung der Zeit: (1.) die Zeit verläuft 'objektiv'. Wirklich ist nur die Gegenwart, die vergangenen Ereignisse sind nichts (nicht wirklich).*

> *(2.) die 'Einstein-Kantische Auffassung': das Vergehen der Zeit besteht in der Änderung unseres Gesichtspunkts, die vergangenen Ereignisse sind ebenso wirklich wie die gegenwärtigen".* [p. 23; This is just an excerpt from the remark].

Bem[erkung] (Phys[ik]): Zwei Auffassungen der vierdimensionalen Welt. Entweder 1. als etwas starr Existierendes [oder] 2. mit einer dreidimensionalen Ebene, die sich darin 'bewegt' (oder überhaupt nur dreidimensional). [p. 10].

English (translation by the present author):

Remark (Philosophy): Which is the correct interpretation of time: (1.) Time passes 'objectively'. Only the present is real; events in the past are nothing ([no longer] real).

(2.) The 'Einstein-Kant interpretation': The passing of time consists in the changing of our [the 'observer's'] point of view; past events are just as real as present events. [p. 23; This is just an excerpt from the remark].

Remark (Physics): Two interpretations of the four-dimensional world. Either (1.) as something which exists fixedly, [or] (2.) with a three-dimensional plane, which 'moves' within [that world] (or generally just three-dimensional)." [p. 10].

The last remark would seem to indicate that he was already considering a cosmos which would move (e.g., rotate) *as a whole*, around three years before he began planning his paper on the philosophy of time (Kant-Einstein), which in turn led to his discovery of the 'Gödel Universe'.

However, what rekindled his immediate interest in GRT, cosmology and time and spurred him to action in 1946 is clear: Paul Arthur Schilpp, for whom Gödel had already written a contribution to the volume on Bertrand Russell in the '*Library of Living Philosophers*' series [Gödel (1944)], was now planning a volume on Einstein, which would fulfil a twofold purpose: to describe Einstein's philosophy for a general audience; and to serve as a *Festschrift* for Einstein's 70th birthday, on March 14th, 1949—Schilpp had wisely begun to prepare the volume early, and he visited the IAS in May, 1946, asking Gödel to contribute a paper to the volume. Gödel of course immediately agreed—and he initially (in a letter to Schilpp a few weeks later) suggested a brief contribution of around 3 pages on Kant's philosophy of time—the lack of an 'objective time'—and the notion of time in General Relativity. Schilpp encouraged him to submit a longer contribution; and in the event, he did, on a much more unique topic: He solved Einstein's field equations for a *rotating universe*, now known as '*Gödel's Universe*', which incidentally (in its original, stationary version) contains closed-loop, time-like worldlines, which at least suggest the possibility of time travel into the past. It also excludes the possibility of a 'universal time', relegating time to the realm of the subjective, as Kant believed (but of course for different reasons).

In fact, Gödel conceived this topic only several months after Schilpp's visit and his original agreement to submit an article.

And how did Gödel arrive at the idea of a *rotating universe*? [Not such a simple proposition; rotation is normally observed relative to some 'fixed frame'. Newton dealt with this notion, and he mentioned the 'fixed stars' as a reference frame for rotation. Later, after the advent of the Maxwell theory, a stationary 'ether' was postulated as the propagation medium for electromagnetic waves, and it would provide a 'fixed frame' for the entire universe. There is even a 'Lorentz ether theory', an alternative to SRT, which holds fast to a 'rest frame'—in contrast to SRT, where there is no privileged frame which is 'at rest' throughout the universe. Einstein for some time held Ernst Mach's idea that the greater portion of the matter in the universe provides the reference frame for rotation—which he called 'Mach's Principle'—to be one of the foundational concepts of General Relativity (along with the Equivalence Principle and the Principle of Relativity)]. But if the entire universe, i.e., spacetime itself, is *rotating*, what is the reference frame for the rotation? Mach's Principle fails completely in that case.

The answer to the question of how Gödel arrived at his topic is unclear. *Éric Audureau*,[17] who has evaluated some of Gödel's unpublished notes on philosophy (the '*Max Phil*' notes in the *Collected Works*, Vol. III), believes, considering the above history of Gödel's interest in GR and time, that Gödel himself arrived at the topic, with or without encouragement in his regular conversations with Einstein. He goes so far as to state[18] that the alternative explanation offered by *Wolfgang Rindler*, among others, that Gödel was influenced by a short publication and/or a letter to Einstein from *George Gamow* in the autumn of 1946, is not relevant: "*Wolfgang Rindler's suggestion [Rindler 2012: 189] that Gamow's letter to* Nature *inspired to Gödel the idea of rotating universes* should be disregarded".

That alternate explanation maintains that the rotating cosmos was suggested to Gödel (indirectly) by George Gamow, who—precisely in 1946, when planning for the Einstein *Festschrift* had just begun—himself became interested in the possibility of a rotating universe (perhaps our own?). He, in turn, had thought of that idea because of the fact that most of the observable objects in our universe (the Solar System, stars, galaxies…) are rotating, many of them in the same sense. This means that the universe may possess an enormous overall angular momentum, and Gamow asked the question as to whether the *entire universe* is rotating. He recognized that in terms of General Relativity, this would imply an *anisotropic solution* of the Einstein equations.[19] His thoughts were published in a brief note in *Nature* on October 19th, 1946.

Just how Gödel became aware of Gamow's note is uncertain. He was assiduous in following the literature, and no doubt would have seen it sooner or later on his own; he was certainly aware of it by the time that he gave his IAS lecture on the Gödel Universe [Gödel (1949c), called (*1949b) in the *Collected Works*, Vol. III]. But his attention may have been called to Gamow's ideas even before the latter's note appeared, through one of Gödel's many conversations with Einstein—it is known[20] that Gamow had written to Einstein about his idea, in September 1946, shortly before his note was published in *Nature*. He asked Einstein's opinion, and Einstein may well have mentioned Gamow's letter, and his own reply, to Gödel, who thus became aware of Gamow's proposal. Either way, Gödel was certainly aware of Gamow's initiative *vis-á-vis* a rotating universe, probably when he started working seriously on his contribution to the Einstein *Festschrift*. It may have supported Gödel's own notion that our universe must be rotating, which he maintained for the rest of his life (both Audureau and Montgomery Link, also in [Crocco & Engelen, eds. (2016)], quote Gödel's telephone call to Freeman Dyson in 1976, less than two years before his death, to inquire about observational evidence for rotation of the universe; "*and when Dyson informed him there was none, he was unwilling to accept the conclusion*" [quoted from Dawson (1997), p. 182]).

Given that Gamow was an interesting character in his own right, we include a thumbnail biography: **George Gamow** (1904–1968), born Георгий Антонович Гамов (*Georgiy Antonovich Gamov*) in Odessa (then part of the Russian Empire, later in the USSR and Ukraine, now spelled 'Odesa'), was a theoretical physicist and cosmologist who was educated in Russia and who emigrated to the USA via Belgium and France, after attending the 7th Solvay Conference in Brussels in 1933. He obtained his education first in Odessa and then at the University of Leningrad, initially working with Alexander Friedmann (known for his early solution of the Einstein equations, leading to an 'expanding universe' cosmology).

Gamow worked as a graduate student in Göttingen, Copenhagen and Cambridge (see Fig. 15.5) on nuclear models and on cosmology before receiving his doctorate in Leningrad. He is known for suggesting quantum tunneling as the mechanism of the alpha decays of atomic nuclei, as well as for developing the liquid-drop model of nuclear structure, and for describing an important mode of nuclear beta decays ('Gamow-Teller' decays). He helped to build the first cyclotron in the USSR. After moving to the USA in 1934, he worked at George Washington University (GWU), collaborating with Edward Teller, Ralph Alpher and Mario Schenberg, among others. Gamow's sense of humor was legendary—he recruited Hans Bethe (from Cornell

University in Ithaca, NY) to be a co-author of a paper with Alpher, so that its authors were 'Alpher, Bethe, and Gamow'. With the Brazilian physicist Mario Schenberg, he wrote a paper on the rapid cooling of supernova remnants by neutrino emission, which they called the '*Urca Process*', after the *Urca* district in Rio de Janeiro, where at that time a casino operated—since 'heat was dissipated from the stellar remnant rapidly, like the money passing out of the hands of gamblers in Urca, by fast exchange'.

Later in his career, he was well known for writing numerous popular books on astrophysics and cosmology [e.g. '*The Birth and Death of the Sun*' (1940/52), '*One, two, three … Infinity*' (1947/61), '*Gravity*' (1962), and the '*Mr. Tompkins*' series (1938–1967)]. He left GWU in 1954 to work at UC Berkeley for a time, then spent the remainder of his career at the University of Colorado, Boulder, from 1956 until his early death at age 64. In his later years, he also worked on deciphering the genetic code by which information is stored in DNA and RNA in terms of the ordering of the four 'genetic' bases.

Thus, we can see that Gödel's idea for his contribution to the Einstein *Festschrift* was in any case not simply the result of 'his conversations with Einstein', as is sometimes claimed.[22] Gödel himself denied that they were its source.[23] In his own words (as told to Hao Wang in 1975–76),

My work on rotating universes was not stimulated by my close association with Einstein. It came from my interest in Kant's views. In what was said about Kant and relativity theory, one only saw the difference, nobody saw the agreement

Fig. 15.5 John Cockcroft and George Gamow (*right*) at the Cavendish Laboratory, Cambridge/UK, in 1931. Photo by K. T. Bainbridge, Cavendish Laboratory Photography Collection[21]

of the two. What is more important is the nature of time. In relativity there is no passage of time, it is coordinated with space. There is no such analogy in ordinary thinking…

Once he had conceived the idea of finding a solution to Einstein's equations for a rotating universe, he set to work. Schilpp, understandably nervous about whether Gödel's contribution would be ready in time for the publication of his book, inquired several times about his progress on the article.

At the same time Gödel, although fascinated by the subject of his planned article, was also busy with more worldly matters. As Dawson[24] reports, Adele had become very concerned about the fate of her family in Vienna following the end of WWII (and in particular after the difficult winter of 1945/46), and she was determined to travel there and see about them in person. Gödel, in turn, supported her project because he thought it necessary for her 'peace of mind' (i.e., her mental health). But there were the usual problems in acquiring visas, booking passage etc., especially in the immediate post-war period. Furthermore, the apartment that they still owned in Vienna had been damaged, and that required on-the-spot attention. Kurt himself was however not prepared to travel to Vienna, not even to accompany Adele. Her trip was postponed several times, and she finally left New York only in May of 1947. She stayed in Vienna for seven months, leaving Kurt to his own devices in Princeton. One might see this as her 'revenge' for his 8 months' absence in 1938/39, just after their marriage, although the pressing need to look after her mother and to straighten out their affairs in Austria were certainly sufficient reasons for her trip.

Kurt took advantage of his time alone to work intensively on his article for the Einstein *Festschrift*. His friend Oskar Morgenstern 'looked after him' in the first months of Adele's absence, but then he himself went to Vienna— among other things, he visited Kurt's mother Marianne and his brother Rudolf there, to some extent as a surrogate for a visit by Kurt. Marianne had left Brünn in 1944, evidently to visit Rudolf in Vienna, and, due to bombing and other military activity, agreed to stay there for the rest of the war. This was very fortunate, in retrospect, since the remaining German inhabitants of Brünn/Brno were not treated well by the Czech majority in the country after the withdrawal of German forces near the end of the war. They were in fact driven out of town on a forced march to Austria at the end of May 1945, a distance of around 60 km (about 38 mi.) from Brünn to the Austrian border, and then forced to return since Austria would not admit them. A considerable number (nearly 2000) did not survive; the incident is often referred to as the 'Brno death march'. Given the arrogant and merciless treatment of the

Czech population during the Nazi occupation, this was understandable, but still not excusable. A wrong cannot be righted by another wrong. Marianne, having left Brünn the previous year, was spared this ordeal. She had however missed the opportunity to sell her villa for an appropriate price and in the end was given only a small compensation, around 10% of its value, after the war.

By the end of September 1947, Kurt wrote to his mother that his article was practically finished. In the event, he produced five drafts, rejecting each one in turn, and finally submitted his essay to Schilpp in March 1949, just in time for Einstein's birthday. He had discovered, by September 1947, a world without 'objective time', where no unique simultaneity can be defined. The possibility of defining what he called a 'natural' cosmic time exists only in a non-rotating cosmos. This was a result which was very relevant to his original topic, the relation of GRT to Kant's notion of time.

Gödel's Cosmology, and his Philosophy of Time

Wolfgang Rindler, in his (2009), recounts the story told by Gödel in his lecture to the IAS in May of 1949. Gödel (and Rindler) begin by considering the Newtonian description of a rotating universe, which is a good approximation to the GR result and is intuitively clear. Our starting point[25] is Poisson's equation for the gravitational potential (compare Table 15.1):

$$\nabla^2 \varphi_G = 4\pi G \rho_M. \tag{15.6}$$

Since we are looking for a solution for a *rotating universe*, we must have an axis of rotation, which can be chosen to be the z axis, and we expect cylindrical symmetry around that axis. Then a reasonable solution for the potential φ_G might be

$$\varphi_G = \pi G \rho_M \left(x^2 + y^2 \right) = \pi G \rho_M r^2, \tag{15.7}$$

corresponding to a field pointing radially toward the axis of rotation. The relevant gravitational force points *toward* the axis (we consider a mass point m and take its value to be 1 in the following):

$$\boldsymbol{F}_G = -\nabla \varphi_G = -2\pi G \rho_M \boldsymbol{r} \tag{15.8}$$

where r is the radius vector (pointing outward from the rotational axis). This force must be balanced by the *centrifugal force* $F_{Ctr} = \Omega^2\, r$, where Ω is the rotational angular velocity (2π times the number of rotations in a given time). This balance condition is satisfied when

$$\Omega^2 = 2\pi G \rho_M. \tag{15.9}$$

We now have a 'model universe' which rotates around the z-axis with the angular velocity Ω. As Rindler points out, our model does *not* violate the cosmological principle (that the universe looks everywhere the same), since it is the *entire spacetime* that is rotating, not anything particular *within* the universe—the galaxies are stationary within the rotating spacetime, and the gravitational force F_G is not detectable anywhere within the universe (it is exactly cancelled everywhere by F_{Ctr}). However, as in any rotating system, there is also a *Coriolis force* which acts on any *moving* body within the system:

$$F_{Cor} = 2v \times \Omega \tag{15.10}$$

where v is the (vector) velocity of the moving object, and Ω is the (vector) angular velocity of the rotation (pointing along the axis of rotation). This 'cross product' or 'vector product' is itself a vector which is perpendicular to both v and Ω. From an observational point of view, each galaxy appears to be at rest on the axis of rotation of the universe which is rotating rigidly at the angular velocity Ω around it. The free motion of any moving objects in this universe, if their initial motion is within a plane perpendicular to the axis of rotation ($z = $ const.), is *circular*; or, if it has an initial component perpendicular to that plane, it is a *helix* (with respect to the rotating spacetime). In the former case, there is no force along the z axis, and within the $z = $ const. plane, the only force (F_{Cor}) is perpendicular to the velocity v, so that the velocity has a constant magnitude but is continually deflected, giving a circular orbit of radius $r = v/2\Omega$. If, for example, the object is moving horizontally outward from the axis of rotation ($v//r$), the Coriolis force will point upward, and the circular motion of the object (with angular velocity ω) will be counter-clockwise as seen along the z axis, while the rotation of the universe itself is clockwise ($\Omega//z$). In general, the angular velocity of such local motions due to the Coriolis force will be opposite to Ω, and it will have the magnitude $\omega = 2\Omega$. [If the object is *not* moving relative to the spacetime, it will be carried along by the rotating spacetime, orbiting the axis of rotation with angular velocity $\omega = \Omega$ as seen by an 'outside observer' (outside the universe!). If it *is* moving with a 'horizontal' velocity v, it will orbit in the opposite direction with an equal but opposite ω, and thus has the relative angular velocity -2Ω w.r.t. the spacetime. It is not hard to show that a similar conclusion holds for *any* velocity $\boldsymbol{v}$ within the model universe].

At this point, Rindler mentions that Gödel used a negative value of the cosmological constant Λ in his relativistic calculations, but he did *not* include that in his Newtonian model. Including it, we would have—instead of Eq. (15.6)—the following 'extended Poisson equation':

$$\nabla^2 \varphi_G + \Lambda c^2 = 4\pi G \rho_M, \tag{15.11}$$

where c is the speed of light and φ_G is now a *joint* potential for gravity and the Λ force. Gödel's relativistic theory requires namely that $\Lambda = -4\pi G \rho_M / c^2$. Putting this value into Eq. (15.11), we find that the solution for the joint potential becomes

$$\varphi_G = 2\pi G \rho_M \left(x^2 + y^2 \right), \tag{15.12}$$

so that for Ω, we find instead of (15.9):

$$\Omega^2 = 4\pi G \rho_M. \tag{15.13}$$

This agrees exactly with Gödel's relativistic result [he himself was satisfied with Eq. (15.9) as a "good approximation" from the Newtonian model— cf. Gödel (1949c)]. As Rindler points out, however, the Newtonian model here does not require the above *negative* value of Λ, in contrast to the situation with Einstein's 1917 static solution which *does*—also in the Newtonian approximation—require a *positive* value of Λ to counteract the contracting effect of gravitation. Thus, the origin of the negative Λ for Gödel's rotating universe remains somewhat mysterious.

Rindler now goes on in his Section 5 to consider stationary metrics, and the form of Gödel's metric for his rotating universe in GRT. This is the next step toward reconstructing Gödel's formulation of his relativistic model. Rindler defines 'stationary' to mean that there is no *evolution* of a universe (expansion or contraction), and 'static' to mean that there is no *rotation* (of the universe as a whole). [Those terms are used rather interchangeably in the general literature!].

Rindler considers a (three-dimensional, rigid, but curved) lattice of points within the spacetime which is subject to a permanent gravitational field, stationary but not static, so that there is a continual rotation of the "compass of inertia" at each lattice point. Each lattice point traces out a worldline in four-dimensional spacetime, and in a *static* spacetime, there would be a three-dimensional 'hypersurface' (i.e., a three-dimensional subspace) perpendicular to the worldlines at each point. Those hypersurfaces represent the possibility of defining simultaneous events in the static spacetime (and thus permit the definition of an 'objective time' in *static* GRT). When *rotation is allowed*, however, the worldlines are *twisted* (by the rotation) and there are no longer unique sets of perpendicular hyperspaces, but rather infinitely many sets, giving many different 'simultaneities'. This applies to Gödel's Universe.

Rindler gives a general expression for the *interval* in a stationary (not static) spacetime in GRT:

$$ds^2 = e^{2\Phi/c^2}\left(cdt - c^{-2}w_i dx^i\right)^2 - h_{ij}dx^i dx^j, \tag{15.14}$$

which contains *two* potentials, a scalar potential Φ and a vector potential w [given as components in Eq. (15.14)]. The metric tensor for the spacelike part is denoted here by h, and the x^i, x^j are again generalized (space) coordinates; the indices i, j take on the values 1, 2, 3, corresponding to the three space dimensions. The metric h is called the 'canonical' metric. The potentials and the metric tensor are functions of the space coordinates only; the time t is here an 'adapted' global time. We again see the analogy with electromagnetism, where a scalar (electric) and a vector (magnetic) potential occur. The relation for the gravitational field g is analogous to that in the Newtonian model: $-\nabla\Phi = g$. The rotational velocity (as a vector) is given by a corresponding expression (compare the formula for deriving the magnetic field from the vector potential in electromagnetism):

$$\mathbf{\Omega} = (1/2c)e^{2\Phi/c^2}(\nabla \times w). \tag{15.15}$$

Rindler (2009) now introduces *gauge symmetries*, a topic that is also well known from electromagnetic theory. Since the relevant quantities are all obtained by taking derivatives with respect to the coordinates, they are insensitive (invariant) with respect to shifts of the *zero points* of the potentials and the coordinates, e.g., to adding functions which are well-behaved w.r.t. the coordinates. [These, again, are familiar from electromagnetic theory (and from everyday practice with electricity—the important quantity for the design of electric circuits is the *voltage*, i.e., the potential *difference*, and not the absolute values of potentials). The 'zero point' of potential of a Faraday cage might be 1,000,000 V relative to ground, but a worker inside can measure small potential differences of millivolts or less, independently of the potential drop to *ground*, taken as the practical zero of electric potential on the Earth's surface]. As we saw from the Newtonian model, the effective global gravitational field is everywhere zero in Gödel's universe, since it is precisely compensated by the centrifugal force at every lattice point. Thus g = 0 and we can take $\Phi = 0$. We can also use a gauge symmetry to transform the zero points of the coordinates, according to

$$x^i \rightarrow x^{i'} = g\left(x^i\right) \tag{15.16a}$$

and

$$t \rightarrow t' = t + f\left(x^i\right) \tag{15.16b}$$

where f and g are arbitrary well-behaved functions of the x^i.

This has the following effects on the potentials Φ and w_i:

$$\Phi \rightarrow \Phi', \quad w_i \rightarrow w_i' = w_i + c^3 \partial f / \partial x^i. \tag{15.17}$$

Given the metric h of the lattice and the gravitational field **g** and rotation rate Ω at every lattice point, the interval (15.14) is uniquely determined up to a gauge transformation. Rindler now considers the construction of the lattice for Gödel's universe, consisting of a set of parallel lines which serve as local axes of rotation, and two-dimensional spaces orthogonal to them (which must be homogeneous, as Gödel's universe is homogeneous, containing a uniform massive 'dust' or an ideal fluid; but it may be curved, as usual in GRT). He then lists possible curvatures, k/a^2 [a is a 'scale factor' that determines the scale of the spacetime; it defines a 'radius of curvature'; k is a 'curvature index' which can be $k = 1$ ('spherical'), $k = 0$ ('flat'), or $k = -1$ ('hyperbolic')], and this gives the corresponding metric for the *two-dimensional* subspaces of constant curvature:

$$dl^2 = a^2\left(dr^2 + \Sigma^2 d\phi^2\right), \tag{15.18}$$

using *polar* coordinates, r = radius and ϕ = azimuthal angle. Σ is a function which depends on the type of curvature [$\Sigma = \sin r$ for $k = 1$; $\Sigma = r$ for $k = 0$; $\Sigma = \sinh r$ for $k = -1$]. Measured ('ruler') distance from the axis of rotation is given by ar, and ϕ is the angle around the axis. Taking z as the third coordinate, along the axis of rotation, we find for the interval of a three-dimensional lattice:

$$dl^2 = a^2\left(dr^2 + \Sigma^2 d\phi^2\right) + dz^2. \qquad (15.19)$$

We thus now have *cylindrical* coordinates, symmetrical around the rotational axis (so the four-dimensional metric must be independent of ϕ; and, due to homogeneity in the r and z directions, also independent of those coordinates). Gauge symmetries are again used to set $\Phi = 0$, and the vector potential w can have no z component (its *curl*, the angular velocity of rotation $\boldsymbol{\Omega}$, points along the z direction), while its r component can be reduced to zero by a gauge transformation. We then find the following expression for the full four-dimensional interval:

$$ds^2 = \left[dt - w(r)d\phi\right]^2 + a^2\left[dr^2 + \Sigma^2 d\phi^2\right] - dz^2 \qquad (15.20)$$

where the units have been chosen to make $c = 1$. By a rescaling of the coordinates ($t \to at$, $w \to aw$, $z \to az$), Gödel finally converts this metric to the 'conformal' form:

$$ds^2 = a^2\left\{\left[dt - w(r)d\phi\right]^2 - \left[dr^2 + \Sigma^2 d\phi^2 + dz^2\right]\right\}. \qquad (15.21)$$

Rindler, following Gödel, now proceeds to derive various properties of this model,[26] finding the rotational angular velocity to be $\Omega = w'/2a\Sigma$. From the condition that $\Omega = $ const., it follows that $w' \propto \Sigma$. Using gauge symmetry, one can now define a 'geometrically preferred' global time for a particular location (e.g., a particular galaxy), choosing a gauge that makes the $t = $ const. hypersurface ('surface of simultaneity') cut the worldline of that location orthogonally, so that t coincides with the local (inertial) time of that location. Such 'time slices' are different at every locality (e.g., for two different galaxies), as they are in SRT for different inertial observers. This is in contrast to 'standard' GRT (e.g., the Friedmann cosmology, the first solution to the field equations to show an expanding universe, where all the galaxies' worldlines cut the 'canonical' time slices orthogonally, permitting the definition of a true *global time*). In Gödel's universe, the worldlines are twisted due to its rotation, so such a condition cannot be satisfied. This is just what Gödel was looking for in his comparison to Kant's view of time.

In his next Section 6, Rindler discusses the question of whether the above model is indeed compatible with Einstein's field equations. We leave off the details here, and refer the interested reader to Rindler (2009) and to Gödel (1949b). Rindler considers the energy–stress tensor T in Gödel's model, and the fact that the full metric $g_{\mu\nu}$, according to (15.21), has a simple structure, with only two nonzero off-diagonal elements, g_{42} and g_{24}. (Rindler uses the modern convention, μ, $\nu = 1,2,3,4$). Also, the energy–stress tensor T has only two nonvanishing off-diagonal components, namely T_{24} and T_{42}, and two diagonal components, T_{22} and T_{44}. Using a computer algorithm to construct the Ricci tensor $R_{\mu\nu}$ in accord with $g_{\mu\nu}$, he finds that the Einstein tensor $G_{\mu\nu} = R_{\mu\nu} - (1/2)\,Rg_{\mu\nu}$ also has a simple structure, with 10 of its off-diagonal elements equal to zero; only G_{24} and G_{42} are nonzero. They, and the diagonal elements, are straightforward functions of w and its derivatives and contain two constants S and C, both derived from Σ, and thus dependent on the curvature parameter k. All the elements of G are independent of the scale factor a. Inserting these results into the Einstein field Eq. (15.3), one finds a relation for the cosmological term:

$$a^2\Lambda = -(1/4)\left(w'^2/S^2\right) \tag{15.22}$$

which shows that it cannot simply be 'left off', and that it must be *negative* to make Gödel's model compatible with GRT. Combining this with the equation for the diagonal component G_{33}, which contains the same term as the right-hand side of (15.22), yields $a^2\Lambda = k/2$, so that, in order to guarantee that $a^2\Lambda$ is negative, we must have $k = -1$, leading to $\Lambda = -1/2a^2$. This is the origin of the negative, nonzero value of the cosmological constant Λ in Gödel's universe (its mathematical, but not its *physical* origin!). With the value of k settled, the remaining terms can be evaluated, giving values for w, w' and the matter density ρ_M (found to be $\rho_M = 1/(8\pi Ga^2)$). There are at this point still two field equations left, but—*mirabile dictu*—they are in fact also fulfilled by the above values of Λ, w, and ρ_M, so that Gödel's model is indeed compatible with GRT!

Stability is an important property of a cosmos which is discussed by Rindler, but not mentioned by Gödel in his (1949c) lecture. In a *stable* cosmos, small changes in the parameters (ρ_M, Λ, radius r_0) cause no drastic changes in the structure and form of that cosmos; it returns to its previous equilibrium or to a nearby one. This is *not* the case for Einstein's stationary, static cosmos of 1917, as was realized only somewhat later: it has a 'fine-tuned' balance between the 'vacuum pressure' represented by Λ and the gravitational attraction of the matter. The former increases as r, while the latter decreases as $1/r^2$. The smallest change in the scale of r will thus destabilize the whole cosmos, resulting in a catastrophic collapse or expansion. In the case of Gödel's universe, in contrast, the value of Λ is uncritical; it must be nonzero and negative for the model to satisfy the field equations, but since its magnitude depends only on the scale factor a, while the Einstein tensor is independent of a, small changes will not destabilize the cosmos. Rindler (2009) gives an argument based on the approximate Newtonian model to demonstrate that variations of ρ_M or the scale factor will also cause no destabilization; there is a balance between gravitation, centrifugal force (which is of course lacking in the Einstein solution) and the Λ force, leading to a restoring force rather than destabilization. See Barrow & Tsagas (2004) for a more complete discussion.

Light Cones, Time Travel, and Geodesics in Gödel's Universe

The title of this subsection is the same as that of Rindler's (2009) Section 7, and it is very appropriate. *Light cones* are an heuristic device which greatly facilitates visualizing and understanding the propagation of massive and massless particles in a relativistic cosmos, and for testing whether causality is maintained within it. They simplify the four-dimensional spacetimes of SRT and GRT by considering only a *two-dimensional* space, with time as the third coordinate axis, taken to be vertical (the usual z-axis). The origin is a point where an 'observer' can imagine themselves to be located, and is the zero of time (i.e., the 'present' moment). It is at the apex of two cones, one spreading out into the future (the green cone in Fig. 15.6), the other stretching back into the past (the lilac-blue cone in the figure). The inclination of a line starting at the origin (the red dot in the figure) and propagating into the future (positive values of time) defines the speed of a particle—be it a 'mass point' or a massless particle such as a photon—along a direction in the horizontal plane, the two-dimensional (x, y) space of this model. The greater the inclination, the higher the speed. By choosing units in which $c = 1$, we make an inclination of $45°$ correspond to the velocity of light, and the surfaces of the cones are therefore termed 'light-like'. All light beams emitted from the origin propagate along those surfaces. Massive objects must move more slowly, and they thus propagate *inside* the cones, in the 'time-like' region. Outside the cones, there is a 'forbidden zone' (the 'space-like' region), where no particles are allowed to propagate and no events can be communicated by light beams, according to the 'speed limit' imposed by relativity.

[Note that only one horizontal axis (the y-axis) is shown in the figure; the x-axis would project out of the plane of the page toward the reader]. *Causality*

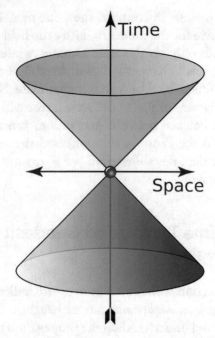

Fig. 15.6 A light cone, with a two-dimensional space (axes *x, y*) in the horizontal plane and *time* as the vertical (*z-*) axis. The green cone shows the *future* of any particle emitted from the origin (red dot); the lilac-blue cone is the *past*, and particles coming from a previous time must move within it or on its surface. Image: a drawing by Incnis Mrsi, June 2008[27]

extends from below to above in the figure: the cause of an event must lie at an earlier time than the event itself: effects occur (perhaps immediately) *after* their causes. In four-dimensional Minkowski space, the expression for the relativistic 'speed limit' is given by:

$$dx^2 + dy^2 + dz^2 \leq c^2 dt^2. \tag{15.23}$$

The full interval in Minkowski space is given by (15.4): $ds^2 = -c^2 dt^2 + dx^2 + dy^2 + dz^2$, and from (15.23), it must be ≤ 0 to obey the 'speed limit' (with the 'signature' $(- + + \ +)$ used above; we could instead use $(+ - - -)$, making $ds^2 \geq 0$). The local equation of the light cone is given by $dx^2 + dy^2 + dz^2 = c^2 dt^2$, making $ds^2 = 0$, and this is the general condition for light-like propagation. The interval ds^2 is the squared distance between nearby points (events or observers) in Minkowski space, and it has the same value for every observer (corresponding to the constant, universal value of *c*). Unlike the squared distance in Euclidean space (cf. Fig. 15.4), which is always positive, ds^2 in Minkowski space can be positive, negative, or zero, the latter value pertaining to the propagation of light (or any other massless particle).

The curved Riemannian spacetime of GRT is *locally flat* (like the surface of the Earth, which is globally curved but locally flat and Euclidean), and thus can be approximated locally by Minkowskian spaces. In a (global) Minkowski space, all the light cones have their axes parallel and worldlines thread through them, always staying within (or on) the cones. But in the curved Riemannian space of GRT, worldlines can also be curved, leading to a situation as shown in Fig. 15.7, right-hand side.

Thus, it would be theoretically possible within GRT that a worldline could curve back upon itself, forming a closed loop in time—endangering causality and possibly permitting 'time travel', with the associated paradoxes. This possibility had occurred to Einstein, but he held it to be unphysical, and expected that some 'limit' would prevent such a universe from existing, just as the 'speed limit' in SRT prevents its lack of simultaneity from leading to a violation of causality.

Rindler (2009) quotes Einstein[28]—already in 1914!—on possible closed-loop worldlines (they are implied by e.g., the anti-de Sitter solution to

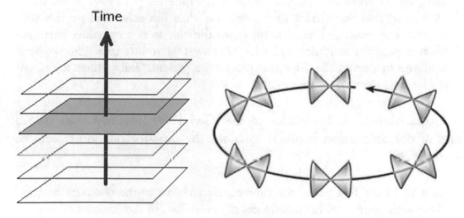

Fig. 15.7 *Left*: a 'stack' of planes corresponding to t = const., i.e., planes of *simultaneity*, all perpendicular to the time axis. This situation is called 'presentism' by philosophers. Each intersection point of a plane with the time axis is a 'present moment' for some observer, and these succeed each other in an orderly progression, giving a time axis which is a straight line, a series of points (the continuum) along which time proceeds at a uniform pace for all observers. *Right*: the situation in Gödel's Universe, where light cones determine the allowed velocities and causality, and worldlines are curved and even closed timelike loops. The significant limiting condition is that the worldlines always remain *within* (for massive particles), or *tangent to* (for photons) the light cones. Images: *Left*, Creative Commons, WikiMedia. *Right*, Creative Commons[31]

the field equations[29] and by Lanczos' solution (1924);[30] but Einstein had recognized their possibility even before completing GRT!):

> This conflicts strongly with my physical intuition. But I am unable to prove that the theory excludes the occurrence of such orbits.

'Such orbits' were inherent in several cosmological models based on GRT, but were not recognized earlier. Gödel was the first to point them out explicitly, rather late in the game (and during the period when GRT was still effectively 'becalmed', between about 1930 and the mid-1950s, which probably limited attention to his publications in the beginning).

Carlo Rovelli, in his book '*The Order of Time*' [Rovelli (2018)], makes the following comments about 'closed-loop timelike worldlines':

> The structure of the *[light]* cones can even be such that, advancing always towards the future, one can return to the same point in spacetime … In this way, a continuous trajectory towards the future returns to the originating event, to where it began. The first to realize this was Kurt Gödel, the great twentieth-century logician who was Einstein's last friend, accompanying him on walks along the streets of Princeton … More than a hundred years have passed since we learned that the 'present of the universe' does not exist. And yet this still continues to confound us and still seems difficult to conceptualize. Every so often a physicist mutinies and tries to show that it isn't true. Philosophers continue to discuss the disappearance of the present. Today, there are often conferences devoted to the subject.

Gödel himself, in his lecture to the ICM 1950 [text in Gödel (1952)], says of the connection between rotation, the time-metric, and closed-loop worldlines:

> … it is, in the first place, the time-metric (relative to the observers moving along with matter) which determines the behaviour of the compass of inertia. In fact, *a necessary and sufficient condition for a spatially homogeneous universe to rotate is that the local simultaneity of the observers moving along with matter be not integrable* (i.e., to not define a simultaneity in the large). This property of the time-metric in rotating universes is closely connected with the possibility of closed time-like lines.

Geodesics are a special kind of worldline, one which represents the shortest distance between two points in spacetime. Massless particles, like photons, follow geodesic paths, as do free-falling (force-free) massive particles. But the closed-loop worldlines leading to 'time travel' and problems with causality in Gödel's universe are *not* geodesics. In order to force a light beam to traverse such a worldline, it would have to be equipped with many mirrors to deflect the beam and keep it on the path. Likewise, a massive object could be made to follow such a worldline only by exerting large forces and producing corresponding accelerations, e.g., with a rocket engine. Note that the curvature of such a loop-like worldline is limited by the requirement that the paths of objects in spacetime remain *within* (or on) the light cones.

Gödel didn't claim that the *geodesics* in his model would form closed loops. In fact, he apparently believed that the extreme practical difficulties of forcing a massive particle or a photon to follow such a closed-loop worldline would act as a limiting factor to prevent—FAPP—any paradoxes or violations of causality from occurring in practice in his universe. That probably did not satisfy Einstein, however, who would have wanted a limit *in principle*, comparable to the 'speed limit' in SRT, which emerges from the theory itself, rather than simply a practical limitation based on energy requirements. Gödel made some estimates of the energy required for sending a massive object around such a loop, and the literature abounds in examples[32] ('if the Earth were sent around a loop in Gödel's Universe, converting some of its mass to energy and ejecting some more underway in order to produce the necessary accelerations, it would have shrunk to a ball of only a few hundred meters in diameter by the time it arrived back at its starting point in its distant past') Gödel apparently didn't make a similar calculation for a light beam traveling around a loop,[33] thereby 'sending a message into the past'; but the practical problems of setting up a system of mirrors around an orbit many light years in diameter make that alternative also effectively impossible to carry out. Rindler discusses these aspects of Gödel's universe in some detail in his (2009), at the end of Section 7. See below, section on '*The Philosophy of Time*'.

Some years after Gödel's publications on his Universe, the famous astrophysicist *Subrahmanyan Chandrasekhar*, together with a colleague from the University of Chicago, James P. Wright, mistakenly assumed that Gödel had in fact claimed *closed-loop geodesics* in his model, and wrote an article[34] 'refuting' him (one is reminded of Zermelo's reaction to his first incompleteness paper). In fact, there was nothing to refute, but no-one seems to have noticed that for some time afterwards. Gödel himself was not informed of their paper nor offered a chance to reply to it (even though by that time he was himself a long-standing member of the National Academy of Sciences, which published the paper in its house journal, PNAS). The 'refutation' was itself refuted only after another 8 years, when a philosopher of science, *Howard Stein*, at the time a professor at Case Western Reserve University in Cleveland, OH, noticed the discrepancy and submitted a paper correcting

Chandrasekhar and Wright's 'correction'. It was at first rejected by the editors of *Philosophy of Science*, probably in deference to Chandrasekhar's great prestige, but after Stein wrote to Gödel and received his assurances that he had never suggested that the closed loop orbits would be geodesics, the paper was accepted [Stein (1970)], and the matter was cleared up; however not without some damage to Gödel's reputation.

An example: In 1963, there was a small conference on '*The Nature of Time*' at Cornell University, not far from Princeton; it was organized by the (originally Viennese) astrophysicists *Hermann Bondi* and *Thomas Gold*.[35] Surprisingly, Gödel did not attend. Either he was not invited (although the organizers certainly knew of his work, and had probably met him previously), or else he declined the invitation. Whether he was aware of Chandrasekhar's objections to his work by this time is not clear. But it is also quite possible that the organizers of the meeting were impressed by Chandrasekhar's article— he *was* invited to the meeting and gave a talk there, of similar content to his (1961) article in PNAS—and they may have discounted Gödel's work as 'wrong' or irrelevant due to Chandrasekhar's mistaken criticisms. [One of Chandrasekhar's objections to the 'Gödel Universe' as the latter introduced it in his 1949 lecture at the IAS (which Chandrasekhar attended) might have been that it was *stationary*, although it was well known by that time that the observable universe is *expanding*. In his later lecture to the international mathematics conference [ICM 1950; see Gödel (1952)], Gödel indeed introduced another rotating universe which *does* exhibit expansion—but no closed-loop worldlines!].

At the end of his exposition of Gödel's 1949 lecture, Rindler briefly recounts the work of *Kornel Lanczos*, published in the early period of GRT, in 1924. He came very close to anticipating Gödel's work on cosmology, 25 years before Gödel. Gödel was evidently not aware of Lanczos' work, and it indeed attracted little attention at the time.

Kornel Lanczos (whose first name was later changed to Cornelius) was born in Hungary in 1893, and studied mathematics and physics in Budapest and Szeged, where he received his doctorate in 1921 with a thesis on relativity theory. He emigrated to Germany, and worked in Freiburg and Frankfurt am Main, there as an assistant to Erwin Madelung. After his *Habilitation* in Frankfurt in 1927, Lanczos went on a fellowship for two years to Berlin and worked as there as assistant to Einstein. Back in Frankfurt, he decided to leave Germany after the Nazi power grab in 1933, and emigrated to the USA, where he initially was on the faculty of Purdue University in Indiana. He later worked in industry and for the National Bureau of Standards at its branch in Los Angeles, where he was a colleague of Olga Taussky-Todd and her husband John Todd. He was considered to be one of '*the Martians*' (see Chap. 8, section on 'The Königsberg Conference'). During the McCarthy era, he left the USA and became a collaborator of Erwin Schrödinger in Dublin, Ireland.

In his later career, Lanczos concentrated on quantum mechanics and numerical mathematics, but in the period 1920–1930 he worked in relativity and cosmology, and it was then that he found and published his solution to the Einstein field equations. It exhibits cylindrical symmetry, and it has much in common with Gödel's solution. Lanczos however failed to realize that it describes a *rotating* world, and also that it could contain *closed-loop worldlines*. His work received little notice at the time, but it was rediscovered 13 years later by *Willem Jacob van Stockum*[36] [van Stockum (1937)], who noticed the rotation but not the loops. But it was only with Gödel's work in 1946–50 that this topic was clearly formulated, and Gödel worked quite independently of Lanczos and van Stockum.

The Philosophy of Time

Gödel's lecture to the *International Congress of Mathematicians* (ICM) in Cambridge/MA in 1950, later published in the Proceedings of the conference [Gödel (1952)], was his last published statement on his GRT work, although he maintained an interest in the topic throughout the rest of his life. *John Wheeler*, perhaps the leading exponent of gravitation and relativistic astrophysics theory (at least in the USA) from 1955–1975, made the following remarks about Gödel's Universe in his autobiography (with Kenneth Ford)[37] (referring to a visit to Gödel in his office at the IAS, with Charles Misner and Kip Thorne, probably in the late 1960s or early 1970s, while they were working on their iconic book, '*Gravitation*'):

> … he asked us what we were going to say about his theory of a 'rotating universe'. (The term doesn't mean literally that the universe as a whole is rotating—for what would it rotate relative to? In Gödel's theory, individual galaxies rotate more in one direction than another—just as the hands of clocks on the wall rotate more in one direction than another.) 'Nothing', we had to confess. This response distressed him. He was so passionately interested in the subject and so desperate for facts and figures, it turned out, that he had taken down the great Hubble atlas of the galaxies, lined up a ruler on each galactic image to estimate the galaxy's axis of rotation, and compiled statistics of the orientations. He found no preferred sense of rotation.

The topic claimed Gödel's full attention for at least 4 years, from 1946–1950, but he published only three works in connection with it: his article for the Einstein *Festschrift* [Gödel (1949a)], the article in *Reviews of Modern Physics* which contained a more detailed exposition of his solution of the field equations [Gödel (1949b), denoted as (1949) in the *Collected Works (CW)*, Vol. II], and the 1950 lecture quoted above [Gödel (1952)]; of course, his lecture to the IAS in May 1949 is also now available in text form [Gödel (1949c), (*1949b) in *CW* III], as are several 'intermediate versions' of his paper for the Einstein book,[38] and numerous relevant remarks in his Philosophical Maxims (the '*Max. Phil*' quoted previously, found in his papers after his death). Nevertheless, there can be little doubt that he considered his 'excursion' into theoretical physics to be essentially a *philosophical* project, and to a considerable extent, philosophy was the *Leitmotiv* of much of his life and career.[39] Rindler (2009) says of Gödel's cosmology project,

> Gödel tells us in his Einstein-Festschrift contribution that he was motivated by sympathy for Kant's philosophy of time to invent his model universe. It was to serve as the first counterexample on the cosmic scale to the "objective" view of time, which treats time as an infinity of layers of "now" coming into existence successively. [Compare Fig. 15.7, left side.]

Wheeler even suggests that Gödel's interest in his rotating universe had deep-seated personal roots:[37]

> I first heard him speak of it in 1949 at a seventieth birthday celebration for Einstein. In a universe with an overall rotation, he concluded, there could exist world lines (spacetime histories) that closed up in loops. In such a universe, one could, in principle, live one's life over and over again. I had to conclude that Gödel's passionate concern for his own health, so openly visible, was matched by an equally passionate, if less visible, wish to defy death and live again.

It is also informative to quote another story from Wheeler's autobiography about Gödel's interest in the rotating universe:[40]

> ...some time after his *[Gödel's]* death, I found myself chatting across a lunch table with a young man visiting from New York *[just who this was is not certain; most likely it was* Charles Parsons, *a co-editor of Volume III of Gödel's* Collected Works. *He was teaching at Columbia University in NY at the time in question].*

> 'What brought you to Princeton?, I asked.'

> 'I am editing the papers of Kurt Gödel', he answered.

So, I couldn't help asking, 'Have you come across any work related to the rotation of galaxies?'

'Interesting question', he responded. 'Recently I found a great pile of sheets filled with numbers, totally unlike the rest of his papers. It took me some time to figure out what they were all about. They were the foundation of his statistical analysis of galactic rotation.'

In 2021, the '*Kurt-Gödel-Freundeskreis-Berlin*' ('*Berlin Circle of Friends of Kurt Gödel*') held its second essay competition on a topic related to Gödel's life and work. The proposed topic was "What does it mean for our world-view, if we assume—with Gödel—the nonexistence of time?" The first prize was awarded to Prof. *Reinhard Kahle*, professor for the theory and history of the sciences at the University of Tübingen; the second prize went to Prof. *Claus Kiefer*, professor of theoretical physics (specialty: quantum gravity) at the University of Cologne.

Their winning essays can be seen on the website of the *Freundeskreis*,[41] along with other activities and documentation by the group.

The essay by Claus Kiefer begins with a quotation from the German author Wilhelm Busch, from a letter to Maria Anderson, dated May 2nd, 1875:

Certainty can be found only in mathematics. But unfortunately, it only scratches the surface of things. Whoever has experienced a deep astonishment about the world will want more. They will philosophize...

This probably comes close to describing Kurt Gödel's feelings about the knowability of the world. His 'first love', as we noted in earlier chapters, was mathematics, but philosophy followed soon after. He came seriously to physics somewhat later, and he described his path to Hao Wang, who wrote in his essay '*Time in Philosophy and in Physics: From Kant and Einstein to Gödel*':[42]

Gödel's interests in physics and in Kant's philosophy began early. At the age of 16 he read some of Kant's work; at about the same time, he evaluated Goethe's dispute with Newton, and decided to side with Newton. The interest in Newton's work must have played a part in his choice to specialize in physics, from 1924 to 1926, when he was 18 to 20. An indication of his deep concern with the philosophical aspect of physics was the fact that he requested from the library, on 26.1.25, Kant's book of 1786 on the foundations of natural science...

There has been a great deal of interest in Gödel's Universe since the publication of his results over seventy years ago. It began slowly, perhaps in part due to the fact that relativity was decidedly 'out of fashion' at the time of his publications. [This however changed soon after, due to the efforts of *John A. Wheeler* in Princeton, as well as *Bryce* and *Cécile DeWitt* in North Carolina and later Texas, in the USA; *Dennis Sciama* in Cambridge and Oxford, later London, UK; and of *Yakov B. Zel'dovitch* and *Igor D. Novikov* in the USSR, among others. They all founded 'schools' of relativity, gravitation, cosmology and astrophysics research, which produced an active and very productive second generation, including Charles Misner, Kip Thorne, Carmen Molina-Paris, Christopher Isham, Stephen Hawking, Roger Penrose, and Rashid Sunyaev—and many others. This development is sometimes referred to as the 'Renaissance of Relativity', and it led to a 'Golden Age' (around 1955–75; some would limit it to the decade of the 1960s)]. Gödel's relativity work inspired numerous projects in theoretical physics (see Chap. 19), but in addition, interest in its *philosophical* aspects has been growing steadily over the past 40 years, and as mentioned by Carlo Rovelli (see the quote in the section on '*Light Cones, Time Travel...*' above), that topic now inspires whole conferences.

It has also inspired dramatizations and exaggerations, much like Gödel's earlier work on incompleteness. It seems that his work was fated to be, on the one hand, *misunderstood*; and on the other, to be *overly dramatized* (and generalized). His Incompleteness Theorems did not signal the end of meaningful mathematics research (and certainly not of mathematics practice!), and Gödel's Universe did not signal the end of time, any more than Heisenberg's Uncertainty Principles signaled the end of physics, although some would have us believe that. But Gödel's Universe has stimulated much philosophical debate and discussion on a topic which has been of great interest to philosophers for thousands of years: the meaning and origin of *time*. Gödel himself was struck by the apparent prescience of Immanuel Kant, and quoted passages from Kant's works which seemed to anticipate the view of time implied by relativity theory. He also quoted passages in Leibniz's works which seemed to presage many developments in modern mathematics.

Claus Kiefer, in his essay for the Berlin *Freundeskreis* [Kiefer (2021)], gives a compact summary of the significance of Gödel's Universe for *physics*:

> "Gödel's solution is especially significant because of its consequences for the *physical concept* of time. That naturally concerns on the one hand the possibility of the existence of worldlines along which one could travel back into one's own past. On the other hand, the presence of rotation leads to the situation that we can no longer maintain the *illusion* of a global time, not

even approximately. To be sure, this relativization of the concept of time is one of the principal consequences of Einstein's theory; but nevertheless, we can get around that consequence if we hold fast to the cosmological solutions with expansion and without rotation as the only relevant solutions *[of the field equations]*. In that case, there is indeed a distinguished time *t*, which holds for all the observers moving at the same velocity, and which can be used to speak objectively of *simultaneity*." [Translation by the present author, emphasis added].

But what of its significance for *philosophy*? The recent essays by Reinhard Kahle and Claus Kiefer give a modern view of that topic, which is in the meantime the subject of a vast literature. It is impossible within this brief chapter to summarize all the views represented in that literature, so we will content ourselves with some highlights mentioned by those two authors, one a historian of science, the other a theoretical physicist with (inevitably) a philosophical bent—and of course including some points noted by earlier biographers of Gödel, in particular Hao Wang and Palle Yourgrau.

Reinhard Kahle mentions two additions to footnotes in Gödel's (1949a), his article for the Einstein *Festschrift*, which he appended to its German translation (1956), and which were thus overlooked by some authors.[43] One concerns the relevance of the 2nd Law of Thermodynamics (the '*entropy principle*') as a possible 'limiting factor' forbidding closed-loop worldlines and time travel in his universe; Gödel states in his addendum that '*he had determined himself that it is in fact compatible with his universe*' (footnote 14). He also added a longer additional text to his footnote 11, decisive for distinguishing between physical *principles* and *contingent properties* (such as the high energy cost of time travel in his universe):

"What was previously *[only]* a *practical* difficulty in microphysics is today, as a result of the uncertainty relations, an *impossibility in principle*; and the same could one day occur not *[only]* for a 'too small', but for a 'too large'." [Translation by the present author; emphasis added.]

Gödel of course is referring to limitations *in principle* of the precision of measurements within the 'nanoworld' due to quantum–mechanical uncertainty, and suggesting that similar limitations might someday be found for the macroworld of cosmology, preventing the occurrence of a *real* universe like that in his model, or—if it did occur as a real universe—preventing time travel and the resulting paradoxes and causality violations within it. While Gödel himself never claimed that his model represents 'our' universe, not even the later versions which included expansion, he seems to have kept a secret hope that it just *might*, and he thus assiduously followed any indications from cosmological observations which could have provided evidence for that, as we saw above. Furthermore, he points out that the appearance of a 'universal time' in static GRT cosmologies is only a result of the particular organization of matter in them, and thus of a *contingent* property and not a principle; so that even if his rotating model does not apply to the real world, the universal time found in some other (static) cosmologies is merely a coincidence and not a *quality* of the world.

Kahle also emphasizes the *specific context* of Gödel's speculations about the existence of time (this is again reminiscent of the discussions regarding his incompleteness results, where the specific context to which they apply is also often forgotten, whether intentionally or simply by negligence). In the case of Gödel's Universe, that context is what Gödel refers to as *idealistic philosophy*, which casts doubts on the existence of an objective time that would be valid for all observers, throughout the universe and irrespective of their locations or proximity to large masses (e.g., galaxies).

Yourgrau (2005) discusses this context in some detail in his section '*What Gödel means by Time*'.[44] He quotes *John M. Ellis McTaggart*, a philosopher active in the early twentieth century, a member of the 'British idealist' school and a Hegelian. McTaggart is also mentioned in a footnote by Gödel in his (1949a), his contribution to the Einstein *Festschrift*. Gödel cites McTaggart as an example of what was contradictory in the philosophical view of the *essence of time*, even before relativity. Yourgrau goes into more detail, describing McTaggart's two notions[45] of time: a geometrical one, similar to that of classical physics, in which time is a fixed sequence of events, like the points along a line in Euclidean space ('the B-series'); and a more intuitive one ('the A-series'), in which events are characterized by their status as 'present' (and therefore *existing*, in a real sense) or 'past' and 'future' (*no longer* existing, or *yet to exist*). 'A-series time' is *not* geometrical, and thus cannot rely on our intuitive understanding of geometry; it is *dynamic*, constantly changing. It is related to, but not a part of '*presentism*', the idea that only what is present, what exists *now*, truly exists, making an ontological statement. The contrasting view is *eternalism*, which holds that 'everything' exists forever, as in the 'block universe'. McTaggart's two 'series' have been the subject of extended linguistic debates, and the emphasis on 'nowness' in the A-series has been called anthropocentric (and nonphysical).

That both views of time can coexist in the human mind is unquestioned; but the relevance of the A-series to physical theories is dubious. McTaggart's 'proof' of the unreality of time was in fact based on his rejection of the A-series, due to its self-contradictory nature. Gödel, with his rotating universe, showed that solutions to the field equations of GRT exist which do not admit of a straightforward B-series view of time, which, to be considered 'global' or 'universal' would require simultaneity; otherwise, the fixed order of events inherent in the B-series is lost. [This was no surprise to Einstein, who had noted that aspect of SRT much earlier (see above; and compare his comments on the essays in the 1949 *Festschrift*[46]).] And, of course, there is no room at all in GRT for the 'A-series': how could its 'flow' be included in the time *t* that is contained in the spacetime metric?

As Yourgrau points out, the A-series is characterized by several unique aspects, and it is the essence of what intuitively distinguishes *time* from *space* (apart from the purely 'physical' factor '*ic*' in the metric of Minkoswki space; what Yourgrau would call the *formal* distinction of time from space). Those aspects are: (i) the *time* of the 'A-series' is *dynamic*, it 'flows' from moment to moment; (ii) it has a 'privileged moment', its *now*; (iii) it makes an onto-logical distinction between different moments (only what is present in the *now* moment truly *exists*), and this makes *simultaneity* an absolute, since it determines existence; (iv) it is *asymmetric* (the past is fixed, while the future is completely open). Yourgrau quotes Gödel to the effect that intuitive time—as it was known before relativity—was essentially a series of *now* moments, as represented by the stack of simultaneity planes in Fig. 15.7 (left), i.e., presen-tism—or even McTaggart's 'A-series' (which McTaggart himself rejected in his (1908) and later).[47] Gödel's rotating-universe solution thus demonstrated that *formally* (i.e., within GRT), that intuitive concept of time could not be represented (just as 'truth' cannot be represented on the formal, syntac-tical level, but only on the intuitive, semantic level: Gödel/Tarski). This is an example of Yourgrau/Wang's "dialectic of the formal and the intuitive", which Yourgrau raises to the level of 'the Gödel program'. An interesting approach, and it certainly unifies Gödel's work in different areas throughout his life. Hao Wang and John Dawson also searched for, and found, parallels between Gödel's proofs in formal logic, his results in set theory, and his excur-sion into GRT. Whether this justifies identifying it as a conscious 'program' which Gödel followed throughout his academic life, is another question. That notion has been praised e.g., by Mark van Atten, who in his (2015) wrote: "*A particularly attractive topic is what has been called by Palle Yourgrau (2005, p. 75) 'the Gödel program'. While Hilbert focused on the formal aspects of mathe-matics, and Brouwer on intuitive aspects, Gödel focused on the interplay between these two aspects …*". Yourgrau's idea of a *Gödel program* in this sense is also suggested by Wang's comment[48] in his (1987):

"… the contrast between formal systems and the axiomatic method as Gödel appears to understand them may be seen as a fundamental aspect of what might be called Gödel's dialectic of the formal and the intuitive … While Gödel is best known for his definitive results on the limitations of formalization, his primary interest is not in what has been formalized but in clarifying our larger intuitions". [Emphasis added.]

Apart from that, the correctness of GRT as a whole and of Gödel's solution in particular, while it rules out the validity of the 'A-series' view of time within *formal theory*, certainly does not rule it out of human experience. Just as Gödel's incompleteness proofs placed limits on formal mathematical logic, but not necessarily on intuitive mathematics, his rotating universe demonstrates that under certain circumstances, the formal notion of time is not compatible with the intuitive. And, of course, Gödel was searching for *concepts*, in his Platonist sense; ideas which are innate in the world, and not just human constructions or apprehensions, which he calls *notions*. It may well be that he consigned the intuitive sense of time (the 'A-series', if one will) to the dustheap of mere *notions*, and he raised the formal definition (something like the 'B-series', but lacking global simultaneity) to the status of a *concept*. His solution for a rotating universe (and his recognition of its implications, something which his predecessors, Lanczos and van Stockum, had failed to do) were, as Einstein acknowledged, a major achievement and a significant contribution to GRT, in any case. And the fact that a sort of 'universal time' can be defined in *static* GRT cosmologies is only a coincidence of the distribution of matter in those solutions; it is 'accidental' and not *in principle*.

Claus Kiefer closes his Section 2, '*Time and Reality*', the heart of his essay, with the words,

The non-existence of time follows uniquely from the possibility of a rotating universe. We owe this insight to Gödel. A consistent picture of our world must take it into account.

Hao Wang begins *his* essay on Gödel's notion of the intuitive view of time in philosophy and physics [Wang (1995)] with the statement that,

Both physical and mental processes take place in time—'that mysterious and seemingly contradictory being, which, on the other hand, seems to form the basis of the world's and our own existence' *[Gödel (1990), p. 202[49]]*.

Wang goes on to state [(1995), p. 216],

By way of this spatial analogue of time, we are able to represent space and time mathematically in such a manner that much of what is in our intuitive conception of space and time is preserved. Given the fact that time is primarily a frame of our inner sense, it is remarkable that this mathematization, through spatialization or externalization, of time—a highly precise but inflexible way of giving form to experience—has turned out to connect our inner and outer senses so well as to agree so completely with our observations of the external world.

Here, he is speaking of Kant's idea of intuitive time (as a geometrical analogue: 'linear time'). Kant of course emphasized that this intuitive, geometrical conception of time is a result of our particular perceptions, and not a true reflection of 'physical reality'. Considering the need for a new conception of time in light of the results of relativity theory, Wang says,

> …The usual emphasis is on the *conflict* between these parts *[of Kant's philosophy]* and the philosophical implications of the new physics. In contrast, Gödel chooses to uncover and argue for a surprising *similarity*, in some respects, between relativity theory and Kant's doctrine about time and space. [Emphasis added. Compare the Gödel quote above in the section 'Gödel's Universe–Relativity and Cosmology'.]

Wang closes his essay on Gödel and time with a list of observations about time made by Gödel in various conversations (mostly by telephone) in 1975/76, during the last decade of his life. One of them, his 'Q5', from Nov. 25th, 1975, is particularly telling (in it, Gödel refers to his study of the philosophy of *Edmund Husserl* (1859–1938), a German philosopher who developed *phenomenology* as a branch of philosophy. Husserl was born in *Prossnitz* (now *Prostějov*) in the Olomouc region of Moravia, only about 40 km from Brünn/Brno, and spent the later years of his career at the University of Freiburg, where he was the principal mentor of Martin Heidegger. See Chap. 16, last section). Gödel was very interested in Husserl's philosophy in the later years of his life, and read the latter's work from 1959 until shortly before his death in 1978:

> …As we present time to ourselves it simply does not agree with fact. To call time subjective is just a euphemism for this failure. Problems remain. One problem is to describe how we arrive at time … Another problem is the relation of our concept of time to real time. The real idea behind time is causation: the time structure of the world is just its causal structure.

Wang's (1995) summary of those conversations characterizes Gödel's attitude toward 'time' near the end of his life:

> Indeed, according to Gödel's general philosophical position, objective reality includes both the physical and the conceptual worlds, of which we can know better and better. In particular, he believes, I think, that there is a sharp concept which corresponds to our vague intuitive concept of time—only we have so far not yet found the right perspective to perceive it clearly.

This can be compared to Gödel's remarks on the successfully precise definition of '*computability*' due to Turing. He had not found a similarly precise definition for *time*.

Some authors have taken a very defensive stance with respect to Gödel (both with respect to his mathematical-logical results and to his GRT work). An example of the latter is *Palli Yourgrau*, who has written three books [his (1991); (1999), a revised version of (1991); and (2005), a more popular work that has biographical character] on *Gödel's Universe*. He suggests that there is (or was) a widespread 'conspiracy of silence', or at least a consensus, to suppress and ignore Gödel's Universe.[50] While it is true that Gödel's publications received little attention in the decade following their appearance, that was equally true of other publications on GRT and SRT during that period, assuming they were even able to be published at all (compare Chap. 19). This changed with the 'renaissance' of relativity, but in Gödel's case that also led to the mistaken criticism by Chandrasekhar and Wright. That however was most likely due—at least in part—to Gödel's own rather dense style of writing, which led to misinterpretations and false claims of 'errors', not only in his writings on the Gödel Universe. The 'conspiracy of silence' regarding SRT/GRT was incidentally not just an American phenomenon; it was certainly present in the UK, and in Continental Europe as well. To some extent, it was simply the result of the ascendency of quantum mechanics and the 'Shut up and calculate!' mentality. For a critique of Yourgrau's defensive narrative, compare Kahle (2021), Stachel (2007), and Hentschel (2005).

Kurt Gödel: *Mathematician, Philosopher, Physicist*

Gödel was all of those things, and more. He was on the one hand a thoroughly rational person, who wanted to explain *all the world* and all of human experience in a *rational* manner, and he found a safe haven in mathematics, where his rationality could express itself most effectively. Alonzo Church considered Gödel to be the most rational person he had ever known.[51] Of his rationality, Hao Wang[52] offers the following opinion:

> To be rational and reasonable are both to be agreeable to reason... Yet being reasonable seems to include more readily a flexibility (or even a weakness) associated with being sensible, sane, moderate, or not excessive. Some of Gödel's applications of the principle that everything has a reason appear not to be reasonable, even though they may perhaps still be termed rational. The extent of his distrust of doctors would seem to be neither reasonable nor even rational. If it could perhaps be argued that his apparent belief in the feasibility of religious metaphysics as an exact theory is rational, one is more strongly inclined to consider it not reasonable ... being 'fanatically rational' is, for me, no longer rational.

And of course in his personal life, Gödel was often highly *irrational*. He suffered from imaginary illnesses, he was anxious about many commonplace things, he took unnecessary medicines and avoided necessary ones, he was more than once seriously undernourished due to his strange and self-destructive eating habits (cf. Chap. 9). Yourgrau (2021) casts some doubts on Wang's opinion as quoted above, but he asks many questions rather than giving answers, a rather sneaky way to discredit an opinion without taking any responsibility for one's own.

None of this detracts from Gödel's great achievements in the three fields named above; on the contrary, the fact that he was able to carry out those achievements in spite of the handicaps thrown in his path and the difficulties of his everyday life makes them even more admirable. That he was often misunderstood was an additional difficulty, which likewise does not detract from his achievements. Rebecca Goldstein believes that he suffered greatly under that misunderstanding, especially on the part of philosophers (and most especially on the part of Ludwig Wittgenstein, a philosopher with whom Gödel by his own account had very little to do). Palle Yourgrau also charges that Gödel, as a philosopher, was consistently and almost criminally ignored by the 'mainstream' of contemporary philosophy, perhaps even the victim of a 'conspiracy of silence'. It is certainly true that the great wave of attention to Gödel's life and work arose only after his death, and it crested around the time of his centennial in 2006, although it has by no means ebbed away in the ensuing 16 years. Let us close this chapter with the conclusion that his 'excursion' into physics was in reality a case of '*to thine own self be true*', and that it was a consummate expression of Gödel the *philosopher* rather than Gödel as a mathematician or a physicist.

Notes

1. Photo: by Paul Ehrenfest (1921), from Wiki Media, public domain; cropped and enhanced. Reused from: https://commons.wikimedia.org/wiki/File:Einstein_en_Lorentz.jpg.
2. See also the recently-published book '*Einstein in Bohemia*', by Michael D. Gordin, Princeton University Press, Feb. 2020; ISBN 978–0-691–17,737-3.
3. Photo: courtesy of the copyright holder, *Astrophysikalisches Institut Potsdam*, September 21st, 2005. Reused from: https://commons.wikimedia.org/wiki/File:Einsteinturm_7443a.jpg.
4. Quoted in Graf-Grossmann (2018), p. 83, from the obituary of Grossmann written by Kollros for the *Neue Helvitische Gesellschaft*, Geneva, 1937. Original in French.
5. *ibid.*, p. 87.
6. Misner, Thorne & Wheeler (1973):'*Gravitation*'. Charles W. Misner, Kip S. Thorne, and John A. Wheeler, Princeton University Press (1973), reissued 2017. ISBN: 978-1-400-88, 909–9.
7. Rindler (2001/2006): '*Relativity: Special, General and Cosmological*', Oxford University Press (2001), reissued 2006. ISBN: 978-0-198-50, 835–9.
8. Photo: by Vysotsky, July 13th, 2016. Licensed under Creative Commons Attribution-Share Alike International 4.0 license; reused from: https://commons.wikimedia.org/wiki/File:EinsteinLeiden4.jpg.
9. Einstein used old-fashioned German *Fraktur* capitals for tensors, but they are hard to recognize for many modern readers. He also used standard (Roman) capitals in his writings e.g. in French.
10. Gödel called this a 'linear element' in his publications.
11. Drawing: from WikiCommons, courtesy of the copyright holder. File modified. Reused from: https://commons.wikimedia.org/wiki/File:Diagonal_uhlopricka.jpg#mw-jump-to-license.
12. See e.g., Liddle (2006).
13. For more information on this topic, see the introduction by David B. Malament to the text of Gödel's 1949 lecture at the IAS, in the *Collected Works*, Vol. III.
14. Wang (1987).
15. See Menger (1994), in particular the section on Gödel, Point 6.
16. In the *Collected Works*, Vol. III, quoted for example by Stein and by Malament (1995) (in that volume), and by Audureau (2016).
17. Compare e.g. the article by Audureau in [Crocco & Engelen, eds. (2016)], '*On the relevance of cosmology for the understanding of Gödel's philosophy*', [Audureau (2016)], Open Edition, online 2021.
18. In Audureau (2016), footnote [17].
19. Gamow (1946).
20. See Jung (2006).

21. Photo: by K.T. Bainbridge, courtesy *AIP Emilio Segrè Visual Archives*, Bainbridge Collection; cropped and enhanced. Reused from: https://reposi tory.aip.org/islandora/object/nbla%3A294,395. Original source: Cavendish Laboratory. Photography Collection, Dept. of Physics, University of Cambridge/UK. No known copyright; fair use.

22. Hintikka (1999), etc.

23. Quoted from Wang (1997), p. 88.

24. Dawson (1997, 2005 edition), pp. 177 ff.

25. We follow the treatment in Rindler (2009). Rindler includes many details and explanations not given by Gödel in his lecture, and his article is of course modernized in terms of notation and terminology as well as methods.

26. Readers who are interested in the mathematical details should consult Rindler (2009) and/or Gödel (1949b).

27. Image: a drawing by Incnis Mrsi, June 2008. Released by the creator to the public domain, see Wiki Media at: https://commons.wikimedia.org/wiki/ File:Light_cone_colour.svg.

28. Quoted from Albert Einstein, '*Die formale Grundlage der allgemeinen Rela- tivitätstheorie*', in: *Sitzungsberichte der Königlich-Preussische Akademie der Wissenschaften*, Vol. XLI (1914), pp. 1030–1085; see p. 1079.

29. Anti de Sitter space: See e.g. Synge (1960), p. 264.

30. Lanczos (1924).

31. Images: *Left*: Creative Commons, Wiki Media. *Right*: Creative Commons Attribution-Share alike 3.0 Unported license. Reused from: https://public ism.info/science/eternity/7.html.

32. See, for example, Natario (2011). See also Pfarr (1981); he gives an estimate for the radius of the earth after returning to its point of origin as 210 m.

33. But see his additions to the footnotes in the German translation of his IAS lecture of May 1949 [in Gödel (1956)].

34. Chandrasckhar & Wright (1961).

35. See Gold & Bondi (1967).

36. See https://www.lorentz.leidenuniv.nl/history/stockum/stockum.html for some information on his history.

37. Wheeler & Ford (1998), pp. 309 ff.

38. Reprinted in the *Collected Works*, Volume III.

39. A conclusion reached in somewhat more detail by Palle Yourgrau, who holds "the dialectic of the formal and the intuitive" to be "the leitmotif of Gödel's lifework". See Yourgrau (2005), p. 124, and Wang (1987), p. 201.

40. Wheeler & Ford, *op. cit.*, pp. 310 and 311/312.

41. See https://kurtgoedel.de/.

42. The essay was published in the journal *Synthese* in 1995; cf. Wang (1995).

43. See Gödel (1956), footnotes 11 and 14.

44. Yourgrau (2005), pp. 124 ff.

45. Cf. McTaggart (1908), and the article on McTaggart by Kris McDaniel (2020) in the *Stanford Encyclopedia of Philosophy*.

46. Einstein (1949): Einstein's remarks on the essays appearing in the *Festschrift* volume, Schilpp (1949), pp. 663–688.
47. Compare the quote by Wolfgang Rindler (2009) at the beginning of this section.
48. Wang (1987), p. 201.
49. Gödel (1990) refers to the *Collected Works*, Volume II.
50. This is reminiscent of Gödel's own belief in a conspiracy to suppress the ideas of Leibniz, which he imagined to be active right up to modern times and even in—for example—American university libraries. His friend Oskar Morgenstern at first scoffed at his obsession, but then he was startled by examples of missing books and manuscripts that Gödel showed him. They carried on a long, unsuccessful campaign to archive copies of Leibniz's papers in Princeton.
51. See the review of Budiansky (2021) by Palle Yourgrau, online at https://kurtgoedel.de/review-essay-by-palle-yourgrau/#_edn7. In his Ref. [viii], he attributes this quote to C. A. Anderson (1998): '*Alonzo Church's Contributions to Philosophy and Intensional Logic*', in: *The Bulletin of Symbolic Logic*, Volume 4, No. 2 (June 1998), p. 129.
52. Wang, *op. cit.*, p. 225.

16

The 1950s. *Disciples—Recognition.*

The Disciples of the Master

Beginning in the mid-1940s, following World War II, Gödel's fame started to spread among a younger generation of mathematicians and logicians. By that time, his activities in mathematical logic had decreased, after several years of unsuccessful efforts to prove the independence of the General Continuum Hypothesis from the axioms of ZFC. He turned more and more to philosophy, interrupted for a few years by his excursion into cosmology, as we saw in the previous chapter.

Gödel's increasing fame—albeit within only a small, select group of people: Mathematicians, and within that larger group, mainly those who studied fundamentals of mathematics and mathematical logic—resulted in many inquiries and occasional unsolicited visits by would-be admirers, some of them 'crackpots'. Gödel's friend Einstein was accustomed to that, and in his case it was much more intense, since he was known all over the world to people in many areas of life, and not just to a small group of specialized academics.

Some of Gödel's admirers, many about 20 years younger, turned out to be very interesting people in their own right, and a few real friendships resulted, but often *distant* friendships, of the kind that Gödel could best deal with. We may realistically call them his '*disciples*'.[1] Hao Wang gives a list in the second chapter of his (posthumously published) book, '*A Logical Journey*' [Wang (1997)]; he calls them '*logicians who saw him* [Gödel] *as their master*': William Boone, Paul J. Cohen, Stephen C. Kleene, Georg Kreisel, Abraham

© The Author(s), under exclusive license to Springer Nature
Switzerland AG 2022, corrected publication 2023
W. D. Brewer, *Kurt Gödel*, Springer Biographies,
https://doi.org/10.1007/978-3-031-11309-3_16

Robinson, Dana Scott, Clifford Spector, Gaisi Takeuti, Stanley Tennenbaum, and Wang himself. Verena Huber-Dyson (2005) gives a similar list which in addition includes John Myhill. We will mention some details about each at the appropriate point in our chronology, but we begin with the three who published the most informative articles and books about Gödel:

Hao Wang, a mathematician/philosopher of Chinese origin who first met Gödel in 1949; *Stephen C. Kleene*, Church's Ph.D. student in Princeton when Gödel first went there in 1933/34; and *Georg Kreisel*, a mathematician originally from Austria, who arrived in Princeton via the United Kingdom in 1955. Wang and Kreisel, in particular, played a role in Gödel's later life and work. Gödel had very few doctoral students and never formed a 'school', but—like Wittgenstein[2]—he had *disciples*, and Wang and Kreisel certainly fall into that category. Wang, in particular, accompanied Kurt Gödel on the last 10 years of his life's journey, publishing three books and a number of articles on Gödel's work and on his philosophy.[3]

Hao Wang (1921–1995) was born in Jinan in the *Shandong* Province in East-Central China. He studied mathematics at the *National Southwestern Associated University*, obtaining the BSc in 1943 (this university was an association of several universities which had to move from Peking and Tianjin due to the Sino-Japanese war beginning in 1937. One of them was the *Tsinghua University* in Peking, where Wang was then granted the MA in philosophy in 1945, with high honors). He obtained a Rockefeller Fellowship to study in the US, at Harvard University, and finished his Ph.D. in mathematical logic under the guidance of W.v.O. Quine there in 1948, in the record time of 18 months.

He was then an assistant professor at Harvard, but the following year, he was in Princeton, where he first met Kurt Gödel. He spent 1950/51 in Zurich with Paul Bernays, held a lectureship at Oxford University from 1956, and in the late 1950s he worked at IBM, where he wrote a program to 'mechanically' prove theorems, which he applied to Whitehead and Russell's *Principia* (PM). From 1961, he was professor of Mathematical Logic and Applied Mathematics at Harvard, moving to Rockefeller University in New York in 1967. He remained there until his retirement in 1991. He is noted for developing the *Wang Tiles*, which can serve as analogs to a Turing machine, and which formed the basis for quasicrystals (Nobel prize for Chemistry in 2011 to Dan Schechtman, who first observed quasicrystals as real objects). Wang died in New York in 1995 of a lymphoma, not quite 74 years old.[4]

Wang (Fig. 16.1) was very interested in both the mathematical logic and the philosophy of Kurt Gödel, and especially in the later 1960s and the 1970s, he carried on an active correspondence and had many conversations with Gödel, some in person but mainly by telephone. They formed the basis

Fig. 16.1 Hao Wang, 1970s. Photo: San-You and Jane Hsiao-Ching Wang, private archive. Used here with permission from the Wang family[5]

for three of his books [his (1974), (1987), and (1997)]. His (1974), entitled *'From Mathematics to Philosophy'*, is a broad survey of the development of mathematical logic and its relation to philosophy, which emphasizes Gödel's contributions but also gives an excellent overview and introduction to many other topics. Gödel himself said that it best reflected his own philosophy. The two later volumes were more straightforwardly biographical in nature: *'Reflections on Kurt Gödel'* (1987), written after Gödel's death, and *'A Logical Journey'* (1997), recollections of Wang's own life and career and his relations to Gödel, published after Wang's death in 1995. All three are valuable sources on Gödel's work, and especially on the relationship of his mathematical logic to his philosophy.

Stephen Cole Kleene (1909–1994) was born in Hartford, CT on January 5th, 1909. He studied mathematics at Amherst College, completing his BA in 1930, and then went to Princeton as an instructor. He worked for a second year as a high-school teacher, then had a doctoral fellowship at Princeton University. He completed his Ph.D. thesis under Alonzo Church there in 1933, and in 1934/35 he had a research fellowship at Princeton.[6] It was during the latter stay that he, together with J. Barkley Rosser, wrote and edited the lecture notes from Gödel's lecture series at the IAS in the spring of 1934. We met up with him in Chap. 10 (in the section on 'Gödel's First

Stay at the IAS'; see also Fig. 10.6). He joined the faculty at the University of Wisconsin (Madison) in 1935, but he was back in Princeton as a Visiting Scholar at the IAS in 1939/40, when he founded the branch of mathematical logic known as *recursion theory*.

After a year (1941) back at his *alma mater* Amherst College, Kleene served in the U.S. Navy during WWII. In 1946, he rejoined the faculty at the University of Wisconsin/Madison, becoming full professor in 1948, and remained there until his retirement in 1979. The mathematics library is named in his honor.

In 1976, Kleene wrote a memoir on Kurt Gödel's work [Kleene (1976)], published in the *Journal of Symbolic Logic* (it was based on a Survey Lecture that he delivered to the *Association for Symbolic Logic* in NY on December 29th, 1975). He began with a discussion of [Gödel (1930)], the published version of Gödel's Completeness proof, of which Kleene says, "*Perhaps no theorem in modern logic has been proved more often than Gödel's complete-ness theorem for the first-order predicate calculus. It stands at the focus of a complex of fundamental theorems, which different scholars have approached from various directions*". He points out that Gödel's proof automatically contains the Löwenheim-Skolem theorem (1915–1920) and the compactness theorem (cf. Chap. 7). Kleene then turns to Gödel's (1931), his incompleteness proofs, and gives a brief history of the formalist school (Hilbert et al.). He proceeds to describe the arithmetization of metamathematics, quoting Alfred Tarski [(1933); see Tarski (1936)]:

> ... 'the idea *[of arithmetization]* was developed far more completely and quite independently by Gödel' ... *[Gödel says]*, 'The formulas of a formal system ... in outward appearance are finite sequences of primitive signs ..., and it is easy to state with complete precision which sequences of primitive signs are meaningful formulas and which are not. Similarly, proofs, from a formal point of view, are nothing but finite sequences of formulas (with certain specifiable properties)'.

Kleene goes on to describe Gödel's recursive functions, which he himself (in 1936) renamed 'primitive recursive functions' (PRFs). He continues with the history of incompleteness, citing Gödel's 1934 lectures and his own and Church's (1936) articles, as well as Turing's thesis (1939).

In the last section of his lecture, Kleene describes Gödel's work on set theory, summarizing with the statement,

In an expository paper, '*What is Cantor's continuum problem?*' (1947), Gödel expressed the views that 'the axiom of choice... is, in the present state of knowledge, exactly as well founded as the system of the other axioms' (Footnote 2), but 'one may on good reason suspect that the role of the continuum problem in set theory will be this, that it will finally lead to the discovery of new axioms which will make it possible to disprove Cantor's conjecture' (p. 524). Paul J. Cohen in (1963/64), using a quite different model than Godel's L, showed that the negation of the continuum hypothesis is consistent with ZFC, thus eliminating the first possibility. So $2^{\aleph_0} = \aleph_1$ is undecidable from the axioms of ZFC.

He concludes his lecture with the sentence,

We can be certain of one thing: Gödel's work here, as well as in the demonstration of undecidability phenomena in formal number theory and in a definitive treatment of the completeness of first-order logic, will have remained a landmark whose visibility from afar will have been undiminished by time" *[paraphrasing von Neumann's remarks about Gödel's work on the occasion of the 1st Einstein Award]*.

Just over two years after the lecture by Kleene quoted above, Gödel had passed on. Kleene himself retired three years later. Some time before his retirement, Daniel Dennett coined the saying, published in 1978, that[7] "*Kleeneness is next to Gödelness*". Kleene died in 1994, at 85 years of age. He was in the most relevant sense of the word a *disciple*, perhaps the very first, of Kurt Gödel.

Georg Kreisel (1923–2015) arrived in Gödel's sphere of influence somewhat later than both Kleene and Wang. He was born in Graz, Austria, to a Jewish family who wisely sent him and his brother to live in England before the *Anschluss* in 1938. After completing his schooling, he went to Trinity College, Cambridge, where he was influenced by Ludwig Wittgenstein, also at Trinity; the latter is quoted as saying of Kreisel that he was '... *the most able philosopher he had ever met who was also a mathematician*'—to the surprise of some of his fellow students and of Rush Rhees, a colleague and later Wittgenstein's literary executor.

After completing his BA in 1944, Kreisel was assigned to war work with the Royal Navy, and he studied the effects of waves on the artificial harbors planned for the Normandy landing. In 1946, he returned to Cambridge and worked on his doctorate. The record is unclear about the outcome; some sources say that he obtained the doctorate but left Trinity because he failed to be granted a research fellowship there; others say that he never received the doctorate. He was, in any case, granted a D.Sc. by Cambridge University in 1962 (a *doctor of science* degree, a 'higher doctorate', in some respects analogous to the *Habilitation* or the *docteur d'état*).

He took a position on the faculty of the University of Reading, where he worked intermittently from 1949–1960. This was however probably not a very appropriate position for someone of his talents; Verena Huber-Dyson wrote, in the *Festschrift* for Kreisel's 70th birthday [Odifreddi, ed. (1996)], "*Although he did not come with the explicit intention of staying, and the possibility of return to Europe was open to him, Kreisel did not have strong ties to his position at Reading, a rather puzzling domicile for a person of his peculiar qualities*". This quote refers to his stay at the IAS in Princeton in 1955–57, where he was invited by Gödel, at the urging of Freeman Dyson (who knew Kreisel from undergraduate days at Cambridge). Huber-Dyson was Freeman Dyson's wife at the time, and the mother of his two children, Esther and George (she also had an older daughter, Katarina, from her first marriage (1942–48) to Hans-Georg Haefeli, a Swiss mathematician). She married Dyson in 1950, after going to the IAS on a postdoctoral fellowship, but was divorced from him in 1958, most likely as a result of her relationship with Georg Kreisel during the latter's stay in Princeton.

Another passage from Huber-Dyson's (2005) regarding Kreisel sheds some light on his personality and his relations with Kurt Gödel. She notes that Kreisel was ('*and probably still is*') a complex personality, a brilliant logician, and difficult in personal relations. She then recounts how during his stay as a Fellow at the IAS in Princeton, he became a close 'family friend' of the Gödels at the time (1955–57). Furthermore, they remained in contact through letters and long telephone conversations after Kreisel left Princeton. She says that Kreisel kept up similar contacts ('*philosophical ruminations*') with Hao Wang, but that Wang never became as intimate with the Gödels as did Kreisel, perhaps because the latter's Austrian origin gave him a 'kinship' to both Kurt and Adele on a personal level, in addition to his common interests in mathematics and philosophy with Kurt.

Kreisel returned to Reading in 1957, but was on leave to Stanford in 1958/59, and again in 1960/62, when he was at the University of Paris. In 1962, he moved permanently to Stanford, where he remained on the faculty until his retirement in 1985. Thereafter, he lived in Salzburg for the rest of his life; he died there in 2015. He had effectively broken off his contacts to Gödel around 1969, when he said that he '*couldn't stand to watch Gödel trying to hide his depression*' any longer. This may be seen as an expression of particular sympathy for Gödel, or also one of considerable egoism on the part of Kreisel.

Fig. 16.2 Georg Kreisel, 1974. Photo: by Paul Halmos, reused with permission[9]

Kreisel (cf. Fig. 16.2) had lifelong friendships with the novelist/philosopher *Iris Murdoch*, whom he met during her stay at Cambridge in 1947, and with the molecular biologist *Francis Crick* (they met while doing work for the Royal Navy during WWII). Murdoch corresponded with Kreisel for many years, and he evidently served as a model for some of her fictional characters. Her biographer *Peter Conradi* was quoted as saying,[8] "*For half a century she nonetheless records variously Kreisel's brilliance, wit and sheer 'dotty' solipsistic strangeness, his amoralism, cruelty, ambiguous vanity and obscenity*".

A rather different characterization of Kreisel's personality is given by his former student Henk Barendregt [see Barendregt (1996)] in his essay '*Kreisel, lambda calculus, a windmill and a castle*'. He recounts how Kreisel, at the time a Visiting Scholar in Utrecht, met an English baroness with Dutch connections who was living near Utrecht in a local castle, and how he dealt with her small son during a dinner at the castle (which Barendregt evidently also attended). During the meal, the young son of the baroness began to whine and cry that he did not want to eat the raisins in his rice pudding, and asking his mother to take them out of the pudding. Kreisel exhibited a side of his personality previously unknown to Barendregt, and showed his considerable understanding of human nature, by calmly asking the boy which would be worse for him: that his mother would refuse to remove the raisins, or that she was simply unable to remove them. After some thought, the boy said that it would be worse if she refused to remove them… to which Kreisel replied that she in fact was simply not able to remove them—leaving the now calm and distracted boy no alternative but to eat his pudding as it was.

Kreisel's major contribution to the memory of Kurt Gödel is the long (78 page) memoir of Gödel and his work that he wrote for the Royal Society in 1980 [cf. Kreisel (1980)]. Gödel was elected to Foreign Membership of the R.S. in 1968, two years after Kreisel was selected to be a Fellow of the Royal Society. Verena Huber-Dyson (2005) says of this memoir: "*By far the most illuminating survey of Gödel's work including its methodological and philosophical implications is Kreisel's memoir for the Royal Society*". It follows the same general plan as Stephen Kleene's memoir quoted above [Kleene (1976)], but is much more detailed, giving the historical context of Gödel's work in mathematical logic, the particular characteristics of his proofs and methods, and a survey of their reception and influence up to 1980.

Kreisel opens his (1980) with a remark on Gödel's place in the history of mathematical logic:

Kurt Godel did not invent mathematical logic; his famous work in the thirties settled questions which had been clearly formulated in the preceding quarter of this century. Despite sensational presentations by crackpots, philosophers and journalists... Godel's results have not revolutionized the silent majority's conception of mathematics, let alone its practice... Certainly, those results refuted most elegantly each of the grand foundational 'theories' current at the time, of which Hilbert's, on the place of rules in mathematical reasoning, and those associated with Frege and Russell, on its reduction to universal systems like set theory, were most popular.

Of Gödel himself and of his private life, Kreisel says in his (1980),

[About Gödel's purported 'lack of sensibility' (*i.e.,* empathy)]: "I myself witnessed a degree of understanding, whether intuitive or as a result of reflection, which is exceptional by any standards ..." [About Adele and Kurt]: "It was a revelation to see him relax in her company. She had little formal education, but a real flair for the *mot juste*, which her somewhat critical mother-in-law eventually noticed too, and a knack for amusing and apparently quite spontaneous twists on a familiar ploy... his [Gödel's] aversion, after World War II, to Austrian academic institutions seems out of all proportion, and remained a total puzzle to his family, as documented, for example, by his mother's letter of 28 January 1963 to her brother Karl. He was offered, and refused, sometimes for mind-boggling reasons, membership and later honorary membership of the *Academy of Sciences* in Vienna, and the highest national medal for science and the arts. (He had no chance to refuse an honorary doctorate of the University of Vienna, since it was awarded posthumously.)".

And about Gödel's relations to the other mathematicians at the IAS,

After all, he always stressed the conflict between his views and the *Zeitgeist*, to which, naturally without empirical checks, he supposed his colleagues at the IAS to be subject. He was more disappointed than he let on by his occasional failures to persuade them; but not nearly as much as he would have been had he realized that he was battling a *Zeitgeist* from another time, the early thirties; and then not what it was, but what the *Wiener Kreis* would have wanted it to be.

We will occasionally quote other passages from this remarkable memoir, which (in spite of a number of trivial errors in dates and details of Gödel's early life) is probably, indeed, one of the best characterizations of his life and work.

Gödel's Honors and Awards

We have seen briefly in Chap. 9 that Gödel received considerable attention in the 1950s, including a number of honors and awards. Some of that may have been due to his friends' efforts to cheer him up and set the record straight (in particular, von Neumann and Morgenstern), but it would seem that the (academic) world had suddenly discovered his achievements and had resolved to honor them. He gave an important lecture (on his relativity work) at the *International Mathematicians' Conference* in September, 1950 (cf. Chap. 15), and then, after his ulcer crisis in early 1951, he was co-recipient of the *Einstein Award*.

Dawson [(1997), 2005 edition, p. 194 ff.] tells the story of how Oskar Morgenstern and J. Robert Oppenheimer, at that time director of the IAS, had discussed how to improve Gödel's mental outlook after his illness, and concluded that the time was ripe for him to be named professor at the Institute. However, there was some opposition to be expected from the faculty, and Oppenheimer could not promise to put through Gödel's appointment in the immediate future (in the event, it took two more years). But as an alternative, he proposed that Gödel should be nominated to receive the first Einstein Award. It had been endowed two years earlier by Louis L. Strauss, an influential trustee of the IAS (who later played a role in Oppenheimer's removal from the Atomic Energy Commission), and it included a gold medal and prize money of $15,000. It was to be awarded on Einstein's birthday, March 14th, beginning in 1951 and continuing at three-year intervals.

There were, however, some matters of protocol which had to be settled. The prize committee consisted of Oppenheimer, Weyl, von Neumann and Einstein. Oppenheimer thought it an impropriety to nominate a member of the IAS by a committee from within that Institute, all of whom were known to be friends of Gödel's. And furthermore, *Julian Schwinger* had already been nominated for the prize, a young theoretical physicist from Harvard who had formulated a new version of quantum electrodynamics (QED) (and was later to receive the Nobel prize in 1965, together with Richard P. Feynman and Shin'Ichirō Tomonaga, for that achievement).

But Schwinger's nomination had not yet been announced, so the committee was able to arrange a double award, to Schwinger and to Gödel. Gödel was apparently very happy about the prize and pleased to finally be given an academic honor for his work (this was his first real public appreciation).

The Einstein Award was given at varying intervals from 1951 to 1979 (the year of Einstein's centennial). It had been endowed by the *Louis and Rosa Strauss Memorial Fund* in honor of Einstein's 70th birthday in 1949. The Award was administered by the IAS, which provided the jury from among its faculty.

The first award, to Gödel and Schwinger, was the only one to be presented to two recipients. Later awardees included Richard Feynman, Edward Teller, Willard Libby, Leó Szilárd, Luis Alvarez, John Wheeler, Marshall Rosenbluth, Yu'val Ne'eman, Eugene Wigner, Stephen Hawking, and Tullio Regge. Schwinger, Feynman, Libby, Alvarez, and Wigner were also Nobel prize winners, mostly after they received the Award.

Two remarks made at the award presentation (Fig. 16.3) and celebration have achieved a certain fame: Einstein, as he presented the medal to Gödel, reportedly said to him, "*And here my dear friend, for you—and you don't need it*". And von Neumann, who gave the *laudatio* of Gödel at the banquet luncheon, famously said of Gödel's work that it was "*a landmark which will remain visible far in space and time*" (cf. Chap. 8).

Perhaps in part because of the Einstein Award, Gödel also received two honorary doctorates within the year: a D.Litt. from Yale, during its 250th anniversary ceremony on October 19th, 1951; and an honorary D.Sc. from Harvard, proposed by W.v.O. Quine and conferred in June, 1952. His work was called there "*the most significant mathematical truth of this century*" [cf. Wang (1987), p. 118]. (He later received two other honorary degrees, from Amherst College in 1967 and from Rockefeller University in 1972. Princeton did not offer him one until 1975, and he refused to go to the ceremony to accept it, saying it was 'too late'. The degree is listed in the announcement

Fig. 16.3 Albert Einstein, Louis Strauss (nearly hidden behind Einstein), Kurt Gödel and Julian Schwinger at the presentation of the first Einstein Award, March 14th, 1951 at the Princeton Inn. Photo by Alan Richards (1951), reused with permission[10]

of the ceremony but was never conferred on Gödel). He was also elected to membership of the US *National Academy of Sciences* (in 1955) and the *American Academy of Arts and Sciences* (in 1957). As mentioned in the above quote from Georg Kreisel, he was also offered membership and honorary membership in the Austrian Academy, but refused both (claiming political reasons as an American citizen).

Another signal honor was granted to Gödel in late 1951: He had been invited to deliver the *Gibbs Lecture* at the annual meeting of the *American Mathematical Society* in December, to be held at Brown University in Providence, RI (jointly with the annual meeting of the *Mathematical Association of America*). The Lecture had been established in 1924, in honor of *Josiah Willard Gibbs*, an eminent American mathematical physicist and thermodynamicist. Gödel's title for his talk was '*Some Basic Theorems on the Foundations of Mathematics and Their Philosophical Implications*'. Its text is reconstructed in the *Collected Works*, Vol. III, pp. 304–323, under the citation [*Gödel *1951*]. (Gödel originally intended to publish the text of his lecture in the proceedings of the meeting, in the *Bulletin of the American Mathematical Society*. He wrote letters to representatives of the Society confirming that in May, 1953, and January, 1954, and worked on his revisions for around two years). But he was never satisfied with the results; Dawson[11] speculates that he found his own arguments not very convincing. His revisions were found in his legacy, with many additions and deletions, so that it was impossible to reconstruct exactly what he had said in the original lecture. See the *Collected Works*, Vol. III.

The Gibbs Lecture, 1951

The Gibbs Lecture had a prestigious roster of previous speakers. In 1951, it was held on December 26, a Wednesday and the day after Christmas. Gödel's subject was his Completeness and Incompleteness theorems from 1929/31, and their relevance to mathematics and philosophy. This was his last public lecture, and it received a certain amount of publicity, in the popular press as well as in scientific journals. In the lecture, he emphasized his Platonic view that mathematical truths are innate in the world and are not created or constructed by human efforts. In support of that idea, he pointed out that the creator (or, in Gödel's view, the discoverer) of a new mathematical concept has very little freedom to pattern it at will—its form is determined by the nature of the subject; and is thus *not* just a product of the human imagination.

Furthermore, his Incompleteness results show that 'mathematical intuition' can develop axioms which cannot however be completely captured by a 'mechanical' procedure, so that either the mind '*infinitely surpasses the powers of any finite machine*'; or else it '*is equivalent* to a finite machine', in which case there would be *absolutely* unsolvable mathematical problems (i.e., undecidable in *any* conceivable mathematical proof system). Gödel then asserts that, "*It is this mathematically established fact which seems to me of great philosophical interest. Of course, in this connection it is of great importance that at least this fact is entirely independent of the special standpoint taken toward the foundations of mathematics*". He then turns to '*the philosophical implications*': "*Of course, in consequence of the undeveloped state of philosophy in our days, you must not expect these inferences to be drawn with mathematical rigour*". He continues:

> … if the first alternative holds, this seems to imply that the working of the human mind cannot be reduced to the working of the brain, which to all appearances is a finite machine… On the other hand, the second alternative, where there exist absolutely undecidable mathematical propositions, seems to disprove the view that mathematics is only our own creation; for the creator necessarily knows all properties of his creatures, because they can't have any

others except those he has given to them. So this alternative seems to imply that mathematical objects and facts (or at least something in them) exist objectively and independently of our mental acts and decisions, that is to say… some form or other of Platonism or 'realism' as to the mathematical objects.

He goes on to give several arguments supporting this kind of Platonistic view, '*irrespectively of which alternative holds*'.

In the Notes on Gödel's Gibbs Lecture, written by George Boolos for the *Collected Works*, Vol. III [cf. Gödel (*1951*) there], Boolos concludes his comments by summarizing the situation some 40 years after Gödel's talk, when that volume was published:

> What is surprising here is not the commitment to Platonism, but the sugges-
> tion, which recalls Leibniz's project for a universal characteristic, that there
> could be a mathematically rigorous discussion of these matters, of which the
> correctness of any such view could be a 'result'. Gödel calls Platonism rather
> unpopular among mathematicians; it is probably rather more popular among
> them now, forty years after he gave his lecture, in some measure because of his
> advocacy of it, but perhaps more importantly because every other leading view
> seems to suffer from serious mathematical or philosophical defects. Gödel's idea
> that we shall one day achieve sufficient clarity about the concepts involved in
> *philosophical* discussion of mathematics to be able to prove, mathematically,
> the truth of some position in the philosophy of mathematics, however, appears
> significantly less credible at present than his Platonism.

Herr Professor Dr. Gödel

The year 1953 was a significant one in Gödel's life for several reasons; the most important was his promotion to the position of Professor in the School of Mathematics of the IAS. As we have seen in Chap. 9, this was considered by most of his friends to be long overdue; von Neumann even remarked,[12] "*How can any of us be called professor when Gödel is not?*" On the other hand, Gödel seemed to be satisfied with his previous position as permanent member, which gave him a secure status without too many demands of an administrative or institutional sort.

Once he became professor, his relations to the rest of the faculty changed; he necessarily took on responsibility for joint decisions, the most important of which was the choosing of new fellows and faculty members. Frictions resulting from decisions on those matters had long since caused problems at the IAS (for example, Abraham Flexner's retirement in 1939 was directly provoked by a disagreement with the faculty on hiring policies). Most of the faculty seems to have been inclined to be disputatious toward the IAS administration and its director, while Gödel was ready to accept authority, but at the same time exceedingly meticulous in making any decisions. This combination isolated him from the rest of the faculty, so that not only was he effectively excluded from participating in decision-making (except for his own specific field of mathematical logic, of which he was often the only representative, or only with Hassler Whitney); he was also increasingly excluded from scientific discourse—and he had never been aggressive in seeking it out. There has been some discussion of whether the 'insulated' environment at the IAS was not counterproductive for his scientific development, even before he became professor (which further exacerbated the points of friction with his fellow faculty members).

Certainly, it could not compete with the earlier Vienna days, with the Circle and the Mathematics Colloquium. But those comparatively happy times had come to an end by the mid-1930s, for reasons far outside the control of any of the participants. In addition to the structural problems at the IAS, Gödel was faced with the loss of several of his old friends and discussion partners in the second half of the 1950s, as we shall see.

Gödel's former friend, mentor and colleague *Rudolf Carnap* spent two years at the IAS, from September 1952 until September 1954. They had experienced a period of intensive contact, and a sort of advisorship on the part of Carnap, in the period 1927–1930. Their positions were reversed after the publication of Gödel's important articles on mathematical logic in 1930/31: Gödel was now more the 'advisor' to Carnap on matters of logic. And their close contacts were essentially ended when Carnap went to Prague in 1931, or at the latest after his emigration in 1935. There seems to be little information available about their possible renewed contacts while they were both in Princeton in 1952–54, but it is relatively certain that they did not interact very intensely.

The Carnap Book

Interestingly, just in this period, in the spring of 1953, Paul Schilpp again turned up with another book project for his series on *Living Philosophers*, namely a volume on Carnap. He again approached Gödel, who had indeed made important contributions to the volumes on Russell and on Einstein, and asked him to write a chapter on Carnap, its subject to be '*Carnap and the Ontology of Mathematics*'. Gödel was willing to write a shorter contribution, which he titled '*Some Observations on the Nominalistic View of the*

Nature of Mathematics'. His intention was to prove that mathematics is not *just* syntax. This refers to Carnap's book, written in 1934 [and published in English in 1937 as '*The Logical Syntax of Language*'; see Carnap (1934/37)]. Gödel worked intensively on his contribution to the planned book in 1954 and 1955, as shown by entries in Morgenstern's diary; and he continued working on it through 1958. In 1959, he informed Schilpp that he would not submit it, probably because he was uncertain of Carnap's reaction and/or how convincing his arguments were. The Carnap volume finally appeared in 1963, without Gödel's contribution. But many drafts were later found in Gödel's papers; Gödel himself later regretted in a conversation with Hao Wang that he had 'spent so much time on it'.

Hannes Leitgeb, in the *Stanford Encyclopedia of Philosophy* [Supplement to the article on Rudolf Carnap: *H. Tolerance, Metaphysics, and Meta-Ontology* (2020)] gives a modern assessment of Gödel's critique:

> Much of the recent discussion of the principle of tolerance *[put forth in later editions of Carnap's Logical Syntax]* was inspired by Gödel's critique of it, unpublished until 1995, that was originally intended for the Schilpp (1963) volume on Carnap in the Library of Living Philosophers. Gödel withdrew his paper, but six successive drafts of it were found in his Nachlaß, of which several have now been published [Collected Works, Vol. III, Gödel (*1953/9)]. It appears, from its title ('*Is Mathematics Syntax of Language?*') and exposition, to focus specifically on the 'syntax' thesis, or more generally on the 'linguistic' accounts of the foundations of mathematics deriving from Wittgenstein's *Tractatus* ... However, Gödel himself understood that Carnap had meanwhile left behind that view (in its ' *Logical Syntax*' form), and recent commentary has focused on the principle of tolerance.

Thus, we can see that Gödel's efforts were in the end not in vain, and that he need not have regretted spending time on his contribution, which—long after his own death, and that of Carnap—stimulated considerable research activity in logic and philosophy.

Bernays and the Dielectica Paper

Paul Bernays had been forced to leave Göttingen in 1934, after the Nazi takeover, and he went to Zurich (being a Swiss citizen), where he held a series of lectureships and an associate professorship. He went on sabbatical as a visiting professor to the University of Pennsylvania, not far from Princeton, in 1956, and took the opportunity while there to renew his collegial contacts with Gödel.[13] They maintained contacts thereafter, mostly by correspondence, until shortly before Bernays' death in 1977. Around the same time (1956), Gödel was invited to contribute to a *Festschrift* for Bernays, whose 70th birthday would take place in October of 1958. Gödel took the opportunity to write a text version of his lectures from 1941 (at Princeton and Yale). He submitted the paper to the Swiss journal *Dielectica*[14] (founded by Paul Bernays, Ferdinand Gonseth, and Karl Popper in 1947), to an issue devoted to Bernays' birthday. Gödel's article is summarized briefly by Dawson[15] and at greater length by Hintikka.[16] This article [Gödel (1958)] was the first that Gödel had published in German for a number of years, and it is often referred to as 'the *Dielectica* paper'. Hintikka (1999) makes the surprising remark that '*It is not clear what motivated Gödel to write it*', although its origin was explained by Dawson in his (1997). Its title was '*Über eine bisher noch nicht benützte Erweiterung des finiten Standpunktes*' ('*On a hitherto unused Extension of the Finitary Viewpoint*').

In his article, Gödel gives an interpretation of first-order logic and of arithmetic. He sets up translation rules for converting statements of first-order arithmetic into higher-order statements, restricting all function variables to recursive values. The translation rules [see Hintikka (1999) for a clear summary] at first seem rather arbitrary, but Hintikka (and before him Dana Scott) interpret them as 'rules of a certain game of verification and falsification'; in other words, they suggest a *game-theoretical* interpretation. Hintikka (1999, p. 63) gives an example of such a game in *first-order logic*, giving a set of rules which correspond to the translation rules in Gödel's article (which might be called the 'conjunction rule', the 'disjunction rule', the 'conditional rules', and the 'negation rule'). The game ends when it leads to an atomic sentence or an identity A, which is true or false (corresponding to a win by one of the two players, the 'verifier' or the 'falsifier'). The game in this form corresponds to the standard interpretation or model of a first-order language. Gödel's *extended* version leads to a non-standard interpretation of logic and arithmetic, restricting all the strategy functions to recursive functions (i.e., to *computable* functions). Hintikka notes that Gödel does not mention this game-theoretical interpretation of his extension, although he was no doubt aware of it.[17]

Dawson gives a much briefer summary of the '*Dielectica* paper', noting that it repeats what Gödel had described in more detail in his lectures in 1941, and that it contains an outline of his notion of a computable function of finite order, and how it might provide a constructive (but not finitary) proof of the consistency of number theory—and might be further extended to 'much stronger systems' which could be used to prove the consistency of analysis (a suggestion that he had already made to Bernays in a letter dated Feb. 6th, 1957, and verbally to Kreisel). Kreisel indeed made use of it to extend Gödel's interpretation to analysis. He lectured on that topic at the *Summer Institute for Symbolic Logic*, held at Cornell in the summer of 1957, thus directing the attention of logicians to Gödel's functional interpretations. The *Dielectica* paper has become an iconic Gödel contribution to mathematical logic.

But the story does not end there. Kreisel, after his stay at the IAS (1955–57), became a regular correspondent and interlocutor with Gödel (and Bernays; and to some extent with Wang). They all discovered another, younger Gödel disciple, of the second generation, so to speak: A doctoral student of Stephen Kleene at U. Wisconsin/Madison, where Kleene had by now been on the faculty for more than 20 years (with interruptions). His name was **Clifford Spector**, and we have seen it in the list of 'Gödel disciples' given above. Gödel invited Bernays to come to the IAS as a visitor in 1959/60, and Spector visited there also, to discuss his ideas on the consistency of analysis with both of them. Spector subsequently also went to the IAS for a longer stay (in 1960/61) and wrote up his work on the consistency of analysis while there. Gödel was favorably impressed, and wrote a recommendation for Spector, calling him 'probably *the* best logician of his age group in this country' [in a letter to George Hay, Feb. 23, 1961; quoted by Dawson (1997), p. 207 and Endnote 475]. Spector tragically died unexpectedly in the summer of 1961, the victim of an acute form of leukemia. Kreisel edited his manuscript, which was then published in the proceedings of a *Symposium in Pure Mathematics* [Vol. 5, *American Mathematical Society*, Providence, RI (1962)]; see [Spector (1962)]. Gödel wrote a postscript to the article, and the article itself was reviewed for example by Fernando Ferreira in a memorial volume dedicated to Gerhard Gentzen; see [Ferreira (2012)]. Gödel's postscript is discussed in Volume II of his *Collected Works* [together with Gödel (1947); see pp. 251–253]. Spector was truly a disciple of Kurt Gödel, and he would no doubt have made many other important contributions to mathematical logic if he had lived longer; his work is not forgotten.

Losses

Kurt Gödel had many colleagues and friends who accompanied him along his career and throughout his life, in spite of his shyness and difficulty with social contacts. We have seen that his important mentors from the period 1925–1935 were Hans Hahn, Karl Menger, Rudolf Carnap and Moritz Schlick; but he had effectively lost them by the later 1930s, due to Hahn's and Schlick's early deaths and the emigration of Carnap and Menger. Gödel continued to have contacts with the latter two, but after about 1940, they were never as close as they had been in the heyday of the *Wiener Kreis* and the *Mathematisches Kolloquium*. His personal friends in Vienna (e.g., Feigl, Natkin) were also dispersed by the mid-1930s. In Princeton, his scientific advocates and personal friends were initially Veblen, von Neumann, Morgenstern and Einstein. His closer colleagues from a purely professional point of view also included Church, Turing and Post, but his direct contacts to them were rather limited.

In the decade of the 1950s, two of his 'Princeton friends', Albert Einstein and John von Neumann, both died; Veblen followed in 1960 (but he had been living in retirement in Maine since 1950). Einstein had suffered from an aortal aneurism since the later 1940s, and his condition worsened threateningly in early 1955. At that time, he was 75 years old, and he evidently felt that his life was reaching its natural conclusion. Even many years before, he had emphasized in correspondence with his later wife Elsa that he was not a supporter of heroic medical interventions to prolong life at all costs, and when that situation became acute for him in April of 1955, he acted accordingly.

Einstein had not revealed the seriousness of his condition to Gödel (this was not difficult, since Gödel was so preoccupied with his own health that he tended to minimize the health problems of others in his perceptions). Einstein was intellectually active up to the end[18]; in March 1955 he celebrated his 76th birthday quietly. His household now consisted of himself, his stepdaughter Margot and his secretary Helen Dukas (since the death of his sister Maja in 1951). He wrote a last autobiographical note in March, and it was published in the *Schweizerische Hochschulzeitung* in connection with the centennial of his *alma mater*, the ETH Zurich. On April 13th, he collapsed—the aneurism had ruptured. He was taken to the Princeton Hospital on April 15th, a Friday, and his son Hans Albert came from California (he was professor of hydraulic engineering at UC Berkeley) and was able to spend the weekend with his father. Einstein died early on the morning of Monday, April 18th, 1955. Gödel had not seen him in his last days, not realizing the seriousness of his illness. So, he was surprised and shocked when he learned of Einstein's death, probably on the Monday, and was informed of its cause by one of Einstein's doctors.

Gödel helped to organize Einstein's papers in his office in Fuld Hall, and he attended a memorial service, a concert given by the Institute in Einstein's memory in late 1955. Dawson quotes Gödel as saying that it was '*the first time that he had allowed himself to endure two hours of Bach, Haydn and the like…*'.[19]

A few months after Einstein's death, von Neumann was diagnosed with bone cancer. After several more months, he was partially paralyzed and had to use a wheelchair. He '*entered Walter Reed Hospital in January, 1956, and remained there, with brief exceptions, until his death in February of 1957*'.[20] Gödel's meetings with him had been rare even before his illness. Von Neumann had been involved in computer development during and since WWII, and he led a project to construct an electronic computer at the IAS after the war. Gödel was occupied with other topics, however, and apparently showed little interest in the computer project, although he kept himself informed. In March of 1956, when von Neumann was already seriously ill and confined to the hospital, Gödel wrote to him, suggesting what later became known as the '$P = NP$' problem (which stands for '**P**olynomial time' and '**N**on-deterministic in **P**olynomial time'). It deals with the question of whether the problems solvable for example by a deterministic algorithm on a Turing machine in 'polynomial time' (i.e., 'mechanically', in a time which increases as n or n^2, where n is the number of symbols in a logical proof) are the same as those problems which are solvable with a *non-deterministic* algorithm in polynomial time. This question is now considered to be a major unsolved hypothesis in theoretical computer science, and it testifies to Gödel's instinct for important problems that he recognized it, even though he had not been seriously involved in computer science since the 1930s. Gödel pointed out that a machine which operated 'in polynomial time' would mean that, '*despite the insolvability of the decision problem, mathematicians' reasoning about yes-or-no questions could be completely mechanized*'.[21] Von Neumann was unfortunately too ill to answer Gödel's letter.

We have seen in the preceding sections how Gödel, who had lost several of his important colleagues and friends in the decade 1950–1960, had also acquired several new (or 'reconnected') ones in that same time period. They include Hao Wang, Georg Kreisel, and Paul Bernays, with whom he had frequent contacts over the following 15 years; with Hao Wang right up to his own death in early 1978, with Bernays until about 1975, and with Kreisel to around 1969. And of course, Oskar Morgenstern remained loyal to Gödel for the rest of his own life, and he died less than a half-year before Gödel himself. So, in spite of his losses in the 1950s, Gödel still had professional and personal contacts throughout that period and in the following decade; perhaps even more than during the 1940s.

Gödel's 'Other Disciples'

In the list of 'disciples' at the beginning of this chapter, there are still seven whom we have not yet discussed. Of those, three belong in later chapters, since their entry into Gödel's sphere of influence occurred after the 1950s (although they certainly knew of him earlier). The remaining four are William W. Boone, John Myhill, Gaisi Takeuti, and Dana Scott, and we will give brief summaries of their careers and their relationships to Kurt Gödel.

William Boone[22] (1920–1983) was born in Urbana, IL/USA and was an undergraduate at the University of Cincinnati, working his way through college with part-time jobs as a barman and an accountant. He originally wanted to be a writer, but he switched to mathematics while in college. He did graduate work at Princeton in the group of Alonzo Church, obtaining his Ph.D. in 1952, with a thesis entitled '*Several Simple, Unsolvable Problems of Group Theory Related to the Word Problem*'. He was assistant professor at the Catholic University in Washington, DC in the early 1950s, and a member of the School of Mathematics at the IAS for two years in 1954–56 as a Fulbright Scholar, and again in 1964–66. Boone was on the faculty of the University of Illinois/Urbana from 1958 until his death. He met Gödel during his first stay at the IAS and they had many conversations; Gödel became interested in Boone's work and supported him with recommendations and advice. Two letters from their correspondence are reproduced in the *Collected Works*, Volume IV. Boone's condolence letter to Adele after Kurt's death is also shown there, and we will quote from it in Chap. 18.

John Myhill (1923–1987) was born in Birmingham, UK. After undergraduate studies at Cambridge, he carried out his doctoral work at Harvard under the supervision of W.v.O. Quine, obtaining his Ph.D. in philosophy in 1949. He was an instructor at Vassar College and Temple University, then assistant professor at Yale. He had a Guggenheim Fellowship in mathematics in 1953/54, then took an assistant professorship in philosophy at UC Berkeley, interrupting his work there to go to the IAS as a Fellow in 1957–1959. He met Gödel during that stay. He returned to Berkeley as professor of philosophy in 1959/60, then was on the Stanford faculty from 1960–1963. After another year at the IAS in 1963/64, he spent two years as visiting professor at the University of Illinois before going to SUNY/Buffalo, where he spent the rest of his career. We quote from the obituary notice written by N. D. Goodman and R. E. Vesley[23]:

> His work included important contributions to mathematics, logic, philosophy, computer science, theoretical linguistics, and computer music. The most recent of his 72 publications appears in the current issue of the Journal of *Pure and*

Applied Logic and is co-authored with his student Robert Flagg of Ohio State University. It describes further progress in the project on which Myhill had long been engaged to extend and correct the ideas of Frege so as to provide type-free foundations for mathematics ... Myhill's contributions to mathematics, logic, and philosophy were informed by deep knowledge of historic developments. He was especially interested in Frege, Cantor, and Leibniz, but his learning extended back to Archimedes and Euclid. Myhill had a life-long interest in musical theory, and first wrote in this area in 1955. During his years at Buffalo he maintained an active interest in computer-assisted composition. He taught regularly in the Department of Music. His compositions were heard in performance in Buffalo, on CBC radio, and at Expo '85 in Japan.

Volume IV of Gödel's *Collected Works* contains numerous mentions of Myhill, many of them references provided by Gödel, who was also instrumental in inviting Myhill to the IAS. He is often forgotten as a 'disciple' of Gödel's, but the record is clear that they had many communications and that Gödel supported Myhill's career quite emphatically.

Dana S. Scott (b. 1932) was born in Berkeley, CA/USA, where he attended school and then obtained his BA in mathematics from UC Berkeley in 1954. He went to Princeton for graduate work, after a disagreement with Tarski, and finished his Ph.D. in mathematics under Alonzo Church in 1958. His first academic position was at the University of Chicago (1958–60), then he was on the faculty at the University of California, Berkeley (1960–63), Stanford University (1963–69), Princeton University (1969–72), and Oxford University (1972–81). In 1981, he joined the faculty at Carnegie Mellon University in Pittsburgh, PA, where he was professor of Mathematical Logic, Computer Science, and Philosophy from 1982 until his retirement as professor emeritus in 2003. He probably first met Gödel during his graduate student years at Princeton. There are many mentions of Scott in Gödel's *Collected Works*; he often received corrected or amended papers from Gödel to be passed on to the editors of various journals after a critical reading. We quote from a biography published in the *Encyclopedia Britannica* and written by William L. Hosch in 2021[24]:

Scott's last position, at Carnegie Mellon, gives some inkling of the remarkable diversity of his academic interests. In addition to contributing his seminal work on automata theory, Scott collaborated in the 1970s with the British computer scientist Christopher Strachey to lay the foundations of the mathematical (or denotational) semantics of computer programming languages. The outgrowth of that work led to Scott's introduction of domain theory, providing, in particular, mathematical models for the λ-calculus ... (a formal mathematical-logical system *[first published]* in 1936 by the American logician Alonzo Church), and many other related theories. Scott was the first editor in chief of *Logical Methods in Computer Science*, an online open-access journal founded in 2005.

Gaisi Takeuti (1926–2017) was born in Kanazawa, Japan and obtained his Bachelor's in mathematics in 1947 and his Ph.D. in 1956 from the University of Tokyo. He went to Princeton specifically to study under Kurt Gödel (his first stay at the IAS was for 1-½ years from January 1959 to July 1960). He was on the mathematics faculty at the University of Illinois/Urbana for many years from 1963 until his retirement in 1992. He contributed to a book entitled '*Memoirs of a Proof Theorist: Godel and Other Logicians* ', and he was president of the *Kurt Gödel Society* from 2003–2009. He spent two later sojourns at the IAS, for two years in 1966–1968 and for one year in 1971/72. His long-term goals were to show the consistency of number theory and to clarify the notion of 'set'. After many conversations with Gödel, he published an article entitled '*About Mathematics*' on the foundations of mathematics [Takeuti (1972)]. In it, he says:

> Usually, it is essential for modern mathematics that it supposes an infinite mind and conjectures what it does. By infinite mind, I mean a mind which can investigate infinitely many things by checking one by one. For example, the law of excluded middle holds for an infinite mind because it can check whether $A(x)$ or $\neg A(x)$ one by one. Similarly, it can see $\{x|A(x)\}$ since it can check whether $A(x)$ or not one by one. On the contrary, a human mind is clearly a finite mind.

This is in contrast to Gödel's idea that the human mind (but not the human brain) 'converges to infinity' because it is dynamic and learns from each experience. Takeuti was truly a 'disciple' of Kurt Gödel, one who had sought him out from afar, looking for a master in the Eastern tradition.

Private Life

During the 1950s, as we saw briefly in Chap. 9, the Gödels led a generally harmonious, if somewhat reclusive personal life. They were ensconced in their own house on Linden Lane, where Adele could cultivate her garden and arrange the house as she wished. She made several trips back to Vienna, in 1953 and in 1956—the latter by air, which shocked her mother-in-law. But by then, transatlantic flights were routine, and she wanted to see her own mother, who was ill. She brought her mother back with her, and she lived with the Gödels until her death three years later. Adele made another European trip in October/November of 1959. Kurt's health, after his ulcer problem in early 1951, was generally good (except for a relapse into acute hypochondria in late 1954, when he was convinced that he was having a cardiac crisis and was about to die). That however passed relatively quickly and without requiring drastic treatment. After Kurt's repeated failures to keep his promise to visit Marianne and Rudolf back in Vienna, the two of them made their first trip to the USA in the spring of 1958, and it was evidently a success for all concerned. They returned at two-year intervals thereafter until 1966, when Marianne was too weak to make the trip. Marianne was initially very critical of Adele, but came to see that she was, in spite of all her flaws, good for Kurt. Budiansky (2021) has described this period in Gödel's life in some detail; the 1950s were perhaps the most tranquil decade of his life, if not the most productive.

In 1959, Kurt began reading the later works of *Edmund Husserl* (1859–1938). Husserl was born in a Moravian town only about 40 km (25 mi.) to the northeast of Brünn/Brno, almost exactly 47 years before Gödel's birth. After elementary school in his hometown, he went to a *Realgymnasium* in Vienna and to the *Staatsgymnasium* in Olmütz/Olmouc. He studied mathematics, physics and astronomy at the University of Leipzig, but his interest in philosophy was awakened early. He transferred to Berlin and there met *Tomáš Garrigue Masaryk*, who had studied philosophy under Franz Brentano and later became the first president of Czechoslovakia (and the eponym of the modern university in Brno). Husserl obtained his doctorate in mathematics at the University of Vienna in 1883 and then returned to Berlin as an assistant to Karl Weierstrass. When Weierstrass became ill, Husserl went back to Vienna and changed his topic of interest to philosophy. He was strongly influenced there by Brentano, attending his lectures on philosophy and philosophical psychology. In 1886, he went to Halle, where he completed his *Habilitation* with a thesis on the 'concept of number'. It was later the basis for his first book, '*Philosophie der Arithmetik*'. He served as *Privatdozent* in Halle from 1887 to 1901, when he moved to Göttingen as associate professor after publishing the first volume of his '*Logische Untersuchungen*' (*Logical Investigations*). Starting about 1907, he studied the work of Kant and Descartes, and founded the philosophical school of *Phenomenology*. Somewhat later he lectured on 'internal time consciousness', a topic which greatly interested Gödel (cf. Chap. 15). Husserl founded the journal '*Jahrbuch für Philosophie und Phänomenologische Forschung*' (*Yearbook for Philosophy and Phenomenological Research*) in Freiburg in 1912, and moved to the University of Freiburg as full professor in 1916. His book '*Ideen...*' (1913) can be seen as the real beginning of his school of Phenomenology.

Husserl remained at the University of Freiburg until his retirement in 1928, some months before his 70th birthday. Among his assistants in Freiburg were Edith Stein[25] and Martin Heidegger; the latter dedicated his famous work 'Sein und Zeit' [Being and Time (1927)] to Husserl, although they later fell out. Heidegger became Husserl's successor on the faculty at Freiburg. Husserl's last works turned more toward history and were concerned with Descartes, Hume and Kant. He died in Freiburg in 1938, aged 79. Interestingly, Husserl, like Gödel, wrote many of his manuscripts in the *Gabelsberger* shorthand. His writings and his library were smuggled out of Nazi Germany[26] and archived at the University of Louvain (*K.U. Leuven*) in Belgium, where they (miraculously) survived WWII and have been published as the *Husserliana* series. Gödel was interested only in Husserl's works on phenomenology, and he studied them for the last 18 years of his life, from 1959 to 1977. We noted this briefly at the end of Chap. 15, and will examine their effect on Gödel's later philosophy in more detail in Chap. 18.

Notes

1. Feferman (1998) calls them 'those who had deeper personal relations and were privy to his thoughts…'.
2. Wittgenstein had an almost hypnotic personality, and 'mesmerized' many intelligent people, including those who barely knew him personally (or not at all). This was apparent during the 'Vienna Circle' days, and it continued after he returned to Cambridge in 1929, and even long after his (early) death in 1951. Goldstein (2005) even suggests that a festering disagreement between Gödel and Wittgenstein influenced the former's life and career and was to some extent responsible for his depressions and paranoia; but there is no real evidence for that conclusion, which Gödel himself vehemently denied.
3. Charles Parsons, in his memoir of Hao Wang [Parsons (1996a), p. 64, Footnote 2], limits the period of intensive contacts between Wang and Gödel to September 1967–January 1978.
4. See the memorial notices by Charles Parsons and Gary R. Mar [Parsons (1996)] and [Mar (2017)].
5. Photo: private archive, property of San-You and Jane Hsiao-Ching Wang. Reused from: https://richardzach.org/2016/09/interview-with-hao-wang-and-robin-gandy/ with the permission of Jane H.-C. Wang.
6. Cf. [Kleene & Rosser (1984)].
7. Dennett (1978).
8. Conradi (2001).
9. Photo by Paul Halmos, from the Paul R. Halmos Photograph Collection, Briscoe Center, Uni Texas. Reused with permission from: https://digitalcollections.briscoecenter.org/islandora/search/Georg%20Kreisel?type=dismax.

10. Photo by Alan Richards, 1951. From the Shelby White and Leon Levy Archives Center, IAS Princeton; reused with permission from: https://albert.ias.edu/handle/20.500.12111/1083.
11. Dawson (1997, 2005 edition) , p. 199 and Endnote 457.
12. *ibid.*, p. 201 and Endnote 462.
13. *ibid.*, p. 206.
14. *Dielectica* is now the official organ of the *European Society for Analytic Philosophy*.
15. Dawson, *op. cit.*, p. 207.
16. Hintikka (1999), pp. 61–67.
17. Compare Gödel's manuscript for his (undelivered) lecture to the *American Philosophical Society*, [Gödel (*1961/?*)]
18. See Pais (1982), pp. 466–478.
19. Dawson, *op. cit.*, p. 204.
20. *loq. cit.*
21. Quoted from Gödel's letter of March 20th, 1956 to von Neumann, archived in von Neumann's papers in the Library of Congress. Dawson (1997) points out in his Endnote 472 that the original text (in German) is reproduced in Volume V of Gödel's *Collected Works*.
22. See e.g. the *Collected Works*, Volume IV, p. 325 ff.
23. Goodman and Vesley (1987), N. D. Goodman & R. E. Vesley, Obituary: John R. Myhill (1923–1987), in: *History and Philosophy of Logic*, Vol. 8, No. 2 (1987), 243–244. Online at https://www.tandfonline.com/doi/pdf/10.1080/01445348708837118.
24. Hosch, William L.: '*Dana Scott*'. In: *Encyclopedia Britannica*, 7 Oct. 2021, online at https://www.britannica.com/biography/Dana-Scott. Accessed March 3rd, 2022.
25. The life and career of *Edith Stein* are worthy of a book in their own right. She obtained her doctorate in philosophy under Husserl, but was refused the *Habilitation* in her native Breslau, as well as in Göttingen and Freiburg, since she was a woman. She became an advocate of women's rights and human rights, converted to Catholicism and served as a nun in cloisters in Cologne and in Echt, NL; but she was murdered in Auschwitz in 1942 due to her Jewish origin. She was canonized by the Catholic Church and is honored by the Evangelical (Lutheran) Church on her death day, August 9th.
26. This is another astonishing story: The Franciscan Pater *Hermann Leo van Breda* risked his life to carry the more than 40,000 pages to safety in Leuven/Louvain, where they were archived (fortunately not in the University Library, which was burned in both WWI and WWII. It has been rebuilt according to the original plans and is today a protected monument). Van Breda went on to publish many of Husserl's later works, e.g., as '*Ideen* II' and III.

17

The 1960s. *Fame and Seclusion*

The Gödels began the decade of the 1960s in good form. Marianne and Rudolf came from Vienna for a longer visit of three months (March–May 1960), and Adele made another visit there in November. As we saw in Chap. 9, she returned to find Kurt depressed, and discovered that he had subsisted almost entirely on eggs during her absence. This was a warning sign of another psychiatric crisis, apparently in connection with his hypochondria, which reached alarming proportions early in 1961. However, he made it through that year, which he later called 'one of his three worst' in a letter to Wang, without major interventions, and was able to work much of the time, hiding his depression and other problems from his colleagues and friends as far as possible. He wrote to his mother about his state of health, as usual, but without revealing the depth of his depression. On March 18th, his letter to her mentioned that he was now keeping a 'normal' schedule for sleeping, going to bed in late evenings and getting up fairly early in the mornings, which he found 'more agreeable' than his previous schedule of working late into the night and arising late the next morning.

Kurt Gödel's health during the remainder of the decade, after he overcame his difficult year in 1961, seems to have remained relatively stable, although his psychiatric problems—hypochondria, general paranoia, depression—gradually increased, probably due to a combination of his own advancing age, and that of Adele, who began to have serious health problems that inevitably reflected on Kurt's daily life and outlook; and also to his increasing isolation. Although he was intellectually active in those years, he was more and

© The Author(s), under exclusive license to Springer Nature
Switzerland AG 2022, corrected publication 2023
W. D. Brewer, *Kurt Gödel*, Springer Biographies,
https://doi.org/10.1007/978-3-031-11309-3_17

more distanced from his colleagues, and even moved into an office adjacent to a new library, with an excellent view, but at some distance from the rest of the mathematics faculty.[1] His contacts to the 'outside world' also gradually decreased—he gave no more public lectures, many of his 'disciples' from previous years were now far away, and some—we saw the example of Georg Kreisel—even deliberately broke off contact because of Gödel's increasing tendency to depression and psychological withdrawal; a vicious circle which he could hardly terminate.

An event that intruded on Gödel's privacy and seclusion at the IAS was 'the great row' of 1962, which arose as a dispute between the faculty and the administration about the hiring of new faculty—a sore topic which had already caused trouble in 1939, when the founding Director, Abraham Flexner, was forced to resign because he had hired some new members without consulting the older faculty. In 1962, the mathematics faculty at the IAS wanted to hire a well-known topologist from Princeton University, John Milnor. The controversy involved the question of inter-institutional rivalry between the IAS and the University. The faculty held that hiring should be based on academic and scientific merit and not on 'extraneous' issues. The Board of Trustees cited an earlier agreement, from the Flexner era, prohibiting competitive hiring between the IAS and the University. Oppenheimer, the Institute's Director at the time, reminded the faculty of that agreement before the vote was taken. The Executive Officer of the School of Mathematics, Atle Selberg, believed that this prejudiced the vote and excluded Oppenheimer from faculty meetings of the School from then on. Gödel was incensed by this and refused to attend himself; but his vote still had to be cast, and that was done henceforth by proxy. The details are given in Dawson (1997), pp. 220/221. The result was a smoldering antipathy between the Faculty and the Director/trustees, which Gödel found rather upsetting.

The 'Theological Correspondence'

In his correspondence with his mother Marianne in the summer and autumn of 1961, Kurt Gödel wrote several "philosophical/theological" letters, which we consider here in some more detail, since they reveal his thinking at the time. He was otherwise reticent about such personal topics with his contemporaries, especially about theological and religious topics, since he didn't want to be seen as overly superstitious or spiritualistic (although he did have such tendencies and perhaps had frustrating experiences in trying to communicate them in his younger years). With his mother, he was more open, although their letters dealt mostly with their everyday lives. Typically, he wrote one letter each month, which she would answer, sometimes writing several answers before he wrote the next one. Unfortunately, all of her letters to Kurt were destroyed after his death (apparently by Adele, who wanted to wipe out the evidence of Marianne's negative opinion of her). The letters from Kurt to Marianne were saved by his brother Rudolf and are archived at the *Wienbibliothek*, which has made them available online.

His letters from July through October of 1961 contain the usual amount of everyday news and trivialities, but in addition Kurt, along with expressing his interest in philosophy and his wish to have a basic philosophical library at home, makes an attempt to explain his philosophical/theological beliefs and what he calls the '*theological worldview*' to his mother, apparently with limited success, according to his responses to her comments in her own replies.

In his July 27th (1961) letter, the last two pages are dedicated to that theological worldview, evidently sparked by Marianne's (perhaps rhetorical) question in her previous letter as to whether there would be a '*Wiedersehen*', clearly referring to '*a meeting again after death*'. She was about to turn 82 and no doubt had thoughts of impending death (in the event, she lived for almost precisely 5 more years after that letter). Kurt replies to this question, which he terms '*schwerwiegend*' ('profound' or 'grave') in a positive sense—he believes that if the world is reasonable (*vernünftig*), this would imply that humans, with their '*broad spectrum of possibilities*', would necessarily live on (*in a 'second life'*).

And he also believes that the world *is* indeed reasonable, owing to the great degree of regularity and order which has been observed by science. '*Order is a form of reasonableness*', he asserts. He then discusses the question of what this other life would be like—'*of course, only conjectures*'—citing the scientific result that *this* world had a beginning and will probably have an ending, but we '*one day found ourselves in it, without knowing why and where we are going*'; and the same may occur again in '*another world*'. For '*why should there be only this one world?*'. He cites the Book of Revelations in the New Testament, saying that science confirms the prediction there, that this world will come to an end—but '*God created a new Heaven and a new Earth*'. All of this is reminiscent of John Wheeler's remark (quoted in Chap. 15) that Gödel wanted his life to go on and on; either in a physical sense, as in Gödel's Universe, or in some more spiritual sense, as in his 'theological worldview' expressed here.

In his next letter, dated August 14th, after writing about everyday subjects (Adele's trip to Italy, his eating habits in her absence) and briefly touching on politics (construction of the Berlin Wall was begun the day before he wrote the letter), Kurt returned to theology. Marianne had evidently written that she '*prayed to Creation*', and that was his opening; he interpreted her statement as meaning that '*Nature is beautiful when not spoiled by human activities*', and he asserted that this could be precisely the reason why there must be two worlds: "... *Only humans can arrive at a better existence through learning, that is, they can thus give their lives a deeper meaning. One, and often the* only *method of learning is to make mistakes. And that truly occurs in* this *world often enough...*". He goes on to discuss the problem of why God had not created humans in such a way that they could '*do everything correctly right from the beginning*', a question related to the theodicy problem. And he arrives at the conclusion[2] that the individuality of humans necessarily includes the property of *having to learn through mistakes*:

> ... Thus, each one *[of us]* could say of him/herself: 'Among all the possible creatures, 'I' am *[the one with]* precisely this particular collection of qualities. If, however, among these qualities is the one that I (cannot) do everything correctly right from the beginning, but rather only after *[acquiring]* a certain experience, then it follows that if God had created other beings instead of us, who would not need to learn anything*, then *we* would not be those creatures. That is, we would not exist at all...
>
> [Gödel's footnote in the letter]: *It must naturally be assumed that such (or nearly such) beings do exist somewhere, or will exist.

Kurt goes on to suggest that,

> ...in religion, if not in the churches, there is much more rationality than one usually believes, but we... are bred right from our earliest youth to have a contrary prejudice; in the schools, by their poor instruction in religion, by books and experiences.

He ends the letter on that note.

Gödel's next letter was written on September 12th. He again returns to 'philosophical musings', after asking about the death of Stefan Zweig (his mother had apparently been reading a biography of Zweig); in particular, why Zweig had committed suicide. He then notes that it would give him great pleasure to be given good philosophy books (he mentions two of Kant's '*Critiques*' as examples), since he would enjoy having a philosophical library at home, so that he could use his free hours to continue reading in them. This leads into his reply to Marianne's answer to his August letter:

> ...That you experienced some difficulties in understanding the 'theological' parts of my previous letter is quite understandable and has nothing to do

with your age. I wrote in a very abbreviated form and touched on several rather profound philosophical questions. At first glance, the worldview that I explained to you appears very improbable. But I believe, if one thinks about it in depth, that it proves to be quite possible and reasonable. In particular, one has to imagine that the 'learning' will take place for the most part in the next world...

He goes on to imagine how experiences in this world, which seem inexplicable to us at the time, could become clear (in terms of cause and effect) upon reflecting on them in the (hypothetical) *next* world; how our present experiences could serve as '*raw material*' for learning there. He quotes Schopenhauer on the '*apparent intentionality in the fates of individuals*', and suggests that some essential truths could be just as clearly recognizable to our (then presumably superior) intellects in the next world, as are primitive truths (like $2 \times 2 = 4$) in *this* world, '*whose fallacy is objectively excluded*'. He then apologizes for '*getting too deep onto philosophy*', and wonders if what he has written could be understood by anyone who had not studied philosophy; but noting that the

...present day study of philosophy would not help much in their understanding, since indeed 90% of today's philosophers see their principal function as knocking religion out of the heads of [*their students*], and thus act in a similar way as the ill-conceived churches.

Kurt Gödel's last serious excursion into theology/philosophy in his letters to Marianne occurs in the letter dated October 6th, 1961. After discussing Marianne's (and presumably also Rudolf's) sojourn at *Mondsee* (near Salzburg); Adele's Italian trip and their household help in Princeton, Mary; and comparisons between Germany (where they had also evidently traveled) and Austria; as well as Zweig and Hammerskjöld (the U.N. Secretary-General, who had recently died in an airplane crash on September 18th, 1961), he answers what must have been Marianne's objection to his previous theological ruminations, defending himself against her presumption that they represent a form of occultism—"*What I wrote is nothing other than a descriptive representation and adaptation to our modern ways of thought of certain theological teachings which have been preached for 2000 years; certainly mixed in with a great deal of nonsense...*". He goes on to take to task in particular the Catholic Church for its past dogmas, and to defend his use of reason in theology, as in every other area of life, giving examples of how reason has led us to new knowledge and abilities over the millennia. He considers the atomic theory as a paradigm of how an originally purely theoretical concept could become a major tenet of science, supported by observations and experiments. He summarizes by writing,

> What I call the 'theological worldview' is the idea that the world and everything in it has meaning and reasonability, indeed a good and indubitable meaning. It follows from this that our earthly existence, which considered by itself has at most a very dubious meaning, can now be *[considered as]* a means to an end for another existence. The notion that everything in the world has a meaning is incidentally quite analogous to the principle that everything has a cause, which forms the basis for all of science.

This correspondence reveals a lot about the basis of Gödel's philosophy, which he evidently pursued on the one hand as a guide to and motivation of his mathematical and scientific work, and on the other as a justification for his personal worldview—of which an important part seems to have been the (desired) certainty of a continuation of life in some form after our inevitable departure from *this* world.

Rudolf and Marianne continued with their biennial rhythm of visiting Princeton in 1962 and 1964. By 1966, when Kurt was preparing to celebrate his 60th birthday in April, Marianne was not well—she was 86—and she was, to her great disappointment, unable to make the trip to the USA. Kurt, of course, could have gone to see her one last time; by then, jet aircraft had made the Atlantic crossing safe, fast, and comfortable. And he had other motives to travel to Vienna at the time: his *alma mater* planned a special celebration for his 60th birthday, and he was also offered membership in the Austrian Academy of Sciences, a national medal for art and science, and an honorary professorship at the University. He refused all those honors, in part for rather specious reasons involving his US citizenship, and thus also missed the chance to see his mother again. But he probably had not allowed himself to realize that she would not live much longer, and he still had his fear of going back to Europe. Whether he regretted this after her death in July of 1966 is not known (and whether he ever set foot in an aircraft is not recorded, but it seems unlikely).

In any case, her death seems not to have caused him a serious mental upset. His last letter to Marianne in the online archive at the *Wienbibliothek* in Vienna is dated December 27th, 1965, and was a typical 'Christmas letter', with thanks for gifts and cards and greetings for the New Year. However, Dawson (1997), p. 227, cites a letter of June 18th, 1966.

In the event, Marianne died suddenly of heart failure on Saturday, July 23rd. Coincidentally, Adele had been on a trip to Italy in June and July of that year and stopped by Vienna on her return trip, and she had planned to visit with Rudolf and Marianne on the 26th. As it was, she was only able to attend the funeral (where she caught a persistent case of bronchitis, having been chilled by rain during the graveside ceremony—perhaps the 'last revenge' of Marianne on her unloved daughter-in-law?). Kurt's last letter to Rudolf in the online archive is from June, 1965—except for two in the 1970s, dated 1.02.1972 and 12.12.1975; but Dawson cites another from August 18th, 1966, in which Kurt defended his own absence from the funeral as due to his frail health. He renounced any inheritance from their mother, saying that Rudolf should handle the matter as he wished. We shall return to his last two letters to Rudolf in Chap. 18.

Publications and Writings, Honors

The 1960s were a lively period for Gödel's publication activities. Dawson's annotated list (1983) shows 21 entries for that decade, many more than in the 1940s (6) and 1950s (4). This was a result of Gödel's increasing fame and the demand for translations of his important articles, up to that time available for the most part only in German. Indeed, 15 of the 21 publications listed were translations, most of them into English or Italian, and mainly of his previously published works from the period 1930–33.

The books edited by Martin Davis ['*The Undecidable*' (1965)] and Jean van Heijenoort ['*From Frege to Gödel...*' (1967)] provided a major impetus for those translations, which formed the bases of chapters or subchapters within them. Of the remaining publications, 4 were simply reprints of earlier articles, edited and to some extent extended by Gödel, including a revised and extended version (published in 1964) of his (1947), '*What is Cantor's Continuum Problem?*'; another was the belated publication of his '*Remarks before the Princeton Bicentennial Conference on Problems in Mathematics*'—his lecture held on December 17th, 1946, whose text was published in [Davis, ed. (1965)]. An original contribution was his postscript to the posthumous publication of Clifford Spector's (1962). There, he describes how that paper was written and emphasizes the contributions of Georg Kreisel (without whom it would not have been published, but who gave full credit to Spector and claimed no authorship himself). Gödel also modified Spector's original title for that article. But on the whole, although his publications list grew, he wrote little that was new or original for publication during this decade.

In 1961, Gödel was elected to membership in the *American Philosophical Society*, and, as we saw in Chap. 9, he was invited to give a lecture there: his 'inaugural lecture' as a new member, scheduled for 1963. Typically, he apparently never answered the invitation to give the lecture, although he wrote a manuscript for its text, which was found in his written legacy (reconstructed as [Gödel (*1961/?)] in the *Collected Works*, Volume III). The commentary preceding the reconstructed text, by Dagfinn Føllesdal, is very thorough and presents an interesting picture of Gödel's philosophical views at the time, in particular his interpretation of Husserl's later phenomenology. We will return to it in the following chapter.

In 1968, Gödel was elected to Foreign Membership in the Royal Society. That is the basis for the obituary written by Georg Kreisel (who was himself a Fellow of the Royal Society (FRS) since 1966—see [Kreisel (1980)]. It is notable that he did not refuse this honor by a foreign scientific society on the grounds of his US citizenship, the reason that he had given for not accepting a similar membership in the Austrian Academy two years earlier.

Later Disciples: Cohen and Tennenbaum

Two more of Gödel's 'disciples' from the list in the previous chapter belong properly to the 1960s, since their connection to Kurt Gödel began in that period. They are *Paul J. Cohen* (1934–2007) and *Stanley Tennenbaum* (1927–2005).

Cohen is famous for his proof that the Continuum Hypothesis (CH) is independent of the axioms of ZFC set theory; Gödel had already shown that it is (relatively) *consistent* with the axioms of ZFC, and Cohen showed that its negation ¬CH is also consistent. He introduced a completely new method, called 'forcing', to demonstrate this. He carried on a lively correspondence with Gödel before publishing his results, which we will examine below, and he later was awarded the Fields Medal for them, on which Gödel congratulated him very sincerely.

Tennenbaum's best-known contribution to mathematical logic is the '*Tennenbaum theorem*', which shows that no countable nonstandard model of Peano arithmetic (PA) can be recursive—but he is remembered equally for his efforts to improve American education at various levels. He was a very unusual personality, as we shall see. In fact, Cohen and Tennenbaum were friends from their graduate school days in Chicago; Cohen even lived with the Tennenbaums for some time.[4]

Paul Joseph Cohen was born on April 2nd, 1934 in Long Branch, NJ/USA, of Polish parents who had immigrated to the USA. He grew up in Brooklyn, NY and was raised by his mother after the age of 9 when his parents separated. He showed an early interest in mathematics, and after three years at Brooklyn College, he was admitted to graduate school at the University of Chicago (at that time still a rather unconventional institution[3]) before receiving his Bachelor's degree. He initially studied number theory, and obtained his Masters degree (1954) under André Weil in that field; but he became interested in logic through his friends and later worked mainly in that area. This period in his life is characterized in a quote from a commemorative article by his fellow student Anil Nerode:[4]

> As a graduate student Cohen's connection with logic were his friendships with a lively group of students who became logicians; Michael Morley, Anil Nerode, Bill Howard, Ray Smullyan, and Stanley Tennenbaum. For a while he lived in Tennenbaum's house and absorbed logic by osmosis, for there were no courses in logic in the Chicago mathematics department.

Cohen (Fig. 17.1) accepted an instructor position at the University of Rochester in 1957, before the official conferral of his PhD in 1958. He spent the following academic year (1958/59) at MIT, then went to the School of Mathematics of the IAS Princeton, where he spent almost two years (September 1959–May 1961). He undoubtedly met Gödel during this period, but their major interactions took place after Cohen had taken a faculty position as assistant professor at Stanford University in mid-1961. By that time, he had published important results in mathematics, with his 'Factorization in Group Algebras' (1959) and his solution of the Littlewood Conjecture (1960). He was an invited speaker at the 1962 ICM in Stockholm (where he also met his future wife, Christina Karls), and he was promoted to a tenured associate professorship at Stanford that same year. In 1962/63, he developed his method of 'forcing' and used it to show the independence of the Axiom of Choice (AC) and the Continuum Hypothesis (CH) from the axioms of ZF set theory, solving the problem on which Gödel had worked for more than five years in 1939–1944. That he can also be considered to have been a 'disciple' of Gödel's can be seen from his later memoir[6] of how he developed the 'forcing' method (quoted here from an Internet summary[7]):

> Cohen explains how he came to the idea of forcing from reading Kurt Gödel's 'The Consistency of the Continuum Hypothesis', a book consisting of notes of a course given at the Institute for Advanced Study in 1938-39. The continuum hypothesis problem was the first of David Hilbert's famous 23 problems delivered to the Second International Congress of Mathematicians in Paris in 1900. Hilbert's famous speech *The Problems of Mathematics* challenged (and

today still challenges) mathematicians to solve these fundamental questions and Cohen has the distinction of solving Problem 1.

The lectures underlying Gödel's book [Gödel (1940)][8] were those that he gave in November/December of 1938 at the IAS. Whether Cohen [and Gödel] indeed have 'the distinction of solving [Hilbert's] Problem 1' is still controversial, and it is listed as 'undecided' in various summaries.[9]

Cohen wrote to Gödel in April of 1963, suggesting that they meet when he was next at the IAS, to discuss his proof that ZF is consistent with ¬CH and ¬AC (—and they are thus independent of the axioms of ZF, since Gödel had already established the relative consistency of the CH/GCH and the AC). Cohen had prepared a manuscript for PNAS, but was uncertain as to whether he should submit it, writing again[10] to Gödel, "*In short, what I am trying to say, is that I feel that only you, with your preeminent position in the field, can give the 'stamp of approval,' which I would so much desire... If I have overstepped any bounds, please excuse me. I can only say that I feel under a great nervous strain...*". Gödel was sympathetic and spent some time and effort to

Fig. 17.1 Paul Joseph Cohen, in 1966, when he received the Fields Medal at age 32. Photo: from [Albers, Alexanderson, & Reid (1986)], re-used with permission from Springer-Verlag[5]

check Cohen's manuscript, which he found to be correct. It is a great credit to Gödel that he (like Frege when confronted with Russell's paradox; cf. Chap. 6 and backnote [18] there) was not envious or disturbed by Cohen's completing the solution of the problem that he had worked on for so many years, but rather was elated to see the successful proof. Budiansky (2021) describes the situation[11] as reflected in their correspondence of May/June 1963:

It was a striking reflection of his *[Gödel's]* great generosity of spirit and encouragement of others, even under circumstances that would have left many in his hypercompetitive field bitterly disappointed at having been beaten by a young upstart to a result he himself had labored on for years with success just out of reach. He never lost the deep aesthetic enjoyment of mathematics that he had absorbed in his heady youthful days in Vienna.

Gödel wrote back to Cohen on June 5th and again on June 20th, reassuring him that his proof was in order and expressing his own pleasure on reading it, and encouraging him to publish it as soon as possible. It was indeed published in PNAS, as two papers in Volume 50, No. 6 (December 1963) and in Volume 51, No. 1 (January 1964). Gödel compared his pleasure at reading Cohen's proof to that of seeing '*a really good play*'.

In 1966, Cohen received the *Fields Medal*, often called 'The Nobel Prize for mathematics', for his work on the independence of the CH and the AC from the axioms of ZF set theory. In an obituary written shortly after Cohen's death in 2007, his work of 44 years earlier was given a place of honor in his career[12]:

An 'enduring and powerful product' of Cohen's work on the continuum hypothesis, and one that has been used by 'countless mathematicians' is known as 'forcing', and it is used to construct mathematical models to test a given hypothesis for truth or falsehood.

Cohen gave a lecture the year before his death, describing his solution to the problem of the continuum hypothesis, at the 2006 *Gödel Centennial Conference* in Vienna.[13] There, he said,

A point of view which the author *[Cohen]* feels may eventually come to be accepted is that CH is obviously false. The main reason one accepts the axiom of infinity is probably that we feel it absurd to think that the process of adding only one set at a time can exhaust the entire universe. Similarly with the higher axioms of infinity. Now $\aleph_1$ is the cardinality of the set of countable ordinals, and this is merely a special and the simplest way of generating a higher cardinal. The set C (the continuum) is, in contrast, generated by a totally new and more powerful principle, namely the power set axiom. It is unreasonable to expect that any description of a larger cardinal which attempts to build up that cardinal from ideas deriving from the replacement axiom can ever reach C.

Thus C is greater than $\aleph_1$, $\aleph_\omega$, $\aleph_\alpha$, where $\alpha = \aleph_\omega$, etc. This point of view regards C as an incredibly rich set given to us by one bold new axiom, which can never be approached by any piecemeal process of construction. Perhaps later generations will see the problem more clearly and express themselves more eloquently.

Here, he agrees with Gödel, who also doubted that the CH would turn out to be true, but who expected that powerful new axioms for set theory would be found which could be used to verify that conjecture.

[In fact, Gödel's position on the CH and the GCH was complex and dynamic: he changed his view of their possible validity at various times. See for example Floyd and Kanamori (2006), or Kanamori (2007)]. In the 1930s, when he set out to solve Hilbert's 1st Problem, Gödel evidently believed that the CH and the GCH were valid. He was able to obtain only a relative consistency proof for CH and GCH. Kanamori [(2007), pp. 176/177] quotes from [Gödel (1947)] on the state of affairs at that time:

Of the three possibilities in axiomatic set theory, that CH could be demonstrable, provable, or undecidable, Gödel [(1947), p. 519] regarded the third as 'most likely' and so advocated the search for a proof to establish *[that]* Con(ZF) implies Con(ZFC + ¬ CH) to complement his own relative consistency result with L. However, Gödel stressed that this would not 'settle the question definitively' and turned to the possibility of new axioms ... Gödel (1947) concluded by forwarding the remarkable opinion that CH 'will turn out to be wrong' since it has as paradoxical consequences the existence of 'thin' (in various senses he articulated) sets of reals of the power of the continuum.

Later, Gödel even claimed that Cantor's original CH was wrong[14]:

In [Gödel (1970a)] [(*1970a) in the *CW*, Vol. III], entitled '*Some considerations leading to the probable conclusion that the true power of the continuum* is $\aleph_2$', Gödel claimed to establish $2^{\aleph_0} = \aleph_2$ ... To modern eyes, there is an affecting, quixotic grandeur to this reaching back to primordial beginnings of set theory to charge the windmill once again ... As set theory was to develop after Gödel, there would be a circling back, with deep and penetrating arguments from strong large-cardinal hypotheses that, after all, lead to $2^{\aleph_0} = \aleph_2$.

Cohen's major work would not have existed without Gödel, and both of them knew that.

Stanley Tennenbaum had a most unusual, unconventional and free-wheeling personality (Fig. 17.2). He was born on April 11th, 1927 in Cincinnati, OH/USA. He decided while in high school to go to Chicago and was admitted as an undergraduate (at 16!) to the (then very liberal) University of Chicago (cf. backnote 3). He quickly showed his talent for mathematics, and he completed his BPh (Bachelor of Philosophy) in 1946 (incorrectly cited as a PhD on the IAS Princeton School of Mathematics website listing former members, visitors etc.).

Tennenbaum continued at Chicago, doing research for a doctorate, and became a member of a clique of graduate students interested in mathematical logic (but without faculty members who worked in that field; see above: the memoir by Anil Nerode[4] on Paul Cohen's student days at Chicago). Tennenbaum married and became the father of a son, Jonathan Tennenbaum, and developed several mathematical ideas which he published already as a graduate student. He however later left Chicago without completing his PhD, and lived for a time in Michigan, where Jonathan acquired his early memories of his father's teaching methods and was allowed to stay out of school when he got into trouble.

Fig. 17.2 Stanley Tennenbaum, mid 1960s. Photo: private archive.[15] From the family archive of Jonathan Tennenbaum

Still later (in 1965), Stanley Tennenbaum obtained a tenured position at the University of Rochester, most unusual for someone who had no PhD or equivalent (but he had by then established his reputation as a talented mathematician). He showed the independence of the *Suslin Hypothesis* from the axioms of ZFC set theory (in part in collaboration with Robert Solovay) and he gave a new proof of the irrationality of $\sqrt{2}$.

In the early 1960s, Tennenbaum became aware of the general decline in American schools (which, in his opinion, were not teaching their pupils/students to *think*). He was especially critical of the 'new math' which became fashionable around that time, and he wanted to counter what he saw as a dangerous downward trend by founding a university based on principles similar to those of Hutchins' Chicago (or perhaps those underlying Black Mountain College—cf. Chap. 13). He tried to accomplish this in Rochester, with the support of some faculty (among them *Robert Marshak*, chairman of the Physics Department). But he ran afoul of the conservative university administration, and finally resigned (to avoid being dismissed on trumped-up charges of mental incompetence, etc.). This led to the breakup of his marriage, and to an extended court battle over custody/responsibility for his two younger children, around 1968.

This was followed by a longer period when he became a sort of academic nomad, taking temporary positions as instructor or visiting scholar, and tried to carry out his 'new university' project by founding a completely new institution, looking for support among rich philanthropists. This was equally unsuccessful. In the midst of this period, he went to the IAS, inspired by Gödel's works, and became a friend and disciple of Gödel (in his own words). His first official visit to the IAS was as a member (probably a Fellow) in the first half of 1967. He returned as an official visitor from May through August of 1972; but he apparently spent much more time there 'unofficially', sometimes sleeping in an office in Fuld Hall (see the Epilogue, Chap. 20).

A great deal of material about Tennenbaum's life and career is contained in the texts from a memorial logic conference held in April 2006 at CUNY/NY, shortly after his death in 2005, and organized by Juliette Kennedy and Roman Kossak. Their book, '*Set Theory, Arithmetic, and Foundations of Mathematics*' [Kennedy & Kossak (2011)], contains many of the contributions to that conference, and others can be found as pdfs on the Internet. Another volume which contains information about Tennenbaum's life and work is Siobhan Roberts' '*Genius at Play. The Curious Mind of John Horton Conway*' [Roberts (2015)]. A contribution intended for the 2006 conference but not delivered there, and not published elsewhere, was written by his son Jonathan Tennenbaum, and is available on a website started in Stanley Tennenbaum's memory by his former student at Rochester, Rob Tully [see Tully (2015)]. Tully is in the meantime deceased (2019), but the site is maintained by 'The Estate of Rob Tully'. Jonathan Tennenbaum's memoir is entitled '*Stan and Education*' and can be found in the *Contributions* section of the Tully website. There, he characterizes Stanley Tennenbaum's relationship with Kurt Gödel. He points out that his father's involvement with the IAS Princeton was a result of his admiration of and interest in Kurt Gödel and his work, which he had admired 'from afar' long before his first visit to the IAS in 1967. And he states his opinion that Stanley Tennenbaum was the '*leading disciple*' and '*defender*' of Gödel during the latter's last 10 years. He also reports that his father had friendships with many people at the IAS, all across the range of faculty, visitors and employees there.

Thus, Stanley Tennenbaum was a 'Gödel disciple' by his own admission and by choice. They met around 1967 when Tennenbaum went to the IAS specifically because of Gödel's presence there, and they maintained contact throughout most of Gödel's final 10 years.

Gerald Sacks was also what might be called a 'Gödel fan', not really a disciple. He went to the IAS for a postdoc year in 1961, after completing his PhD at Cornell under the guidance of J. Barkley Rosser, and no doubt met Gödel then. He made several important advances in recursion theory, and held a joint professorship at MIT and Harvard from 1972 until his retirement in 2006/2012. He was again at the IAS for a year in 1974/75, and had more contacts with Gödel at that time. Sacks died in 2019. Famously, he gave a lecture entitled '*Reflections on Gödel*' at the University of Pennsylvania's Mathematics Department in April, 2007, which is available online as an audio file. Various Gödel biographers have quoted from it, notably Budiansky (2021, p. 105), and Yourgrau (2021), in his review of Budiansky's book.

Particularly piquant is Sacks' story about Gödel's response when he questioned the philosophy of Leibniz, a favorite philosopher of Gödel's for many years.[16] (He was apparently rather forward with Gödel, in the brash manner of a fresh PhD who is out to conquer the world). Gödel, by then a *devoté* of Husserl, answered in an unexpected way, saying that Leibniz was wrong with his monad theory. And he continued by saying that Leibniz was wrong about *everything*—but that's just as hard as being *right* about everything. Sacks was unsure as to whether this was a joke, an ironic remark, or a serious affirmation on the part of Gödel.

This is reminiscent of Gödel's earlier (1933) remark about Platonism,[17] somewhat ambiguous because of his placement of a comma, where he seems to be repudiating his early philosophical idol, Plato. It is indeed difficult to be certain as to whether Gödel was really describing his rejection of

Leibniz's philosophy, or simply joking (in a sober/serious way, most likely meant ironically) in his answer to Gerald Sacks' question.

Another source of information about Gödel's thoughts on philosophy in his later years are the '*Conversations with Gödel*', recorded by **Sue Walker Toledo** from 1972–75 [see Toledo (2011)]; their conversations were motivated by Stanley Tennenbaum and published in Kennedy & Kossak (2011). They are considered in detail there and cited in van Atten & Kennedy (2003, 2009) as well as by Kennedy (2021),[18] and we will return to them in the following chapter, where we take up Gödel's *philosophy* in retrospect. An amusing parallel is suggested in Kennedy & Kossak (2011)[19]: on the one hand, we have the ancient triad *Plato/Phaedo/Socrates* (referring to Plato's famous dialogue '*Phaedo*', in which a student of Socrates, Phaedo of Elis, relates the last words of Socrates to a Pythagorian philosopher, Echecrates. Of course, Plato is in fact relating the story). On the other hand, we have the modern triad *Tenenbaum/Toledo/Gödel*, where Toledo and Gödel hold dialogues, recording some of the later thoughts of Gödel, the whole thing orchestrated by Tennenbaum.

Toledo was in any case not a 'disciple' of Gödel's, but possibly a 'fan', and definitely an interested observer.

In the final years of the decade of the 1960s, Adele's health—she was 70 in 1969—and Gödel's (mental) health were both in decline. Adele had collapsed during a trip to Italy in 1965, although she soon recovered, apparently without lasting effects; and she entered a hospital for tests in 1968, but the reasons for them are unknown. Gödel lists a number of her possible illnesses in his last two letters to Rudolf, but they are in the end only speculation. It was noted by Morgenstern that Gödel looked very bad by spring, 1968, and it is clear that his hypochondria had intensified in the last years of the decade. He took a sudden turn for the worse in early 1970, as we saw in Chap. 9; more details will be given in the following chapter.

Notes

1. He moved into his new office in 1967, according to Budiansky (2021), p. 270; Dawson (1997), p. 229, puts the date of his move in 1965. The office was in a glass-front building adjacent to the new Social Science Library: its

Mathematics Wing. The buildings, connected by a glassed-in bridge, were completed in 1965.

2. Quoted from Kurt Gödel's letter to his mother Marianne, dated 14.08.1961, and later letters. Translation by the author.

3. This was soon after the era of Robert Maynard Hutchins' presidency of the University of Chicago (1929–1951). Hutchins attempted an almost utopian experiment: a liberal, classical education based on 100 'great books', early admission, and comprehensive examinations. The experiment was ended (mainly for lack of outside funding) after Hutchins left Chicago.

4. Anil Nerode, '*In Memoriam: Paul J Cohen 1934–2007*', in the *Bulletin of Symbolic Logic*, Vol. 15, No. 4 (2009), pp. 439–440.

5. Photo from the book [Albers, Alexanderson, & Reid (1986)], with permission from Springer-Verlag. Reused from: https://www.mathunion.org/filead min/IMU/Prizes/Fields/1966/index.html.

6. P. J. Cohen, '*The Discovery of Forcing*', in: *Rocky Mountain Journal of Mathethematics*, Volume 32, No. 4 (2002), pp. 1071–1100.

7. MacTutor Biography of Paul J. Cohen, online at: https://mathshistory.st-and rews.ac.uk/Biographies/Cohen/. Consulted on March 6th, 2022.

8. Gödel's 1940 book (Item 11 in Appendix A) was based on the notes from his Nov./Dec. 1938 lectures at the IAS, taken by George Brown. It was released as No. 3 in the *Annals of Mathematics Studies* series, published by Princeton University Press (in 1940).

9. See Backnote [13] in Chap. 8.

10. From Cohen's letter to Gödel, dated May 6th, 1963. Also quoted in Budiansky (2021), p. 263.

11. Budiansky (2021), p. 264.

12. Jeremy Pearce, '*Paul J. Cohen, Mathematics Trailblazer, Dies at 72*'. In *The New York Times*, April 2nd, 2007.

13. Cf. the Paul Cohen lecture video, six parts, from the *Gödel Centennial*, Vienna 2006 (on YouTube).

14. Quoted by, and from, Kanamori (2007), pp. 181–183.

15. Photo from the private archive of Jonathan Tennenbaum; photographer unknown, public domain. Reused with permission and thanks to Jonathan Tennenbaum.

16. But note that Sacks met Gödel *after* the latter's 'turn' away from Leibniz and toward Husserl. See e.g., van Atten & Kennedy (2003).

17. The remark was included in Gödel's lecture on the foundations of mathematics at a meeting of the AMS in late 1933 in Cambridge, MA. It is quoted by Dawson (1997), p. 100, by Feferman (1995), and by van Atten & Kennedy (2003), p. 430. We quoted it in Chap. 10, and again in Chap. 12.

18. Juliette Kennedy transcribed the handwritten notes of Sue Toledo which were the record of those conversations, probably in the early 2000s, i.e., some 25 years after they took place.
19. Kennedy & Kossak (2011), p. xiii, describing the chapter 'Stanley Tennenbaum's Socrates' by Curtis Franks.

18

Gödel's Last Years—*Philosophy, Set Theory, Logic*

The decade of the 1970s began badly for Kurt Gödel, as we have seen. (See Chap. 9, section 'The Beginning of the End'). Dawson (1997) calls this period '*Withdrawal*'. Gödel's psychiatric difficulties worsened gradually over the second half of the 1960s, and they reached a crisis in January of 1970. His friend Oskar Morgenstern drove him to the hospital on January 23rd, 1970.[1] He was showing symptoms of heart trouble as well as diabetes. He was however released after only four days, and was, for once, not displeased with his doctors. Nevertheless, a few days later, Morgenstern discovered that Gödel was taking (self-prescribed) *digitalis*, instead of the *Isordil*[2] prescribed by his doctor, which had seemed effective, at least for treating his symptoms. Shortly thereafter, Gödel's paranoia worsened dramatically, and he expressed his belief that his doctors were liars, and that the medical literature was full of errors, accidental or intentional.

Morgenstern tried to rationalize this behavior as an 'act', a dramatization on the part of Gödel; but evidently, he was truly suffering from paranoia and really believed what he was fearing and saying, including his own impending death. During this period in early 1970, he experienced 'ups and downs', almost as if he had become manic/depressive (bipolar syndrome), in addition to his other problems. Morgenstern, in his diary, follows Gödel's mood swings through mid-April, when he seemed to be in very bad condition and was in and out of treatment.

Kurt Gödel was convinced that he was about to die, and entrusted various manuscripts to Morgenstern for posthumous publication (and also to Dana

© The Author(s), under exclusive license to Springer Nature
Switzerland AG 2022, corrected publication 2023
W. D. Brewer, *Kurt Gödel*, Springer Biographies,
https://doi.org/10.1007/978-3-031-11309-3_18

Scott, as Morgenstern later learned). Gödel himself said at this point that he *'felt himself to be under hypnosis, compelled to do the opposite of what he knew to be right'*.[3] His brother Rudolf arrived in Princeton in early April, and apparently was in favor of psychiatric treatment for Kurt, which only increased the latter's paranoia. Adele was evidently also not very happy with Rudolf's presence, although both were pursuing the same goal: Kurt's recovery. Kurt later mentioned to Hao Wang that 1970, along with 1961 and 1936, was among the 'three worst years' in his life in terms of his health.

The account in Morgenstern's diary is interrupted for some months at that point, as he was traveling and no longer had contact to Gödel. When he arrived back in Princeton in August, he found, to his surprise, that Gödel had made a 'miraculous' recovery and seemed quite well. He had, indeed, written drafts for three articles, one on his *'Ontological Proof '* of the existence of God, one on the *'True power of the continuum C'*, and another (labeled as the second version (*II. Fassung*) of the previous one) on a proof of the Continuum Hypothesis. He sent the *'True power'* paper to Alfred Tarski to get his opinion before submitting one of them to PNAS for publication. The drafts were later found to contain errors.

As Gödel mentioned to Tarski in a letter written later that year (**1970c*), by which time he had a *third* version of his article on the CH, he had been taking psychopharmaceuticals, and they may have contributed to his 'mood swings' before the dosage was properly adjusted; and also to the confusion in his manuscripts written that year. The 'miracle' of Gödel's rapid recovery evidently consisted in his having submitted to treatment, something he usually resisted; in this case with psychoactive drugs, which did their job. His letter to Tarski remained unsent, and the article(s) were not published.[4]

However, a whole series of re-printings of earlier articles and of translations (in particular into Italian and Portuguese) are listed by Dawson (1983) as publications in the decade of the 1970s. All together, there are 14 entries, of which 9 were translations, the rest reprints. Among his unpublished manuscripts, the *Collected Works*, Vol. III, lists in that decade only the four mentioned above (as **1970*, **1970a*, **1970b*, and **1970c*; the latter is his (unsent) letter to Tarski about the CH manuscripts).

Gödel's Ontological Proof

The first of these manuscripts, [*Gödel *1970*], is Gödel's *'Ontological Proof '* of the existence of God. According to *Robert M. Adams*,[5] Gödel showed it to Dana Scott after he had written it, and gave him a copy to publish in case of his own death. But he later told Morgenstern that he hesitated to publish it because he did not want to be thought to be a deist—but that on the contrary, it was simply a logical exercise to demonstrate the *possibility* of such a proof. It has in the meantime acquired a certain following and has been the subject of attempts to carry it out in a 'mechanical' manner.[6] The steps involved in the proof itself are given in the *Collected Works* (Vol. III, p. 403 ff.) and are also available online.[7] Dana Scott gave a seminar lecture at Princeton University on Gödel's ontological proof in the autumn of 1970, and the text of his lecture, along with Gödel's manuscript, was published somewhat later [after Gödel's death—see Scott (1987)].

The Gödel proof itself dates back to 1941; a single page dated to that year was found in Gödel's legacy, containing many of the ideas used in the proof, and notes about it were also found in his unpublished notebook '*Phil XIV*' (1946–1955; from the second series of '*MaxPhil*' notebooks, see below). The pages referring to the proof were apparently written at Asbury Park/NJ between late August and early October of 1954, where Gödel was vacationing during that period.

Predecessors in formulating 'ontological proofs' of God's existence were of course *St. Anselm* of Canterbury (1033–1109) and *René Descartes* (1596–1650), among others—but Gödel's interest apparently stemmed from his reading of *Leibniz's* works, which he pursued intensively from the later 1930s until at least 1956. Leibniz dealt with Descartes' proof and improved upon it.[5] Adams discusses the history and details of Leibniz's writings on the proof, and demonstrates the parallels to it in Gödel's version, which is expressed in terms of symbolic logic. It is an example of 'modal logic', using modal qualifiers such as *possibly* or *necessarily*. Adams[5] notes that "*Gödel resembles Leibniz in making the ontological proof proceed by way of the conditional thesis that if the divine existence is so much as* possible, *then it is* actual, *and indeed* necessary" [emphasis added]. But he also points out that "*Gödel shows no clear influence of Leibniz's fullest argument for the thesis, which turns on a rather different conception of 'essence' from Gödel's*".

Since the publication of the *Collected Works* (Vol. III, 1995), there has been increased interest in Gödel's proof,[8] and there is now a voluminous literature concerning it, including articles by computer scientists interested in 'automating' the proof (cf. backnote [7] and the references given there). We have here another example of a work initiated by Gödel but not published during his lifetime, about whose publication he himself had ambiguous feelings, which has much later had a considerable effect on scholarship and research in philosophy, mathematical logic and computer science. A proof of the existence of God is of course an especially emotionally-charged topic, quite apart from its formal-logical aspects, and is more likely to be picked up and amplified by the popular press[6] than some of his more arcane writings.

The Continuum Hypothesis

The other three writings from 1970 [the articles Gödel (*1970 a-b*) and his unsent letter to Tarski, (*1970c*)] are all concerned with the Continuum Hypothesis and represent a series of attempts on Gödel's part to find a definitive proof of the power of the continuum. They are reproduced in Volume III of the *Collected Works* (pp. 420 ff.) and introduced there by *Robert M. Solovay* (pp. 405–420). They are flawed, as mentioned above. In those papers, Gödel follows his own program for proceeding from the post-Cohen state of affairs to the proof (or disproof) of the CH/GCH, i.e., by introducing new axioms into ZFC set theory. He proposes four such new axioms, discussed by Solovay in Sect. 2 of his Introductory Notes, and interested readers can find the details there and in the more recent literature, e.g., Kanamori (2007). According to Solovay, Axiom 3 was '*misstated in *1970a*', and Axiom 4 was only hinted at there, so that Solovay, with help from *Stevo Todorčević* and *Gaisi Takeuti*, reconstructed it for his Notes. In his next section, Solovay outlines the proof itself, which begins by demonstrating that $2^{\aleph_0} \geq \aleph_2$. This can be shown unproblematically from Axioms 3 and 4. The next step is the demonstration that $2^{\aleph_0} \leq \aleph_2$. Solovay says that he is '*unable to follow this part of the proof*'. Gödel claimed that it follows from Axioms 1–3, which Solovay however contradicts. In this part of his proof, Gödel also uses Axiom 4 and the '*rectangular conjecture*',[9] and shows that the latter is implied by Axioms 1 and 2 (called the *square axioms*). Gödel later specifically repudiated this part of the proof in his letter to Tarski (*1970c*).

Solovay's Sect. 4 deals with the origin of Axioms 3 and 4, while Sect. 5 expands on the rectangular conjecture and on bounds for the continuum. In his (long) Sect. 6, he describes a model based on Cohen's forcing method which illuminates Gödel's claimed proof. He concludes it by stating that Axioms 1–3 imply no bound on the size of the continuum, while he finds Axiom 4 to be dubious. He finally asserts that the *square axioms* do *not* entail the rectangular conjectures within ZFC (in contradiction to Gödel's original claim).

Solovay's Sect. 7 lists the open questions remaining concerning Gödel's 1970 proofs: (1) Whether Axioms 1–4 are in fact consistent with [the standard axioms of] ZFC; and (2) Whether, specifically, Axioms 3 and 4 are consistent with ZFC. In his 'Final Remarks' (Sect. 8), Solovay determines that Gödel *did not prove* that $2^{\aleph_0} = \aleph_2$ follows from his Axioms 1–4. The questions of the relative consistency of those axioms and whether such a proof could be found by other methods are left open. He questions the plausibility of Axioms 3 and 4, in particular, and also finds Axiom 1 unconvincing. The problem of how Gödel could have made such mistakes he attributes to the latter's feelings of urgency due to his (presumed) impending death, combined with the effects of his illness (and perhaps to drugs that he was taking, as Gödel himself suggested).

We have seen in Chap. 9 how the rest of the 1970s played out for Kurt Gödel. His health remained stable for a few more years after his recovery from his psychiatric problems in 1970, apart from his prostate difficulties which reached a crisis in 1974; but that was resolved, albeit in a less than optimal way. He retired officially from the IAS in July, 1976, a few months after his 70th birthday. Adele's health problems worsened around 1975, and she required several hospital stays in 1975 and 1976.

Kurt's last letters to his brother Rudi were written in 1972 and 1975. Both are rather brief; the first one, dated February 1st, 1972, is a birthday greeting from Kurt and Adele for Rudi's 70th birthday on February 7th of that year. Their general good wishes are accompanied by the hope that his health will soon improve. They had heard from Oskar Morgenstern, who had evidently visited Rudolf in Vienna shortly before, that Rudi was '*already in a good humor*' and '*not looking all that bad*'. They included a '*little present*' with the letter, and hope that he will be pleased with it. In a P. S., Kurt adds that all is well with them, that Adele's condition (presumably her health) is unchanged, and that she thanks [Rudolf] heartily for his nice letter of December 19th (1971).

The letter from 1975, dated December 12th, is an acknowledgement with thanks of a letter sent by Rudolf shortly before. He had announced the death of 'Eva', and Kurt expresses their sorrow at hearing of her passing, mentioning that '*she was not really very old*'. He writes rather plaintively that '*it is the same here*' [in Princeton], and that '*One after another, they are all passing*'. He sends their wishes to Rudi (and *Gitti*, a nickname for Brigitte) for the '*best possible holidays*'. Although communication between the two brothers was minimal in the last years of Kurt's life, he clearly made an effort to be cordial to Rudolf, if not really affectionate.

Gödel and Abraham Robinson

From the list of 'Gödel's disciples' given in Chap. 16, one name still remains to be explored: *Abraham Robinson*. He was a talented logician (and aeronautical engineer!) who came into Gödel's sphere rather late in both their lives, and left it again soon due to his early death in 1974. See Macintyre (1977), or O'Connor & Robertson (2000) for brief summaries of Robinson's life and works.

Abraham Robinson (1918–1974) was born Abraham *Robinsohn* on October 6th, 1918 in *Waldenburg*, the second largest city in German Lower Silesia. (It is now the Polish city of *Wałbrzych* in the province of Lower Silesia. Its Polish name is derived from its German Silesian-dialect name, '*Walmbrich*'). His father died shortly before his birth, and he and his older brother Saul were raised by their mother Hedwig, a teacher. They were Jewish, inclined to Zionism, and after the Nazi takeover of Germany in 1933, they emigrated to Jerusalem, at that time in the British Protectorate of Palestine. Abraham completed school there and studied at the Hebrew University (one of his teachers was *Abraham Fraenkel*, the 'F' in ZFC set theory). He graduated in 1939 with honors, receiving a fellowship for graduate studies at the *Sorbonne* in Paris.

His matriculation there coincided closely with the beginning of WWII, and his first 8 months as a graduate student took place during the '*drôle de guerre*' ('phony war'), the period of waiting between the declaration of war on September 1st, 1939, and the German invasion of France in May of 1940. On June 14th, Paris was occupied by German forces. Robinson (who adopted that spelling of his family name that same year) was able to escape and made his way by train and foot to Bordeaux, where he boarded a small boat that took him to England. There, he joined the *Free French Air Force*, but was soon delegated to the British *Royal Aircraft Establishment* at Farnborough, where he spent the war learning about, and designing, airfoils for the wings of military aircraft. By 1945, he had a double vocation as aeronautical engineer and as mathematical logician. He spent some time in postwar Germany as a Scientific Officer and then became a lecturer at the College of Aeronautics in Cranfield, UK.

His interest in mathematics continued, however, and he received an MS from the Hebrew University in 1946. This enabled him to begin doctoral research in model theory and the metamathematics of algebraic systems at *Birkbeck College*, University of London. Robinson received his doctorate there in 1949, and he accepted his first senior academic position at the University of Toronto, Canada, in 1951.

This was the beginning of a period of repeated moves to new positions and new countries: In 1957, he went to Jerusalem to become Fraenkel's successor at the Hebrew University; in 1962, he moved to UCLA in California as professor of mathematics and philosophy; in 1967 he went to Yale University, where he became Sterling Professor of Mathematics in 1971. From January to April, 1973, he was a Visiting Professor at the IAS in Princeton, and it was certainly during that visit that he came into closer contact with Kurt Gödel. They clearly knew of each other for some time before, and had probably met at the latest after Robinson moved to Yale in 1967. It seems somewhat misplaced to call him a 'disciple' of Gödel's; he was the second-oldest in the list, but unlike Stephen Kleene, the oldest, he did not meet Gödel early in his career. And yet, they apparently had a relatively close relationship during the few years until Robinson's early death from cancer in April 1974, just a year after his visit to the IAS.

Macintyre (1977) says of Robinson: "[he] *was a gentleman, unfailingly courteous, with inexhaustible enthusiasm. He took modest pleasure in his many honors. He was much respected for his willingness to listen, and for the sincerity of his advice"*. He also states that *"Gödel has stressed that Robinson more than any other brought logic closer to mathematics as traditionally understood"*. In summarizing Robinson's career, Macintyre says, *"From 1968 to 1970 he was President of the Association for Symbolic Logic, and on retiring in 1970 he delivered a challenging series of problems to the world's logicians"*. This latter lecture is reminiscent of Hilbert's 1900 list of problems, and both were delivered at ICM conferences—Hilbert's in Paris, and Robinson's exactly 70 years later, in Nice, France. The photo of Robinson in Fig. 18.1 was taken at the latter meeting.

Robinson was one of the founders of Model Theory in the 1950s. Dawson (1997) cites its founders as '*A.I. Mal'cev, Leon Henkin, Abraham Robinson, and Alfred Tarski'*. It has often been asked why Gödel himself did not participate in, or anticipate, these developments, given that it was his work of 1930/31 that stimulated them and formed their basis; the answer is usually claimed to lie in his *philosophy* (but this is a chicken-and-egg situation: Did his philosophy shape and limit what he was willing to do in mathematical logic, or was there a basic limitation in his personality that led to that philosophy and to the accompanying limitations in his work? We will return to this question below in the section on *Gödel's Philosophy*).

Of Gödel's work in the early 1970s, Dawson (1997) says, "*He persevered with research efforts, but as time went on, he lost touch with what other logicians were doing. There were a few exceptions, notably Abraham Robinson and Hao Wang, with whom his contacts actually increased in the years 1968–74…*".

Fig. 18.1 Abraham Robinson, taken at the ICM 1970 in Nice, France. Photo by Konrad Jacobs. *Oberwolfach Photo* Collection (MFO)[10]

Dawson also recounts[11] the following incident which bears on Gödel's relation to Robinson:

> In March of 1973 *[during his stay at the IAS]* Abraham Robinson spoke at the institute on his work on nonstandard analysis… at the end of the talk Gödel took the opportunity to express his opinion of the significance of Robinson's achievement. Nonstandard analysis was not, he said, '*a fad of mathematical logicians*' but was destined to become '*the analysis of the future*'. Indeed, he predicted, '*in coming centuries it will be considered a great oddity…that the first exact theory of infinitesimals was developed 300 years after the invention of the differential calculus*'.

Robinson's early death from pancreatic cancer a year later brought his inter-actions with Gödel to an abrupt end. By that time (April 1974), Gödel was in any case suffering from increasing health problems of his own; and those of his wife Adele, who was essential to his well-being in everyday life, were also on the increase, so that it is questionable how much more they could have interacted even had Robinson lived longer. But Robinson had been Gödel's 'connection to the outside world' for around five years, and Robinson, in turn, was certainly among those who, in the words of Hao Wang (1997, p. 33), "*saw him* [Gödel] *as their master*".

Gödel's Philosophy

It is clear from many sources that philosophy played a dominant role in Kurt Gödel's intellectual life and even in his mathematical logic. Gödel himself made many statements about his philosophy and left a large quantity of unpublished writings on that topic—in particular his '*MaxPhil*' notebooks,[12] begun in 1934 and continuing in two series until around 1956. A number of authors have given analyses of his philosophical positions and how they affected his work on logic, set theory and cosmology.

Of course, we cannot even begin to survey this vast literature here; instead, we will try to emphasize a limited number of salient characteristics and keywords which typify his philosophical positions, and to draw some preliminary conclusions about the relationship between (his) philosophy and the mathematical and scientific work for which he is most famous. From the outset, however, we can say with some certainty that Gödel without philosophy is unthinkable; it was a guideline for his work in all other areas and even for his behavior, to a considerable extent.

As we saw in previous chapters, Kurt Gödel was first introduced to philosophy during his last two years of high school (1922–24), when he was required to take courses in that subject. He was assigned reading of *Immanuel Kant*'s works—no doubt parts of the *Critiques*—and this aroused his philosophical vein, which remained active for the rest of his life. In his second year at university, 1925/26, he attended the survey lectures given by Heinrich Gomperz (History of European Philosophy), and there, his enthusiasm (some would say 'passion') for *Platonism* was aroused. He declared himself to be a Platonist many times during his later life, but this must be regarded with some reservations, since he apparently favored certain aspects of Platonism without embracing it as an all-encompassing philosophy. Still later, he 'discovered' *Gottfried Wilhelm Leibniz*, and beginning in the mid-1930s, he read many of Leibniz's writings, continuing until around 1956. Starting at the end of 1959, he began reading *Edmund Husserl*'s later works, and continued to do so for around 18 years, nearly to the end of his life. There has been speculation that Gödel experienced a kind of epiphany [see e.g., Goldstein

(2005), where his presumed early 'conversion' to Platonism is described in passionate tones, as though it were an *amour fou*]. Gödel himself believed in such life-changing experiences (e.g., in connection with Husserl before and after 1909; see below), but he never indicated that he had experienced one himself, although he was apparently willing and eager to. This was one motivation for his thorough reading of Husserl's later works. Gödel's conversion to Husserl's philosophy has been treated in detail by van Atten & Kennedy (2003) ('*On the Philosophical Development of Kurt Gödel*'), by Tieszen (2011) ('*After Gödel...*') and by Crocco & Engelen, eds. (2016) ('*Kurt Gödel. Philosopher-Scientist*'). This leads us to our first keyword, which I will call '*epiphany*'. Gödel referred to it in connection with Husserl's 'moment of enlightenment'; in a conversation with Sue Toledo in 1972[13]; he describes it as follows:

> There is a certain moment in the life of any real philosopher when he for the first time grasps directly the system of primitive terms and their relationships. This is what had happened to Husserl ... Husserl's philosophy is very different before 1909 from what it is after 1909. At this point he made a fundamental philosophical discovery, which changed his whole philosophical outlook and is even reflected in his style of writing. He describes this as a time of crisis in his life, both intellectual and personal. Both were resolved by his discovery.

Such an experience might be called a 'change of paradigm' in the natural sciences, or simply a 'turn' in philosophy, or a 'transformation'; but Gödel seems to mean something deeper: a life-changing event, almost a mystical experience. Wang (1997), pp. 169–170, calls it a '*sudden illumination*'; Gödel himself refers to a '*psychological crisis*'. He suggests that other philosophers besides Husserl had experienced such a moment, and he mentions specifically Descartes and Schelling (later also Leibniz and Plato). He believes that it is a very individual event, and it cannot be 'taught' by one philosopher to another; but one reason that he studied Husserl's writings so carefully was to find hints of how it comes about. In her (2021), Juliette Kennedy refers to it as a '*second sailing*', quoting a metaphor applied by Seth Bernardete[14] to Socrates. [Remaining in the nautical vein, one could also speak of a '*sea change*'.] Kennedy (2021) also quotes a conversation between Gödel and Hao Wang,[15] in which Gödel specifically denies having such an experience himself (although Kennedy remains skeptical about this point):

> I myself have never had such an experience. For me there is no absolute knowledge: everything goes only by probability. Both Descartes and Schelling

explicitly reported an experience of sudden illumination when they began to see everything in a different light.

Other authors [e.g., Hintikka (2005); his article is tellingly entitled '*What Platonism?* ...'] see Gödel's often-cited Platonism as simply embodying a 'realistic' ('objective', 'actualistic') view of *mathematical concepts*, i.e., the view that such concepts are innate in the universe, and waiting to be discovered through what Gödel called 'mathematical *intuition*'—just as physical, material features of nature can be discovered by using our 'normal' sensory perceptions. (Our second keyword: *mathematical intuition*). This is, in essence, a 'sixth sense', which operates not in the physical, material world like the first five, but rather in the abstract (but real) world of *mathematical concepts*. Talia Leven, in her (2019), discusses the comparison of Gödel's view of intuition with that of Abraham Robinson, before and after Paul Cohen's proof of the independence of the AC and CH from the axioms of ZFC. Robinson seems to have experienced such an illumination, stimulated by Cohen's work, which led him to a reorientation of his philosophy of mathematics, and in particular of his view of mathematical intuition. Whether his 'turn' can be classed as an *epiphany* is however unclear. Gödel referred to Robinson's approach after Cohen as the '*as if*' viewpoint; he continues to think 'as if' the Platonic realism viewpoint were valid for infinite totalities, but no longer believes that it really is.

This view of 'mathematical realism' is often called 'naïve Platonism' or 'objectivism'. Hintikka, in his (1999), also points out the related difference between (Gödel's) '*actualism*', i.e., his concentration on the *actual* world, the 'real' universe in which we live, rather than considering possible, hypothetical concepts that might exist in other worlds (the 'many worlds' approach). [This applies specifically to his early work on logic and set theory; in his cosmological work (1946–50), leading to the 'Gödel Universe', he was clearly more speculative and inclined to a 'many worlds' viewpoint, although he seems to have hoped that his rotating universe might turn out to be *actual*. Compare Chap. 15]. We thus have our third keyword: '*actualism*'.

We might expect some insights on Gödel's own philosophy from the text of his planned inaugural lecture to the *American Philosophical Society*, written after 1961 but never delivered [see [Gödel (*1961/?*)] in the *Collected Works*, Volume III, pp. 374 ff., and the corresponding Notes by Dagfinn Føllesdal (pp. 364–373)]. Føllesdal discusses Gödel's interest in Husserl's phenomenology:

> According to Gödel, there is a certain kind of obviousness that accrues to mathematical statements once their meaning has been clarified... the method for

this clarification of meaning Gödel finds in Husserl's phenomenology. Gödel describes Husserl's method as '*focusing more sharply on the concepts concerned by directing our attention … onto our own acts in the use of these concepts, onto our powers in carrying out our acts, etc*'. It is '*a procedure or technique that should produce in us a new state of consciousness in which we describe in detail the basic concepts we use in our thought, or grasp other basic concepts hitherto unknown to us*' [*p. 8 of the manuscript*].

He quotes Gödel's manuscript on his program of introducing new axioms:

… in the systematic establishment of the axioms of mathematics, new axioms, which do not follow by formal logic from those previously established, again and again become evident. It is not at all excluded by the negative results mentioned earlier [*i.e., the incompleteness theorems*] that nevertheless every clearly posed mathematical yes-or-no question is solvable in this way. For it is just this becoming evident of more and more new axioms on the basis of the meaning of the primitive notions that a machine cannot imitate [*p. 9 of the manuscript*].

Gödel continued to hold his realistic, actualistic view throughout most of his life, and he expressed it for example in his Gibbs Lecture in late 1951; but he was not sufficiently sure of the strength of his arguments to publish the text of that lecture. As we saw in Chap. 16, it was published only posthumously, in Volume III of his *Collected Works*. And, unlike Robinson, he did not essentially change his viewpoint even after the Cohen proofs— but his 'naïve Platonism' became 'constituted platonism' [Tieszen (2011)]. And Gödel believed, before and after Cohen, that any mathematical problem could be solved *rationally* (known in some circles as *epistemic optimism*). This he shared with Hilbert (*non ignorabimus*), as shown by the above quote.

Talia Levin, in her (2019), proposes that the different reactions of Gödel and of Robinson to Cohen's results were due to their different notions[16] of intuition: "*the key to these different responses stems from the meanings that Gödel and Robinson gave to the concept of intuition, as well as to the relationship between epistemology and ontology*".

It has been suggested [e.g. by Parsons (1995)] that Gödel's interest in Husserl's phenomenology was motivated by Cohen's results, but this would not seem to be borne out by the chronology (Gödel began reading Husserl's works seriously in late 1959, as he later told Wang; but Cohen's forcing method and his proof of the independence of AC and CH came to Gödel's attention only in 1963, shortly before Cohen published them).

Robinson, in contrast, was discouraged from such optimism by Cohen's results. He turned to a version of Hilbert's formalism, and in a talk given at the 1973 Logic Colloquium[17] in Bristol, he was rather pessimistic regarding Platonism:

> The incompleteness of Peano arithmetic and the undesirability of the continuum conjecture have not led to a general abandonment of Platonism although it would seem to some, including myself, that they provide evidence against it.

Gödel began what is sometimes called his philosophical '*program*' (Wang) or '*project*' (Kennedy) in conjunction with his study of Husserl's phenomenology. Van Atten and Kennedy (2003, p. 470) term it to be '*phenomenology as a systematic means to combine the two strands of thought he had adopted earlier, his strong realist view of mathematics and the Leibnizian framework that put subjectivity in central position (monadology)*'. He did not see his realist project as being threatened by Cohen's results; in a letter to Church, dated September 9th, 1966, and quoted by van Atten & Kennedy (2003, p. 470), he says:

> You know that I disagree about the philosophical consequences of Cohen's result. In particular I don't think realists need expect any permanent ramifications[18] as long as they are guided, in the choice of the axioms, by mathematical intuition and by other criteria of rationality.[19]

Gödel did, in any case, read Plato's writings as well as those of Kant, Leibniz, Husserl and other philosophers of the modern era. In van Atten and Kennedy (2003, 2009), the conversations between Sue Toledo and Kurt Gödel during Toledo's stay at the IAS in 1972–74 [cf. Chaps. 17 and 20, and Toledo (2011)] are evaluated in relation to his 'turn' toward Husserl in 1959 and his views on mathematical intuition. In a more recent paper (2021), Juliette Kennedy discusses in addition to Gödel's ideas on Husserl also in particular a conversation that he had with Toledo about Plato's dialogue *Euthyphro*, with which Gödel was evidently quite familiar.

A fourth keyword could well be '*limitations*'. Several authors have noted that Gödel introduced a number of novel concepts and ideas, some of them quite revolutionary, but failed to follow up on their implications. For example, he clearly established the difference between formal-syntactical concepts and semantic knowledge, and he could (should?) have been a founder of what is now called *model theory*. He even anticipated Tarski's demonstration that 'truth' belongs only to the latter category (in his reply to Zermelo in October of 1931). But he then dropped the subject, and let Tarski, Church, Kleene, Rosser and Turing continue the developments that he had initiated. There has been much speculation as to how this was a result of his philosophy of mathematics. But perhaps the opposite is true—given the limitations of his personality (his need for *predictability*) as described in Chap. 9—he may simply have required a circumscribed environment in which to work, and his philosophy adjusted itself to that requirement and justified it *ex post facto*.

Finally, there is a persistent tendency to connect Gödel to *logical positivism* (or logical empiricism), due of course to his participation in the Vienna Circle for around three years at a formative stage in his intellectual life. He himself admitted that his attendance of the Circle's meetings stimulated his interest in philosophy (and probably also in mathematical logic); but he insisted that his own philosophy was already oriented toward Platonism, even before he joined the Circle, and that didn't change as a result of his participation. Positivism is—in modern terms—a WYSIWYG philosophy, which emphatically rejects any hints of metaphysics; while Gödel, as we have seen, firmly believed in an abstract world of mathematical concepts, inaccessible to the five 'normal' senses, but reachable through *intuition*.

We can fittingly close this brief survey of Gödel's philosophy of mathematics by quoting Hao Wang [(1987), pp. 192, 221; as quoted indirectly by Charles Parsons (1996a), p. 78]:

> Wang has written that Gödel did not believe he had fulfilled his own aspirations in systematic philosophy. Very often he *[Wang]* also gives the impression that he did not find what he himself was seeking in philosophical work. Both Gödel and Wang struggled with the tension inherent to the enterprise of general philosophizing taking off from a background of much more specialized research.

The Final Year—*January 1977–January 1978*

And then began the fateful year 1977. If Kurt could have continued his previous conversations with Hao Wang late that year or early in 1978, he certainly would have put 1977 at the top of his list of 'worst years'. They did indeed speak, in person on December 17th, 1977, while Kurt was still alone at home—Adele had been in hospitals for nearly six months, and he had refused all offers of help—and again by telephone when he was himself in the hospital on January 11th, 1978, shortly before his death. But he was detached and distant in those last conversations, successfully hiding his real condition from Wang, and revealing little of his desperate state. By January, he had retreated from the world and was simply waiting for the end, which

Fig. 18.2 Kurt Gödel and Dorothy Morgenstern at the Institute garden party on Oct. 7th, 1973: A happy moment near the end of Gödel's life. Compare Chap. 9, last section, and backnote 47 there. *Photo* by A.G. Wightman, IAS archives, Shelby White and Leon Levy Archives Center, Institute for Advanced Study, Princeton, NJ[20]

came on Saturday, January 14th. Figure 18.2 shows Kurt Gödel in 1973 at a garden party given by the IAS director (with Dorothy Morgenstern in the background), when he was still in good health.

As noted in Chap. 16, William Boone, one of Gödel's true disciples, wrote to Adele Gödel two weeks after Kurt's passing, offering his condolences. His letter (dated February 1st, 1978; in *CW*, Vol. IV, p. 326) gives a concise summary of Gödel's effects on many of those people who were lucky enough to have interacted personally with him, and characterizes both his professional career and his personal life:

> I hope you know, dear Mrs. Gödel, how much Professor Gödel did to help me with my work. It was he who was willing to check my work and say it was correct. He made many suggestions as to what I should work on, too. But what was the most wonderful aspect of our relationship for me, is that he became my friend, always willing to talk to me about mathematics or the world in general.

> Certainly, he was the most brilliant person I have ever known. I treasured every moment to speak with him ...

Notes

1. Dawson (1997, 2005 edition), p. 232.
2. *Isordil* is a trade name for isosorbide dinitrate, which (like nitroglycerin) is used to reduce chest pain (*angina pectoris*) in patients with coronary artery disease. This drug class ('nitrates') relaxes and expands the blood vessels serving the heart.
3. Dawson *op. cit.*, p. 234.
4. See the *Introductory Note* to the **1970a-c* manuscripts by Robert M. Solovay in the *Collected Works*, Vol. III, p. 405, on later work stimulated by those articles when they became available after Gödel's death. See also the previous chapter in this book, at the end of the section on Paul Cohen.
5. In the *Collected Works*, Vol. III: Robert M. Adams, '*Introduction to* [Gödel (**1970*)]', pp. 388–402.
6. See for example the *Spiegel* (Germany) article of Sept. 9th, 2013 by Tobias Hürter, '*Mathematiker bestätigen Gottesbeweis*' ('*Mathematicians Verify Proof of God's Existence*'), or the *Tagesspiegel* (Berlin) articles of 17.10.2013 by Christoph Benzmüller, '*Gödels "Gottesbeweis" bestätigt*' ('*Gödel's "Proof of* [the Existence of] *God" Confirmed*') and 12.04.2016 by Luisa Hommerich, '*FU Lehrpreis für Christoph Benzmüller, den "Gottesbeweiser"*' ('*Teaching Prize of the Freie Universität Berlin* [awarded to] *Christoph Benzmüller, the "God-Prover"*'). In all of these articles, a certain degree of journalistic sensationalism reigns, implying that the verification of the logical consistency of Gödel's 'proof ' means that indeed, the existence of God (presumably the Christian God of the New Testament) has been scientifically proven. Gödel must be spinning in his grave. In any case, his fears of sensationalism and misinterpretation should he publish his article (**1970*) can be seen to have been justified.
7. See https://github.com/FormalTheology/GoedelGod. The article documenting the digital verification is archived under http://arxiv.org/abs/1308.4526. (C. Benzmüller and B.W. Paleo, '*Formalization, Mechanization and Automation of Gödel's Proof of God's Existence* '). A number of other references are given there (q.v.).
8. More modern views of Gödel's *Ontological Proof*, including attempts to improve it, are given for example by Sobel (2004), Koons (2005), and Pruss (2009).
9. See J. Brendle, P. Larson, and S. Todorčević, '*Rectangular Axioms, Perfect Set Properties, and Decomposition*', in *Bulletin T.CXXXVII of the Serbian Academy of Sciences and Arts*. Online at jstor, https://www.jstor.org/stable/44095604?seq=1. Cf. [Brendle, Larson & Todorčević (2008)].
10. Photo by Konrad Jacobs. *Oberwolfach* Photo Collection (MFO), released to the public domain. Licensed under Creative Commons Attribution-Share

Alike Germany 2.0 license. Reused from: htpps://opc.mfo.de/detail?photo_id = 3540.

11. Dawson, *op. cit.*, p. 244 and footnote [8] there.

12. Those notebooks were evidently intended as memos to himself and as a storehouse of ideas, mostly not developed. They were certainly not intended for publication, and they were in the main written in his 'private code', the *Gabelsberger* shorthand. They have been painstakingly transcribed and reconstructed since his death, and are now being published, in particular by the '*Kurt-Gödel-Forschungsstelle*' (Kurt Gödel Research Group) of the *Berlin-Brandenburg Academy of Sciences and Humanities*, led by Eva-Maria Engelen, as the *Philosophische Notizbücher* (*Gödel*) [*Gödel's Philosophical Notebooks*; see Engelen (ed.), (2019), (2020)]. Gödel's abbreviation *MaxPhil* can be taken to mean 'Maxims on Philosophy' or 'Maxims and Philosophy'.

13. From Toledo (2011), quoted in van Atten & Kennedy (2003), p. 432, and in Kennedy (2021), p. 875.

14. Bernardete (1989), p. 2. Quoted by Kennedy (2021), p. 874.

15. From Wang (1997), pp. 169–70, undated (but certainly between 1972 and 1977). Quoted by Kennedy, *op. cit.*

16. Levin (2019), p. 442.

17. In a talk entitled '*Concerning Progress in the Philosophy of Mathematics*', given at the Bristol Logic Colloquium (1973), and published in its *Proceedings* (eds. H.E. Rose & J.C. Shepherdson), North-Holland, Amsterdam (1975), pp. 41–52. See Robinson (1975).

18. Gödel at this point inserted a note referring to Church's manuscript from his talk at the ICM 1966, in Moscow, whose text is available in the Conference Proceedings (1968), and of which Gödel apparently had an advance copy in 1966.

19. See van Atten & Kennedy (2003); they give the original reference to Gödel's letter as [Gödel *Nachlass*, folder 1/26, 010,334.36].

20. Photo: Kurt Gödel at the IAS garden party in October, 1973; taken by A.G. Wightman, property of the IAS. From the Shelby White and Leon Levy Archives Center, IAS Princeton, courtesy of the IAS. Reused from https://alb ert.ias.edu/handle/20.500.12111/4234?show=full. Cropped and enhanced.

19

Gödel's Legacy

Here, we refer to 'legacy' in both the *material, physical* sense (the writings, papers, objects,[1] real estate, money and other artifacts left behind by someone who has departed this life), and in the *intellectual, scholarly* sense—how that person may have changed the world of the mind and left traces which can guide later thinkers, perhaps leading them to continue work begun but not finished by the departed.

As far as the *physical* legacy of Kurt Gödel is concerned, it has fortunately been saved in the archives at the Firestone Library of Princeton University, placed there for archival storage after it was left to the IAS by his widow, Adele, who died in 1981. (She apparently destroyed some of the private correspondence in Kurt's papers, in particular the letters from his mother, Marianne, for personal reasons. Whether she destroyed or lost other papers will probably never be known).

A fictional version of the story of how his papers were saved was written by *Yannick Grannec* [see Grannec (2013), and Chap. 5], who recounts the story of a young woman who convinced Adele during her last year to save his *written* legacy and donate it to the IAS. Just how that in fact came about in reality is not clear; the introductions to the *Collected Works*, for example, give no details. However, John Dawson, in his interview for the Oral History Project at the IAS, Princeton [Dawson (2013)], tells the story of how he came to take on the imposing task of cataloging, organizing and making available the writings in Gödel's legacy, both those that were published [which he had already listed in his *Annotated Bibliography*, Dawson (1983)], and the many

© The Author(s), under exclusive license to Springer Nature
Switzerland AG 2022, corrected publication 2023
W. D. Brewer, *Kurt Gödel*, Springer Biographies,
https://doi.org/10.1007/978-3-031-11309-3_19

notes, unpublished manuscripts, unsent letters and other correspondence and papers in the archives.

It is certain that the collation, transcription and translation of Gödel's papers was a labor of love, complicated by the fact that many of his manuscripts and most of his private notes were written in an obsolete form of shorthand, the *Gabelsberger Schnellschrift*, which he had learned while in high school. Fortunately, Dawson's wife Cheryl agreed to learn to read that shorthand, aided by an old textbook found among Gödel's papers and by a photographer of German origin in New York, *Herman Landshoff*, who had learned the shorthand in his youth.

Gödel's *intellectual* legacy is perhaps less tedious, but it is rather complex and is still developing, in part because the physical legacy has not yet been completely made available. Furthermore, interest in his work has been increasing steadily over the years, so the subject is dynamic and ongoing. This is true of both his legacy in logic and set theory, and of his philosophical legacy, as we have seen in previous chapters. He was not really known as a philosopher until the publication of his (1944), the chapter in the *Living Philosophers* volume on Bertrand Russell. His status in that field was cemented in 1961 with his election to membership in the *American Philosophical Society*. But, as we have also seen, he published very little during his life, and only after his many notes, manuscripts and memoranda on philosophy have all been published and digested—in particular all of the '*Max/Phil*' notebooks[2]—will the full scope of his philosophical thought be known.

The history and legacy of Gödel's excursion into general relativity and *cosmology*: the '*Gödel Universe*', is rather different. It attracted little attention in the first 10 years after its publication(s), and what it did attract was in part negative [cf. the paper by Chandrasekhar and Wright (1961)]. And that, as we saw in Chap. 15, was typical of works on relativity and gravitation in that era, still in the 'backwater' period of those fields. Publications on the *philosophy* of the Gödel Universe were also rare in the early years after its appearance, but, as we have seen, there has been an increasing number in the later decades of the twentieth century, and continuing up to the present.

Concerning the *physics* literature, during the 'renaissance' of gravitation and relativity, Gödel's Universe became very well known [despite its not being included in the Wheeler/Misner/Thorne *opus magnum* of 1973, *Gravitation*!]. Here is a brief tale that demonstrates with one example how widespread it in fact later became:

The Brazilian theoretical physicist *Jayme Tiomno* (1920–2011) was born almost exactly 14 years after Kurt Gödel, on April 16th, 1920, in Rio de Janeiro, the son of Russian-Jewish immigrants who had arrived in Brazil

Fig. 19.1 Jayme Tiomno, studying in an office at Princeton University, mid-1949. Photo: Courtesy of the private archives of J. Tiomno

10 years previously. He studied medicine, and then physics, at the old *Universidade do Brasíl* (the predecessor of the modern *Federal University*, UFRJ) and, after an interruption for military service during WWII, he began graduate work under *Mario Schenberg* (1914–1990; cf. Chap. 15, section on George Gamow) at the University of São Paulo (USP). His first project involved a formulation of General Relativity in Minkowski space (it was completed but never published, due to the difficulty of publishing on GRT if one was not associated with Einstein in those days, and to Schenberg's political problems which caused his exile to Europe after 1948).

In 1947, Tiomno received a fellowship from the US Department of State to continue his graduate work at Princeton University (where Schenberg and also Tiomno's colleague *José Leite Lopes* (1918–2006) had earlier done graduate work; the latter received his PhD there in 1946). Tiomno (Fig. 19.1) began work in early 1948 in the research group of John A. Wheeler (at that time still in his 'nuclear and particle physics' phase), and he was given a problem in GRT (a harbinger of Wheeler's 'turn' to GRT, gravitation and cosmology four years later). Tiomno had difficulty with the problem, in spite of his consultation with Einstein, and soon turned to a project involving the pion-muon-electron decay chain which also was of great interest to Wheeler. The GRT problem was solved and published a year later by Leopold Infeld [cf. Infeld & Schild (1949); in the same volume of *Rev. Mod. Phys.* as Gödel's (1949b)].

Tiomno received his MSc in 1949 with Wheeler and then wrote a PhD thesis (on neutrino physics and double beta decay) under the direction of Eugene Wigner (since Wheeler was away from Princeton for about two years beginning in June, 1949). See [Brewer & Tolmasquim (2020)] for more details.

Tiomno's and Gödel's paths crossed two times at an interval of just over 20 years—we have here a case of the '*bizarre threads*' mentioned by Gödel in describing his own path to science (cf. Chap. 3, section on 'Mathematics and Philosophy'). Whether they actually met will remain forever unknown; neither of them mentioned such a meeting in a manner that was preserved for posterity, but it is not unlikely. Tiomno was in Princeton from early 1948 until October of 1950, when he completed his PhD and returned to Brazil—this coincides with the period when Gödel was working intensely on his 'cosmological excursion', the *Gödel Universe*. It is indeed probable that Tiomno heard Gödel's May 1949 talk at the IAS on his new cosmology, given his interest in GRT.

And Tiomno returned to Princeton again just over 20 years later, in early 1971, with a joint appointment to the University and the IAS, invited by John Wheeler and Freeman Dyson, due to his blacklisting in Brazil by the military dictatorship at that time. He spent most of 1971 and the first half of 1972 there, accompanied by his wife, the physicist *Elisa Frota-Pessôa* (Fig. 19.2). He worked with members and collaborators of Wheeler's group, including *Remo Ruffini, Leonard Parker, C. V. Vishveshwara, Jeffrey M. Cohen, Frank Zerilli* and *Robert Wald*, publishing over a dozen articles on relativistic astrophysics, especially 'black holes', during and shortly after his stay. None of those related directly to Gödel's work, however.

Tiomno had led a number of students to participate in the short-lived, utopian project of founding the new *Universidade de Brasília* in 1965, conceived as a liberal, open research university (somewhat like Stanley Tennenbaum's unsuccessful project in the USA around the same time). That project was also quashed by the dictatorship, but Tiomno and Elisa saw to it that their students could continue their studies elsewhere.

One of those students was *Mario Novello* (born 1942), who later (in 1972) received his PhD from the University of Geneva under *Joseph-Maria Jauch*. After a postdoc stay at Oxford, where he caught the 'cosmology bug', Novello returned to the *Brazilian Center for Physics Research* (CBPF) in Rio de Janeiro, an institute founded in the early 1950s by *José Leite Lopes, César Lattes, Jaime Tiomno* and *Elisa Frota-Pessôa*, among others. Novello (Fig. 19.3) set up a group in relativity and cosmology there, together with his student *Ivano*

Fig. 19.2   Jayme Tiomno and Elisa Frota-Pessôa in Princeton, 1971. Photo: Courtesy of the private archives of J. Tiomno and E. Frota-Pessôa

Damião Soares, and they were joined in 1980 by Tiomno, after his black-listing was rescinded by an amnesty. They founded the new Department of Relativity and Particles (DRP) at the CBPF and established a lively research activity.

Novello, in the meantime, had specialized in cosmology, and had studied non-stationary universes which evolve toward Gödel's Universe as a final stage. Tiomno and a Novello student, *Marcelo José Rebouças* (born 1951), became interested in Gödel's cosmology and wrote a series of articles treating variations on its themes.

Rebouças (Fig. 19.4) went on to become an expert on Gödel's cosmology, and he has continued publishing in that area, as well as on other aspects of cosmology and gravitation theory (references to the publications by Tiomno, Novello and Rebouças on aspects of Gödel's Universe are collected in Appendix C at the end of this book). As we can see there, their research activity began in the year of Gödel's death, 1978, and has continued for over 40 years, carried out by three generations of theoretical physicists. The

Fig. 19.3 Jayme Tiomno, Mario Novello, and Elisa Frota-Pessôa in 1991. Photo: Courtesy of the private archives of J. Tiomno and E. Frota-Pessôa

majority (18) of the papers were published by Rebouças as author or co-author; 5 included Tiomno and 3 included Novello (who in fact originated the interest in the topic of Gödel's Universe at the CBPF). This is not the place to give an exhaustive bibliography of work all over the world on the Gödel cosmos, but we provide a brief summary in a backnote.[3]

This is only one example of the interest in Gödel's cosmology, which arose 20 years and more after his original work. Gödel himself died too soon to become aware of this activity, but it would certainly have gratified him if he had known of it. In contrast, his hopes that the actual universe would prove to be rotating have not been fulfilled: all the observational evidence has failed to show indications of the predicted asymmetries.

As we can see, Kurt Gödel's lifework, in spite of his occasional hesitancy and his time-consuming perfectionism in his writing, has led to a whole 'industry' spread over several diverse fields: among others, mathematical logic, set theory, cosmology, theoretical physics, computer science, and philosophy. As pointed out by his friends and colleagues John von Neumann and Steven Cole Kleene, his accomplishments rise far above the norms of the times, and they will be remembered not only in an historical sense as important contributions to science, but also as stimulants for further scholarship, for a long time to come.

Fig. 19.4 Marcelo J. Rebouças, 1980s. Photo: Courtesy of the CBPF, Rio de Janeiro

Notes

1. The usual technical term in the literature for this *written legacy* is '*Nachlass*', the German word which variously means 'legacy', 'inheritance', 'estate', or 'assets', along with several legal/financial meanings (its written form with a '*scharfes s*' (ß) is no longer correct since the German Orthographic Reform of 1996). It is however a word with unambiguous one-word translations, so using the German term in an English-language text seems superfluous and a bit pompous. There are, of course, many words in every language which have *no* one-word equivalents in other languages, for example '*gemütlich*' or '*Schadenfreude*' in German, or '*saudade*' in Portuguese, and usage of the original word even in a translated text is then justified. But this is not the case for *Nachlass*, so I have chosen to use the English term 'legacy' in this book, in both senses—of a material legacy and an intellectual legacy.

2. See the website of the '*Kurt Gödel Research Centre*' at the *Berlin-Brandenburg Academy of Sciences and Humanities*, which is carrying out the project of publishing the '*Max/Phil*' notebooks. Online at: https://www.bbaw.de/en/research/kurt-goedel-forschungsstelle-die-philosophischen-bemerkungen-kurt-goedels-kurt-goedel-research-centre-the-philosophical-remarks-of-kurt-goedel.

3. *Physics* publications on '*Gödel's Universe*': a survey of the *arXiv* preprint server yielded 56 entries over the period 1994–2022, submitted by groups in Brazil, Canada, China, Czech Republic, France, India, Italy, Serbia, the USA. The field is still very active and is quite international.

20

Epilogue

At the time of this writing, it has been just over 45 years since the beginning of Kurt Gödel's difficult final year. His generation—born between about 1890 and 1915 into the 'Old World', which today is only a fading memory—has essentially passed away, and even the generation of his disciples, born for the most part between 1920 and 1935, are almost all gone. Hardly anyone is alive today who knew him personally; and yet, due to the hard work and careful scholarship of those who have cataloged, transcribed and translated thousands of pages from his written legacy, we now probably have a clearer picture of his thought and his opinions than was available to all but a scant few during his lifetime.

—But not, of course, of his intimate personality, his most private thoughts, hopes and fears, insofar as he did not document them in some form that has survived. That knowledge died with Marianne, Adele and Rudi, with Rudolf Carnap, Karl Menger, Olga Taußky-Todd, Oskar Morgenstern, Albert Einstein, Hao Wang and Georg Kreisel, with Stanley Tennenbaum and Abraham Robinson. Those few who may still be living are not accessible.

A rather mysterious figure in the latter category is *Sue Walker Toledo*. Her presence on the Internet is very limited,[1] but we know that she held conversations with Gödel on his work in logic and set theory and on his philosophy (and philosophy in general) over a three-year period from 1972–1975. It is remarkable that she was able to speak to him personally, face to face, repeatedly. Hao Wang, in contrast, whom Gödel had known for some time and certainly trusted, mentions that during his sabbatical year at the IAS (1975/76, Gödel's last year at the IAS before his formal retirement in July,

W. D. Brewer, *Kurt Gödel*, Springer Biographies, https://doi.org/10.1007/978-3-031-11309-3_20

1976), he spoke to Gödel 'in person' only a few times, indeed only once at length. Their numerous other conversations were all by telephone, at Gödel's request. Stanley Tennenbaum, who suggested the Toledo/Gödel conversations, clearly knew what he was doing. Sue Toledo was well known (before her marriage to Domingo Toledo) as Sue Ann Walker to some of the 'Logic Cadets' from the University of Chicago in the 1950s (see below). Another, younger logicist who knew Kurt Gödel in his later years is *Rudy Rucker*; see his blog at https://www.rudyrucker.com/blog/2012/08/01/memories-of-kurt-godel/.

The world has changed in almost incomprehensible ways since Gödel was born in Brünn, the provincial capital of Moravia in the Habsburg realm and the 'Manchester of Austria-Hungary'. Modern Brno is a lively city, with a youthful population, many cafés and restaurants, spontaneous concerts in little alleys—and all the locales where Kurt Gödel lived and studied are still there, mostly very well preserved and renovated. They are all still there in Vienna, as well, many of them recalling his passage with memorial plaques—but the spirit of the days of the *Wiener Kreis* and the *mathematisches Kolloquium* is gone forever. And the small, intimate School of Mathematics at the IAS in Princeton, temporarily housed in the old Fine Hall, where he went in 1933, has become a sleek, modern campus with dozens of members.

In any case, it is certain that Gödel exerted a great fascination for many very intelligent people, in particular young people interested in mathematical logic, long before his fame became widespread. This he shares with Ludwig Wittgenstein, although they otherwise had little in common. Stanley Tennenbaum is a good example: William Howard, in his memoir given at the 2006 meeting at CUNY in honor of Tennenbaum's life and work,[2] recalls him from the early 1950s as a dapper young fellow student at the University of Chicago who was obsessed with Gödel's incompleteness theorem. He recounts that Tennenbaum, like the Ancient Mariner in Coleridge's epic poem, would fasten on any unwary listener and ply him or her with information about incompleteness. He had apparently acquired his interest in Gödel's theory from Carnap, who taught for some time at Chicago. Howard presumes that Carnap hated the theory, since it countered some of his positivistic ideas about determining truth; but in view of the complex relationship between Carnap and Gödel, and the development of the former's ideas during the 1930s and '40s, this seems unlikely. Howard notes that the (few, but enthusiastic) students who were interested in mathematical logic were ironically called the 'Logic Cadets' by some of the faculty (themselves not logicians).

But the interesting comment in this passage is about Stanley Tennenbaum's obsession with Gödel's incompleteness proofs, at the age of 23, already with a Bachelor of Philosophy degree but no real training in mathematical logic. At that time, Gödel was still strictly an insiders' tip.

A similar case is of course that of the (even younger) Douglas Hofstadter, 10 years later, who discovered Gödel through the 1958 book by Nagel & Newman (perhaps one of the first publications to introduce him to a wider audience outside the scholarly literature). He went on to write his own book, '*Gödel, Escher, Bach*' (*GEB*), which probably had more influence in making Gödel's name and work known to the general public than any other single source. And Gregory Chaitin was inspired by the same (1958) book to write his own treatise on Gödel and Turing (cf. Chap. 12).

Verena Huber-Dyson's review[3] of Hofstadter's 1979 book *GEB*, is, in contrast, a complete roasting; it isn't her kind of book, although she must admit that it made the name 'Gödel', previously practically unknown, into almost a household word. But she cannot resist giving a compact summary of Gödel's incompleteness proofs herself. Here, there is a strong element of 'insider knowledge', implying that only those within the inner circle of mathematical logic, those who have been sanctified, initiated, or tempered in the crucible of puzzling out logical proofs themselves can possibly understand, or dare to comment on, the work of other logicians. Or is it simply enthusiasm, as with Tennenbaum?

Returning to the 'Logic Cadets', again referring to William Howard's contributions to the Tully/Tennenbaum site: he lists the logic students as including himself, Tennenbaum, Ray Smullyan, and Anil Nerode, as well as Michael Morley. He mentioned in 2004 to Morley that they had an exciting logic group at Chicago in the 1950s', and Morley replied, '*We had logic students but no logic professors*'.

Concerning the earlier history of Sue Ann (Walker) Toledo, Howard reports that she was later at the IAS in Princeton at the same time as Tennenbaum and himself; she was a research assistant to Hassler Whitney for a mathematics education project, and she had her own office (which he believes to have been Gödel's old office). Stanley Tennenbaum reportedly often slept there, leaving a certain chaos in the office the following morning. They were both good friends with Sue Walker, whom Howard had met through Stanley Tennenbaum at a conference on intuitionism and proof theory at Buffalo University in 1968.

Tennenbaum and Howard were both at the IAS in 1972/73, when Sue Toledo, with her fresh PhD from 1971, went there as a research assistant and began her conversations with Kurt Gödel at Tennenbaum's instigation. This was a clever move, and she was able to hold face-to-face conversations with Gödel over several years, in spite of his deteriorating health at that time. Stanley Tennenbaum apparently also saw Gödel often. Howard mentions the biography of John Horton Conway by Siobhan Roberts,[4] which reports

Tennenbaum's conversations with Gödel and lists some of the wide range of topics that they discussed.

About Stanley Tennenbaum's son, Jonathan, Howard says that he completed his PhD under Errett Bishop, a friend and colleague of his father's, in 1973, and recalls the latter's interest in Bishop's theories. An interesting interaction between Stanley Tennenbaum and Errett Bishop, while Tennenbaum was teaching at New Mexico State University and gave a seminar series on Bishop's 'constructive mathematics', is also described on the Tully/Tennenbaum site, illustrating Tennenbaum's remarkable teaching ability.

So now we have arrived at the end of our long journey, which began in the provincial capital *Brünn* in the mid-nineteenth century, and has taken us to Vienna and Princeton, and to modern logic, computer science, cosmology, and philosophy, 170 years later. That journey has followed the unusual life of an unusual human being, Kurt (Friedrich) Gödel. He combined many seeming opposites: striking brilliance in difficult, abstract fields of thought, with an apparent *naïveté* about many everyday aspects of human existence; daring innovations at the forefront of diverse fields of science and philosophy, with an almost childlike timidity and otherworldliness.

John Dawson ended his pathbreaking biography of Gödel 25 years ago with the remark that Gödel's life was well worthy of being dramatized. In the meantime, a number of plays, films and even dance performances dealing with Kurt Gödel's life and work[5] have been produced.

The German/Austrian writer and playwright *Daniel Kehlmann*, (born 1975), author of the best-selling novel '*Die Vermessung der Welt*' ('*Measuring the World*', 2005/06), along with a dozen other novels and short-story collections as well as several stage plays and a number of film scripts, has written the drama '*Geister in Princeton*' ('*Ghosts of Princeton*').[5,6] It premiered onstage in Graz, Austria in 2011 and at the *Renaissance Theater* in Berlin in 2012. The play follows Kurt and Adele from Vienna to Princeton, and it contains a dialogue between Albert Einstein and Kurt Gödel about a variety of topics, ranging from Einstein's wardrobe to cosmology. Gödel's philosophy of time however plays an important role in their dialogue, not surprising given its connection to relativity theory. And, in somewhat altered form, it provides the background for the whole drama, with events widely separated in 'real' time overlapping and leap-frogging each other along the story line. Time is cyclic, as it might be seen along one of the closed, timelike worldlines in Gödel's Universe (Fig. 20.1).

Fig. 20.1 A scene from '*Geister in Princeton*', *Schauspielhaus* Graz, 2012. It shows the actors Johannes Silberschneider (Gödel's ghost) and Kilian Langner (young Gödel). Photo: © *Schauspielhaus* Graz, by Lupi Spuma, re-used with permission[9]

Kehlmann's drama continues the tradition of Michael Frayn's *Copenhagen*, which also revolved around a dialogue between famous scientists—but that dialogue was real, not fictitious, and could be pinpointed in time; and its subject was extremely serious (the question of whether nuclear weapons could—or should—be constructed, especially for a nation with such a brutal and inhumane ideology as that of Nazi Germany, to which Werner Heisenberg had remained loyal, at least outwardly. That distanced him from his hitherto fatherly friend and mentor Niels Bohr). No one knows what was actually said in that conversation in September 1941, in occupied Denmark, but its consequences were certainly world-shaking.

Kehlmann's Einstein-Gödel dialogue is by comparison harmless and even trivial, but it serves as a foil against which Gödel's contradictory foibles can be displayed. Kehlmann's drama has an almost surrealistic atmosphere, including several versions of Gödel (as a curious young boy, '*der Herr Warum*'; as a young man, meeting Adele in the Vienna *Volkspark*;[7] in middle age, carrying on a dialogue with his friend Albert Einstein; and as a reclusive old man, slowly starving himself in his house on Linden Lane; and finally, Gödel as a kind of ghost, able to observe but not speak to the other personages—a 'silent observer'). Its English-language version was performed by the *Workshop Theater Company* in NY in 2012.[8] Their website describes it in the following words:

> This touching play tells the story of Viennese mathematician and logician Kurt Gödel (1906–1978), who by the age of 24 revolutionized mathematics. After the Nazis took over Austria in 1938, Gödel and his wife Adele emigrated to Princeton where he joined his good friend Albert Einstein at the Institute for Advanced Study. In 'Ghosts of Princeton', a play of facts, fiction and philosophy, Kehlmann follows the giant footsteps of Gödel and Adele on their journey from Vienna to Princeton. Breakthrough thinking, brilliant logic and a self-destructive rationalism characterized Gödel's remarkable history.

It would seem that Kehlmann's play captures much of the personality of Kurt Gödel, not without a certain tendency to historical gossip and sensationalism. It is not very careful with details of dates and events; but as a work of fiction, it doesn't need to be. But as one theater critic put it, "Why Gödel?"

And... *Apostolos Doxiadis* (born 1953), a Greek writer of fiction, stage plays, film scripts and comics,[10] has written '*Incompleteness*' (2005), a play which dramatizes the last 17 days of Gödel's life in the hospital in Princeton (premiered in Athens in 2006; its Greek title translates[11] as 'Seventeenth Night'). In his memoir[12] about the writing of the play, Doxiadis recounts how the idea came to him while he was reading a play by Harold Pinter, and it cost him considerable time and effort to turn it into reality. His own idea was to center his play on fictional conversations between the dying Kurt Gödel and a lady dietician during Gödel's final days at the Princeton Hospital. Doxiadis spent much more time than he had probably planned on the writing of his Gödel play, and suffered bouts of depression, self-doubt and anxiety before it was finally finished. One is tempted to think that Gödel's own psychiatric difficulties are somehow contagious.

Doxiadis sees Gödel as a tragic, paradoxical figure: he firmly believed that every aspect of the world can be understood by rational thinking, but yet he caused his own premature death by behaving in a most irrational way. His mistrust of doctors and medicine and his paranoid fear of poisoning led him to self-inflicted starvation. What happened in detail in the hospital in Princeton during those 17 days will never be known; the dietician with whom Gödel carries on a dialogue in the play is fictitious.

Evidently, Gödel refused to eat the hospital food, and would not allow any other life-saving measures such as intubation or intravenous feeding. Adele was no longer physically or psychologically able to help him back to the world, as she had done during several previous self-starvation episodes. And the doctors didn't try to force him to accept their help; that would have entailed declaring him to be mentally incompetent, and no-one wanted to do that. Thus far, Doxiadis stays with the facts.

But he has a requirement which he imposes on his writing, as he explains in his memoir: he believes it is his artistic duty to place his principal character in the context of that character's own most important work, and to bring the work to the audience's attention and understanding in the process. He finds that Gödel's last days represent a paradox (the *ultra-rationalist* behaving irrationally), which is not only interesting in itself, but is in a greater context representative of a parallel with Gödel's work (in particular, his incompleteness results: his '*great discovery*'), which Doxiadis sees as revealing the failure of a rationalistic viewpoint to completely describe reality.

Here, in my opinion, Doxiadis takes Gödel's incompleteness theorems out of the realm of formal mathematical logic, where they belong, and exalts them to a general and overriding philosophical principle which determines the limits of human knowledge and understanding *per se*. He puts Gödel within a sequence of scholars who made important discoveries in the past 180 years, each one reducing the height of the pedestal upon which humanity had placed itself in previous eras: Marx, Nietzsche, Darwin, Freud, Einstein, Heisenberg ... and Gödel. And while Gödel, in terms of his scholarly accomplishments, definitely also belongs on that list, the long-term effects of his discoveries may not be so drastic for humanity's self-image as those of some of the others on it. It is true—his theorems sidelined the formalistic program for mathematics of the logicists/formalists (Russell, Hilbert) in the early twentieth century (compare [Kreisel (1980)]). But even Gödel himself still believed in Hilbert's optimistic *non ignorabimus* long after the publication of his own incompleteness results. He showed that *truth* could not be verified within formal-syntactic systems, and that they will contain undecidable sentences; but *not* that truth is unknowable in principle.

The notion that the '*ultra-rationalist viewpoint does not give a complete view of reality*' has more to do with the limits of human rationality, even in the most intelligent of human beings; and the fact that humans are never perfectly rational creatures, however much they might want to be (cf. Gödel!), than it does with the incompleteness results of 1930/31. And Gödel's own self-destructive vein had more to do with his personality disorder (as correctly diagnosed on his death certificate) than with his brilliant results in mathematical logic.

Given his premise as stated above, it is no wonder that writing the play turned into a tortuous process for Doxiadis. But I well understand how trying to deal with Gödel's paradoxical life could produce such difficulties, and recommend Doxiadis' memoir (backnote [12]) in that context.

While the writing of this book also had its ups and downs, as is usual with any longer work, the 'downs' were fortunately not so deep as those described by Doxiadis.

I also have learned much along this journey. Rest in Peace, dear Kurt Gödel.

Notes

1. For example the IAS School of Mathematics site listing members and visitors shows that she was a research assistant there from 1972–74, but gives the university where she received her PhD in 1971 as Cornell, instead of Yeshiva University, where she in fact worked under Raymond Smullyan. Indeed, she is listed there under her maiden name (Sue Ann Walker), with the correct institution for her PhD, and as Sue Walker Toledo, with the incorrect institution. She is mentioned in WikiData, but only as a PhD student of Smullyan's, and is otherwise not present on Wikipedia or the other usual sites.

2. Posted on the Rob Tully memorial site for Tennenbaum, at https://stanle ytennenbaumamericanoriginal.com/bill-howard/. A number of other contributions from Howard and several other friends and colleagues of Tennenbaum are also available there (cited in the literature list ('References') as [Tully (2015)]). This site deserves archiving in a more permanent manner! Consulted in February/March 2022.

3. See Huber-Dyson (1981).

4. Cf. Roberts (2015).

5. Among the various dramatic works on Gödel, we might mention the play by *Apostolos Doxiadis*, '*Incompleteness*' (2005); see below. Another play which has received considerable publicity is *Daniel Kehlmann*'s '*Geister in Princeton*' (*Ghosts of Princeton*, 2011), also discussed below. In addition, among other dramas, there is a play by *David Fristrom*, '*Foundations*'

(2013)—see https://www.davidfristromplaywright.com; and a film written and directed by *Martha Goddard*, 'Gödel Incomplete' (2013), a romantic fiction—see: https://www.imdb.com/title/tt2722636/?ref_=tt_urv. See also [van Atten (2015)].

6. Cf. the website https://schauspielhaus-graz.buehnen-graz.com/play-detail/gei ster-in-princeton-ua/ on its premiere in Graz, Austria. Consulted on March 28th, 2022.

7. *sic*. Probably the *Volksgarten* is meant, across from the Parliament and just inside the *Burgring*. They most likely did *not* meet there, although Adele might well have walked through it on her way home from the *Nachtfalter*; cf. Chap. 5.

8. Quoted from http://archive.workshoptheater.org/mainstage/2012/ghosts_in_ princeton (consulted March 28th, 2022).

9. Photo: ©Schauspielhaus Graz, by Lupi Spuma, re-used with permission and thanks, from: https://schauspielhaus-graz.buehnen-graz.com/play-detail/gei ster-in-princeton-ua/.

10. Compare also [Doxiadis & Papadimitriou (2009)], Item 33 in Appendix B.

11. Gödel entered the hospital on December 29th, 1977, and died on the 17th day of his stay there: January 14th, 1978. He spent only 16 nights in the hospital; on the 17th night he had already passed away, and his dialogue partner was free to consider the conversations that they had during his last days.

12. Cf. the pdf of Doxiadis' talk: 'Writing Incompleteness' at the *Mathematics and Culture Conference*, Venice 2004, about the writing of his play. Online at: https://www.apostolosdoxiadis.com/wp-content/uploads/2012/09/Writing-Incom-pleteness.pdf. Consulted March 28th, 2022.

Correction to: Kurt Gödel

Correction to:
W. D. Brewer, *Kurt Gödel*, Springer Biographies,
https://doi.org/10.1007/978-3-031-11309-3

The original version of this book was published with a few fatal errors in equations. These, together with some more minor errors have been corrected in this revised version. The most significant corrections are as follows:

p. 30 'Footnote vi' changed to 'Backnote 12' (last paragraph).

p. 97 Subscripts to 'Aleph' moved from lower left to lower right (5 occurrences, 3rd paragraph).

p. 132 '9th District' changed to '8th District' (4th paragraph).

p. 143 New Backnote inserted, explaining Russell's and Witgenstein's misunderstanding of Gödel's Incompleteness Theorems (referenced at end of 4th paragraph, note added on p. 169).

p. 170 In Backnote 31: unnecessary text 'Bew' removed from line 9 in Backnote.

p. 186 'July' changed to 'June' (last paragraph).

p. 229 'easternmost province' changed to 'eastern province' (1st paragraph, middle).

The updated version of the book can be found at https://doi.org/10.1007/978-3-031-11309-3

© The Author(s), under exclusive license to Springer Nature
Switzerland AG 2023
W. D. Brewer, *Kurt Gödel*, Springer Biographies,
https://doi.org/10.1007/978-3-031-11309-3_21

p. 235 Last sentence in 1st paragraph ("Cantor proved…", with backnote), moved to 2nd paragraph following "… of natural numbers." Third-to-last sentence in 2nd paragraph now reads: "A related theorem (Cantor's Theorem) states that the cardinality…".

p. 326 Symbols for the gravitational field unified to lower-case 'g', italic and boldface, in Table 15.1.

p. 347 'kilometers' changed to 'meters' (2nd paragraph).

p. 361 In Backnote 32, added phrase: 'See also Pfarr (1981); he gives an estimate for the radius of the earth after returning to its point of origin as 210 m'.

p. 400 Corrected lower-right subscripts to 'Aleph$_2$' instead of '0' (2 occurrences, last paragraph).

p. 426 Year '2012' changed to '2011' (last paragraph).

p. 454 Year '2010' changed to '2011' (item 36 at top of page).

p. 470 Additional reference to 'Malament 1984' inserted (after 'Majer, U. (1995)').

p. 471 Additional reference to 'Pfarr 1981' inserted (after 'Parsons, C. (1996b)').

p. 476 Additional references to 'Witgenstein 1922, 1953, 1956' inserted (after 'Whitehead…').

p. 481 Phrase 'cf. Penrose, neural networks, 280' inserted after topic 'Computers—Simulation of thought'.

Appendix A: Publications by Kurt Gödel

This appendix gives a brief overview of the works that Kurt Gödel published during his lifetime. A more extensive list, with comments on each of the publications, was given by John W. Dawson (1983). Still more complete comments are contained in the *Collected Works* of Kurt Gödel, Volumes I and II (1986). The notation of the citations in this list is for the most part that used by Dawson (1983), and the more obscure references were taken from that publication.

1. Gödel (1930): '*Die Vollständigkeit der Axiome des logischen Funktionenkalküls*' ('*The Completeness of the Axioms of the Logical Calculus of Functions*'), In: *Monatshefte für Mathematik und Physik*, Vol. 37 (1930), pp. 349–360.
2. Gödel (1930a): '*Über die Vollständigkeit der Axiome des Logikkalküls*' ('On the Completeness of the Axioms of the Logical Calculus') (Abstract), in *Die Naturwissenschaften*, Vol. 18 (1930), p. 1068.
3. Gödel (1930b): '*Einige metamathematische Resultate über Entscheidungsdefinitheit und Widerspruchsfreiheit*' ('Some Metamathematical Results on Decidability and Consistency'), in: *Anzeiger der Akademie der Wissenschaften in Wien*, Vol. 67 (1930), pp. 214–215.
4. Gödel (1931): '*Über formal unentscheidbare Sätze der Principia Mathematica und verwandter Systeme* I' ('*On Formally Undecidable Theorems in the* Principia Mathematica *and related Systems* I'), in: *Monatshefte für Mathematik und Physik*, Vol. 38 (1931), pp. 173–198.

© The Editor(s) (if applicable) and The Author(s), under exclusive
license to Springer Nature Switzerland AG 2022, corrected publication 2023
W. D. Brewer, *Kurt Gödel*, Springer Biographies,
https://doi.org/10.1007/978-3-031-11309-3

5. Gödel (1931a): '*Diskussion zur Grundlegung der Mathematik*' ('*Discussion of Fundamentals of Mathematics*'), in: *Erkenntnis, Vol. 2 (1931), pp. 147–151 (Gödel's talks at the Königsberg Conference, September 1930).*

6. Gödel (1932): '*Zum intuitionistischen Aussagenkalkül*' ('*On the Intuitionistic Predicate Calculus*'), In: *Anzeiger der Akademie der Wissenschaften in Wien, Mathematisch-naturwissenschaftliche Klasse,* Vol. 69 (1932), pp. 65–66.

7. Gödel (1932a): '*Ein Spezialfall des Entscheidungsproblems der theoretischen Logik*' ('A Special Case of the Decision Problem in Theoretical Logic'), in: *Ergebnisse eines mathematischen Kolloquiums, Heft* 2 (*Gesammelte Mitteilungen,* 1929/30), pp. 27–28.

8. Gödel (1932b): '*Über Vollständigkeit und Widerspruchsfreiheit*' ('*On Completeness and Consistency*'), in: *Ergebnisse eines mathematischen Kolloquiums, Heft* 3 (1930/31), pp. 12–13.

9. Gödel (1932c): '*Eine Eigenschaft der Realisierungen des Aussagenkalküls*' ('A Property of the Realization of the Predicate Calculus'), in: *Ergebnisse eines mathematischen Kolloquiums, Heft* 3 (1930/31), pp. 20–21.

10. Gödel (1933): Untitled remark following W.T. Parry, '*Ein Axiomensystem für eine neue Art von Implikation (analytische Implikation)*' ('An Axiom System for a new Kind of Implication (Analytic Implication)'), in: *Ergebnisse eines mathematischen Kolloquiums, Heft* 4 (1931/32), p. 6.

11. Gödel (1933b): '*Über die metrische Einbettbarkeit der Quadrupel des R_3 in Kugelflächen*' ('On the Metric Embeddability of the Quadruple of R_3 in Spherical Surfaces'), in: *Ergebnisse eines mathematischen Kolloquiums, Heft* 4 (1931/32), pp. 16–17.

12. Gödel (1933c): '*Über die Wald'sche Axiomatik des Zwischenbegriffes*' ('On Wald's Axiomatics of the Intermediate Concept'), in *Ergebnisse eines mathematischen Kolloquiums, Heft* 4 (1931/32), pp. 17–18.

13. Gödel (1933d): '*Zur Axiomatik der elementargeometrischen Verknüpfungsrelationen*' ('On the Axiomatics of the Elementary Geometrical Connection Relations'), in: *Ergebnisse eines mathematischen Kolloquiums, Heft* 4 (1931/32), p. 34.

14. Gödel (1933e): '*Zur intuitionistischen Arithmetik und Zahlentheorie*' ('On Intuitionistic Arithmetic and Number Theory'), in: *Ergebnisse eines mathematischen Kolloquiums, Heft* 4 (for 1931–3, published 1933), pp. 34–38.

15. Gödel (1933f): '*Eine Interpretation des intuitionistischen Aussagenkalküls*' ('An Interpretation of the Intuitionistic Predicate Calculus'), in: *Ergebnisse eines mathematischen Kolloquiums, Heft* 4 (1931/32), pp. 39–40.

16. Gödel (1933g): Reprint of (1932), in: *Ergebnisse eines mathematischen Kolloquiums, Heft* 4 (1931/32), p. 40 (in the *Gesammelte Mitteilungen* for 1931/32).

17. Gödel (1933h): '*Bemerkung über projektive Abbildungen*' ('A Remark on Projective Mappings'), in: *Ergebnisse eines mathematischen Kolloquiums, Heft* 5 (1932/33), p. 1.

18. Gödel (1933i): '*Diskussion über koordinatenlose Differentialgeometrie*' ('Discussion of Coordinate-free Differential Geometry'), in: *Ergebnisse eines mathematischen Kolloquiums, Heft* 5 (1932/33), pp. 25–26 (with K. Menger and A. Wald).

19. Gödel (1933j): '*Zum Entscheidungsproblem des logischen Funktionenkalküls*' ('On the Decision Problem of the Logical Functional Calculus'), in: *Monatshefte für Mathematik und Physik*, Vol. 40 (1933), pp. 433–443.

20. Gödel (1934): *On undecidable propositions of formal mathematical systems*, Notes by S. C. Kleene and Barkley Rosser from lectures at the Institute for Advanced Study, Princeton (1934), Princeton, N.J., 30 pp. Reprinted in [Davis, ed. (1965)].

21. Gödel (R1934a): Review of *Skolem (1933)*, In: *Zentralblatt für Mathematik und ihre Grenzgebiete*, Vol. 7 (1934), pp. 193–194.

22. Gödel (1936): Untitled remark following A. Wald, '*Über die Produktionsgleichungen der ökonomischen Wertlehre*' ('On the Production Equations in Economic Value Theory'), in: *Ergebnisse eines mathematischen Kolloquiums, Heft* 7 (1934/35), p. 6.

23. Gödel (1936a): '*Über die Länge von Beweisen*' ('On the Length of Proofs'), in: *Ergebnisse eines mathematischen Kolloquiums, Heft* 7 (for 1934–5, published 1936), pp. 23–24.

24. Gödel (1938): '*The consistency of the axiom of choice and of the generalized continuum-hypothesis*', in: *Proceedings of the National Academy of Sciences*, Vol. 24 (1938), pp. 556–557.

25. Gödel (1939): '*The consistency of the generalized continuum-hypothesis*' (Abstract), in the *Bulletin of the American Mathematical Society*, Vol. 45 (1939), p. 93.

26. Gödel (1939a): '*Consistency-proof for the generalized continuum-hypothesis*', in: *Proceedings of the National Academy of Sciences*, Vol. 25 (1939), pp. 220–224.

27. Gödel (1940): '*The consistency of the axiom of choice and of the generalized continuum-hypothesis with the axioms of set theory*', Notes by George W. Brown from lectures at the Institute for Advanced Study, Princeton. Autumn term 1938–9, *Annals of Mathematics Studies*, No. 3, Princeton

University Press, Princeton, N.J. (1940) [Reviewed (at the invitation of J. D. Tamarkin) by S. C. Kleene in *Mathematical Reviews*, Vol. 2 (1941), pp. 66–67].

28. Gödel (1944): '*Russell's mathematical logic*', pp. 123–153 in P.A. Schilpp (ed.), *The Philosophy of Bertrand Russell* (*Library of Living Philosophers* series), Northwestern University Press, Evanston, IL (1944).

29. Gödel (1947): '*What is Cantor's continuum problem?*', in: *American Mathematical Monthly*, Vol. 54 (1947), pp. 515–525.

30. Gödel (1948): Russian translation of (1940) by A.A. Markov, in *Uspéhi Matematiceskih Nauk* (n.s.), Vol. 3, No. 1 (1948), pp. 96–149.

31. Gödel (1949a): '*A remark about the relationship between relativity theory and idealistic philosophy*'. Kurt Gödel, in: P.A. Schilpp (ed.): *Albert Einstein. Philosopher—Scientist* (*The Library of Living Philosophers* series, Volume VII). La Salle, IL, pp. 557–562 (Reprint of the first edition from 1949).

32. Gödel (1949b): '*An example of a new type of cosmological solutions of Einstein's field equations of gravitation*'. Kurt Gödel, in: *Reviews of Modern Physics*, Vol. 21 (1949), pp. 447–450.

33. Gödel (1950/52): '*Rotating universes in general relativity theory*'. Kurt Gödel, in: *Proceedings of the 11th International Congress of Mathematicians, 1950*. Volume 1, Cambridge, MA (1952), pp. 175–181.

34. Gödel (1956): '*Eine Bemerkung über die Beziehungen zwischen der Relativitatstheorie und der idealistischen Philosophie*', German translation of (1949a) by Hans Hartmann, pp. 406–412, in '*Albert Einstein als Philosoph und Naturforscher*', W. Kohlhammer, Stuttgart, 1955. Reprinted 1979, Vieweg & Sohn, Wiesbaden. [This translation includes additions by Gödel to Footnotes 11, 13 and 14.]

35. Gödel (1958): '*Über eine bisher noch nicht benutzte Erweiterung des finiten Standpunktes*' ('On a Hitherto Unused Extension of the Finitistic Viewpoint'), in *Dialectica*, Vol. 12 (1958), pp. 280–287.

36. Gödel (1959): Reprint of (1958), pp. 76–83 in *Logica: Studia Paul Bernays Dedicata*, Editions du Griffon, Neuchatel-Suisse, 1959.

37. Gödel (1961): Italian translation of (1931) by Evandro Agazzi, pp. 203–228 in *Introduzione ai problemi dell'assiomatica*, Pubblicazioni dell'Universitá Cattolica del Sacro Cuore, Serie terza, Scienze filosofiche no. 4, Societa Editrice Vita e Pensiero, Milan, 1961.

38. Gödel (1962): '*On Formally Undecidable Propositions of Principia Mathematica and Related Systems*', English translation of (1931) by B. Meltzer, Oliver and Boyd, Edinburgh (1962).

39. Gödel (1962a): Postscript to [Spector (1962)], p. 27.

40. Gödel (1964): Reprint of (1944), pp. 211–232 in [Benacerraf & Putnam, eds. (1964)].

41. Gödel (1965): '*On formally undecidable propositions of Principia Mathematica and related systems, I*', English translation of (1931) by Elliott Mendelson, pp. 4–38 in [Davis, ed. (1965)].

42. Gödel (1965a): '*On undecidable propositions of formal mathematical systems*', pp. 39–74 in [Davis, ed. (1965)].

43. Gödel (1965b): '*On intuitionistic arithmetic and number theory*', English translation of (1933e) by Martin Davis, pp. 75–81 in [Davis, ed. (1965)].

44. Gödel (1965c): '*On the length of proofs*', English translation of (1936a) by Martin Davis, pp. 82–83 in [Davis, ed. (1965)].

45. Gödel (1965d): '*Remarks before the Princeton Bicentennial Conference on Problems in Mathematics*', (December 17, 1946), pp. 84–88 in [Davis, ed. (1965)].

46. Gödel (1967): 'The completeness of the axioms of the functional calculus of logic', English translation of (1930) by Stefan Bauer Mengelberg, pp. 582–591 in [van Heijenoort, ed. (1967)].

47. Gödel (1967a): '*Some metamathematical results on completeness and consistency*', English translation of (1930b) by Stefan Bauer-Mengelberg, pp. 595–596 in [van Heijenoort, ed. (1967)].

48. Gödel (1967b): '*On formally undecidable propositions of Principia Mathematica and related systems I*', English translation of (1931) by Jean van Heijenoort, pp. 596–616 in [van Heijenoort, ed. (1967)]. [With added note by Gödel, August 28, 1963.)

49. Gödel (1967c): 'On completeness and consistency', English translation of (1932b) by Jean van Heijenoort, pp. 616–617 in [van Heijenoort, ed. (1967)] [With addition by Gödel to Footnote 1, May 18, 1966.]

50. Gödel (1967d): Italian translation of (1944) by Francesco Gana, pp. 81–112 in [Cellucci, ed. (1967)].

51. Gödel (1967e): Italian translation of (1947) by Carlo Cellucci, pp. 113–136 in [Cellucci, ed. (1967)].

52. Gödel (1967f): Italian translation of (1965d) by Carlo Cellucci, pp. 137–141 in [Cellucci, ed. (1967)].

53. Gödel (1967g): Russian translation of (1958) by G.E. Minc (with added appendix), pp. 299–310 in A.V. Idél'son and G.E. Minc, eds., in *Matematičéskaá téoriá logičeskogo vyvoda, Matematičéskaá logika i osnovaniá matématiki, Izdatel'stvo 'Nauka'*, Moscow (1967).

54. Gödel (1968): Reprint of (1965d), with a few minor changes in wording, pp. 250–253 in R. Klibansky (ed.), *Contemporary Philosophy, A Survey, I, Logic and Foundations of Mathematics*, La Nuova Italia Editrice (1968).

55. Gödel (1969): '*An interpretation of the intuitionistic sentential logic*', English translation of (1933f) by J. Hintikka and L. Rossi, pp. 128–129 in J. Hintikka, ed., *The Philosophy of Mathematics*, Oxford University Press, London (1969).

56. Gödel (1969a): '*La logique mathematique de Russell*', French translation of (1944) by J.A. Miller and J.C. Milner, pp. 87–107 in *La Formalization, Cahiers pour l'Analyse*, Vol. 10, Editions du seuil, Paris (1969).

Dawson, in his (1983) bibliography, also lists a whole series of reprints [of (1932), (1933f), (1930), (1930b), (1944), and (1939a)], published in 1971, 1972 and 1979, and Italian and Portuguese translations [of (1931a), (1944), (1947), (1931), (1965a), (1933e), (1936a), (1965d), and an English translation of (1958)], published in 1973, 1979 and 1980, as well as a Spanish 'Complete Edition' and an Italian translation of (1958), published in 1981. For details see Dawson (1983). There have of course been further translations and reprints of original works by Gödel in the years since 1983. See the *Collected Works*, Volumes I and II (1986).

Furthermore, in his middle years in Vienna (1931–1935), Gödel was very active in reviewing the articles of other logicians, publishing his reviews in the *Zentralblatt für Mathematik und ihre Grenzgebiete*. Dawson lists 27 such reviews in his (1983), and he suggests that there may be a few others in other publications. For details, see Dawson (1983).

There has also been a continuing effort to make Gödel's original but unpublished works available, particularly his philosophical works which make up a large part of his written legacy. Here are two examples:

Gödel & Rodrigues-Consuegra (1995): *Kurt Gödel: Unpublished Philosophical Essays*. Ed. Francisco Rodrigues-Consuegra. Springer Science & Business Media (1995). ISBN 3-764-3-53104. Introduction by the editor (Rodrigues-Consuegra 1995).

Gödel & Engelen (2019): *Philosophische Notizbücher* (*Gödel*). Ed. Eva-Maria Engelen. DeGruyter, Berlin (Vol. 1, 2019; Vol. 2, 2020) [Edited by the leader of the Gödel Research Group at the *Berlin-Brandenburgische Akademie der Wissenschaften*] (Engelen (ed.) 2019, 2020).

Appendix B: Literature About Kurt Gödel

1. ***Gödel's Proof***. Ernest Nagel and James R. Newman. NYU Press, 2001 (Originally published in 1958. Reissued 2001, 2005, 2008). ISBN 978-0-814-75816-8 [Also available as a paperback and an e-book] [Nagel & Newman (1958–2008)]. See also ***Der Gödel'sche Beweis***. Ernest Nagel, James R. Newman. *Scientia Nova*, Oldenburg 2006. ISBN 3-486 4 52185 (German edition).
2. ***From Mathematics to Philosophy***. Hao Wang. Routledge, 1974. ISBN 978-1-138–68773-8 [Later re-issued; also available as an e-book] [(Wang 1974)].
3. ***Gödel, Escher, Bach***. Douglas R. Hofstadter. Penguin Philosophy Series, 1979 (reprinted 1980). ISBN 0-140 1 79976. GEB was re-issued in 1999. See also ***I am a Strange Loop***. Klett-Kotta, 2008. ISBN 978 3 608 94444-0 [English edition as Kindle or paperback] [(Hofstadter 1979)] and [Hofstadter (2007)].
4. **Brief memoirs/Biographical sketches:** ***Kurt Gödel***.
   Georg Kreisel (1980): Biographical Memoirs, *Royal Society* UK. Online at: https://royalsocietypublishing.org/doi/10.1098/rsbm.1980.0005.
   Stephen C. Kleene (1987): US *National Academy of Sciences* Biographical memoir. See: http://www.nasonline.org/publications/biographical-memoirs/memoir-pdfs/gdel-kurt.pdf.
   J.J. O'Connor & E.F. Robertson (2003): MacTutor Biography, Kurt Gödel. See: https://mathshistory.st-andrews.ac.uk/Biographies/Godel/.

© The Editor(s) (if applicable) and The Author(s), under exclusive
license to Springer Nature Switzerland AG 2022, corrected publication 2023
W. D. Brewer, *Kurt Gödel*, Springer Biographies,
https://doi.org/10.1007/978-3-031-11309-3

Solomon Feferman (2006): IAS Princeton Biography: 'Kurt Gödel: Life, Work, and Legacy' (2006). https://www.ias.edu/kurt-g%C3%B6del-life-work-and-legacy.

Juliette Kennedy (2007): *Stanford Encyclopedia of Philosophy* Biography. https://plato.stanford.edu/entries/goedel/.

5. ***Kurt Gödel: Collected Works***. Eds. Solomon Feferman, John W. Dawson, Jr. et al.
   Vol. I: Publications 1929—1936. Oxford University Press USA, 1986. ISBN 0-195-0-39645-5. *Vol. II: Publications 1938—1974*. Oxford University Press USA, 1986. ISBN 0-195-0-3972-6. *Vol. III: Unpublished Essays and Lectures*. Oxford Univ. Press USA, 1995. ISBN 0-195-1-4720-0.
   Vol. IV: Selected Correspondence, A—G. Clarendon Press, 2003. ISBN 0-198-5-0073-4.
   Vol. V: Selected Correspondence, H—Z. Clarendon Press, 2014. ISBN 0-191-0-0377-8.
   [Feferman et al. (1986–2014)].

6. ***Gödel, Götzen und Computer. Eine Kritik der unreinen Vernunft***. Max Woitschach. Poller, Stuttgart, 1986. ISBN 3-879-5-9294-2 (Polemic against set theory; only a brief section on Gödel) [Woitschach (1986)].

7. ***Reflections on Kurt Gödel***. Hao Wang. MIT Press, 1987. ISBN 978-0-26223-127-5. [Personal recollections of a mathematician. First extended biography] (Wang 1987). See also ***A Logical Journey: From Gödel to Philosophy***. Hao Wang, *Bradford Books*, 1997 (MIT Press). ISBN 0-262-2-61251 [Wang (1997)].

8. ***Who got Einstein's Office?*** Edward Regis. Addison-Wesley, Reading MA (1987). ISBN 0-201-1-2065-8. German edition: ***Einstein, Gödel & Co.: Genialität und Exzentrik***. Ed Regis, Springer, Heidelbeg (2013). ISBN 3-034-8-5272-X [Regis (1987)].

9. ***Gödel Remembered***. Proceedings of a symposium in Salzburg, 10–12 July 1983, edited by Paul Weingartner and Leopold Schmetterer. *History of logic* No. 4, Bibliopolis, Naples (1987) [Contains Rudolf Gödel's memoir of his brother.] [Weingartner & Schmetterer (1987)].

10. ***Gödel's Theorem in Focus***. Stuart Shanker. Psychology Press, 1989. ISBN 0-415-0-4575-4 [Specialized on the Incompleteness theorems] [Shanker (1989)].

11. ***Gödel's Incompleteness Theorems***. Raymond M. Smullyan. Oxford University Press, New York, 1992. ISBN 0-19-5046-72-2 [Introduction

to incompleteness theorems for general mathematicians, philosophers, computer scientists] [Smullyan (1992)].

12. *Shadows of the Mind*. Roger Penrose. Oxford University Press, 1994. ISBN 0-19-8539-78-9 (An interpretation of the implications of Gödel's Incompleteness Theorems.) [Penrose (1994)]. See also R. Penrose: *The Emperor's New Mind*. Oxford University Press, New York, London, 1990. ISBN 0-09-9771-70-5 [Penrose (1990)].

13. *Kurt Gödel: ein mathematischer Mythos*. Werner DePauli-Schimanovich, Peter Weibel. Verlag Holder–Pichler–Tempsky, 1997. ISBN 3-209-0-0865-5 [Based on a film, ORF 1986. Many pictures, short descriptions] [Schimanovich & Weibel (1997)]

14. *Logical Dilemmas: The Life and Work of Kurt Gödel*. John W. Dawson. Taylor & Francis, 1997; reprint CRC Press, 2005. ISBN 1-568 8 1256-6 [By a Gödel expert and historian of mathematics. The definitive Gödel biography. The German edition is ISBN-13: 978-067-1-733-35-3, Transl. J. Kellner, Springer 1999 (Hardcover out of print; paperback only) [Dawson (1997)].

15. *In the Light of Logic*. Solomon Feferman. Oxford University Press, NY, Oxford (1998). ISBN 0-19-5-080-30-0 [Gödel's life summary and his philosophy of mathematics; authoritative. E-book] [Feferman (1998)].

16. *On Gödel* (Wadsworth Philosophers Series). Jaakko Hintikka. Cengage Learning Inc., Independence/KY, 1999. ISBN 978-053-4-575-95-3 [Brief text aimed at philosophy students. Concise, clearly written] [Hintikka (1999)].

17. *Gödel. A Life of Logic*. John L. Casti and Werner DePauli-Schimanovich. Perseus Publishing, 2000. ISBN 0-738-2-0518-4 [Casti & DePauli-Schimanovich (2000)] [Biography for a general audience. No e-book.] See also *Kurt Gödel und die mathematische Logik*, item 23, below.

18. *Gödel—Une revolution en mathématiques*. André Delessert. *Presses polytechniques et universitaires romandes*, Lausanne, 2000. ISBN 2-880-744-49-0 [Delessert (2000)].

19. *Types, Tableaus, and Gödel's God*. Melvin Fitting, Springer Nature, Heidelberg, 2002. ISBN 978-94-010-0411-4 [Mathematical/philosophical analysis in terms of modal type theory] [Fitting (2002)].

20. *Kurt Gödel. Wahrheit und Beweisbarkeit*. Vol. 1: Documents and historical analyses. Eds. E. Köhler et al., öbv et hpt, Vienna, 2002. ISBN 3-209-03834-1.

Vol. 2: Compendium. Eds. B. Buldt et al., öbv et hpt, Vienna, 2002. ISBN 3-209-03835-X [Köhler et al. (2002)].

21. *Kurt Gödel. Logische Paradoxien und mathematische Wahrheit*. Gianbruno Guerrerio. Spektrum Verlag, Heidelberg 2002. ISBN 3-936-2-7804-0 [Apparently out of print. No e-book. Italian edition available] [Guerrerio (2002)].

22. *Gödel*. Pierre Cassou-Noguès. *Les Belles Lettres*, Paris, 2004. ISBN 2-251-7-6040-7. See also *Les Démons de Gödel. Logique et Folie*. *Média Diffusion*, Paris, 2015. ISBN 2-021-0-0934-3 [Kindle (French only). Print version (French/English)] [Cassou-Noguès (2004)].

23. *Kurt Gödel und die mathematische Logik*. Werner DePauli-Schimanovich, Univ. of Vienna, 2005. ISBN 3-854-8-7815-X [Volume 5 of *Europolis*. Part of an extensive scholarly study of Gödel and his works, including 4 other books to which the author (from the Vienna *Gödel-Gesellschaft*) contributed. See the Introduction to this volume.] [Schimanovich (2005)].

24. *A World Without Time: The Forgotten Legacy of Gödel and Einstein*. Palle Yourgrau. Perseus, Basic books, NY, 2005. ISBN 978-0-465-09293-2 (Specialized to Gödel's Universe and the Einstein friendship, aimed at a general audience). See also Palle Yourgrau: *Gödel Meets Einstein: Time Travel in the Gödel Universe*. Open Court Publishing, 1999. ISBN 978-0-812-69408-6 [An expansion on his earlier book: *The Disappearance of Time. Kurt Gödel and the Idealistic Tradition in Philosophy*. Palle Yourgrau, Cambridge University Press, NY, 1991. ISBN 0-521-4-1012-6. Both for professional/philosophical audiences] [Yourgrau (1991), (1999), (2005)].

25. *Asperger Syndrome: A Gift or a Curse?* Victoria Lyons, Michael Fitzgerald (See Chap. VIII on *K. Gödel*). Nova Publishers, 2005. ISBN 1-594-5-4387-9 ["Currently unavailable"] [Lyons & Fitzgerald (2005)].

26. *Gödel's Theorem. An Incomplete Guide to its Use and Abuse*. Torkel Franzén, CRC Press, Taylor & Francis Group, Boca Raton/FL, 2005. ISBN 978-1-56881-238-0 [Mainly about Incompleteness; corrects mistakes of interpretation] [Franzén (2005)].

27. *Incompleteness: The Proof and Paradox of Kurt Gödel (Great Discoveries)*. Rebecca Goldstein, Atlas Books (2005), W.W.Norton (2006). ISBN 978-0-73945744-3 [Specialized to the philosophical significance of the incompleteness theorems] [Goldstein (2005)].

28. *Kurt Gödel 1906—1978. Genealogie*. Jiri Prochazka.
Vol. 1. Brno, 2006. ISBN 8-090 2 29794.
Vol. 2. Brno, 2006. ISBN 8-090 3 47606.

Vol. 3. Brno, 2008. ISBN 978-80 903476-4-9.
Vol. 4. Brno/Princeton, 2008. ISBN 978-80 903476-5-6.
Vol. 5. Brno/Princeton, 2010. ISBN 978-80 903476-9-4.
See also *Historie*. Brno/Wien/Princeton, 2012. ISBN 978-80 903476-2-5 [Prochazka (2006–2012)].

29. *Kurt Gödel—das Album*. Karl Sigmund, John W. Dawson, Kurt Mühlberger, Springer, Heidelberg 2007. ISBN 3-834-8-9189-4 [Introduction for a general audience, many pictures; based on a photo exhibition in Vienna, curated by the authors. Essentially a picture book with accompanying texts] [Sigmund et al. (2007)].

30. *Kurt Gödel: Jahrhundertmathematiker und großer Entdecker*. Rebecca Goldstein. Piper, 2006/7. ISBN 3-492-2-4960-4 (German version of Item 27) [Goldstein (2007)].

31. *Thinking about Gödel and Turing*. Gregory J. Chaitin, World Scientific, Singapore (2007). ISBN 9-812-7-0897-9 [Essays on complexity by a mathematician, 23 papers originally published between 1970 and 2007. "Speculative". No e-book. Hardcover/paperback] [Chaitin (2007)].

32. *An Introduction to Gödel's Theorems*. Peter Smith, Cambridge University Press, 2007. ISBN 1-139-4-6593-7 [Introductory text by a philosopher, specialized to the Incompleteness theorems. New edition August 2020. Cf. also Items 1, 10 and 27] [Smith (2007), (2020)].

33. *Logicomix*: *An Epic Search for Truth*. Apostolos K. Doxiadēs, Christos H. Papadimitriou, Bloomsbury, 2009. ISBN: 978-3-86497-004-7 [Docu-comic/graphic novel about Bertrand Russel's mathematical/philosophical attempts to justify rationalism, with references to Gödel] [Doxiadis & Papadimitriou (2009)].

34. *Kurt Gödel: Essays for his Centennial*. Solomon Feferman, Charles Parsons, Stephen G. Simpson. Cambridge University Press, 2010. ISBN 1-139-4-8775-2 [Volume 33 of *Lecture Notes in Logic*. Mathematical/philosophical essays, historical introduction. No e-book] [Feferman et al. (2010)].

35. *Kurt Gödel and the Foundations of Mathematics: Horizons of Truth*. Mathias Baaz, Christos H. Papadimitriou, Hilary W. Putnam, Dana S, Scott, Charles L. Harper, Jr., Cambridge University Press, 2011. ISBN 1-139-4-9843-6 [Impact of Gödel's work and future developments in Computer Science, AI, Physics, Cosmology, Philosophy, Theology, Mathematics.] [Baaz et al. (2011)].

36. *Is God a Mathematician?* Mario Livio. Simon and Shuster, 2011. ISBN 1-416-5-9443-4 [A history of mathematics by an astrophysicist and publicist. Good summary of Gödel's work, pp. 195 ff.] [Livio (2011)].

37. *Die Gödel'schen Unvollständigkeitssätze: Eine geführte Reise.* Dirk W. Hoffmann, Springer Spektrum, 2013–2017. ISBN 3-827-4-3000-3 [Explication of the incompleteness theorems by a computer scientist.] [Hoffmann (2013), (2017)].

38. *Reminiscences of the Vienna Circle and the Mathematical Colloquium.* Karl Menger, Kluwer (1994). Reprinted in 2013: Eds. L. Golland, B.F. McGuinness, Abe Sklar. Springer Science & Business Media, 2013. ISBN 9-401-1-1102-2 [Memories of Gödel by a friend and colleague from Vienna; see pp. 201 ff.] [Menger et al. (1994), (2013)].

39. *Computability: Turing, Gödel, Church and Beyond.* B. Jack Copeland, Carl J. Posy, Oron Shagrir, eds. MIT Press, 2013. ISBN 978-0-262-0-1899-3 [Discussions of the foundations of computability by an interdisciplinary group.] [Copeland et al. (2013)]; 'CPS'.

40. *La Déesse des petites victoires.* Yannick Grannec. *S.N. Éditions Anne Carrière*, Paris, 2012. ISBN: 978-2-380-82050-8 [The story of the life of Kurt and Adele Gödel as told by Adele. Historical docunovel.] [Grannec (2012)].

41. *Interpreting Gödel: Critical Essays.* Ed. Juliette Kennedy. Cambridge University Press, 2014. ISBN 1-139-9-9175-2 [A collection of essays by philosophers on Gödel's work and legacy.] [Kennedy, ed. (2014)].

42. *Kurt Gödel Philosopher-Scientist.* Gabriella Crocco and Eva-Maria Engelen, eds., *Presse Universitaire de Provence*, OpenEdition Books (2016). Online edition (2021) at: http://books.openedition.org/pup/53500.

43. *Simply Gödel.* Richard Tieszen, *simplycharly*, NYC, 2017. ISBN 978-1-943657-15-5 [A brief summary of Gödel's life and work with simplified explanations of his important contributions] [Tieszen (2017)].

44. *Sie nannten sich der Wiener Kreis. Exaktes Denken am Rand des Untergangs.* Karl Sigmund, Springer Spektrum, 2015. ISBN 978-3-658-08534-6 (Print); ISBN 978-3-658-08535-3 (e-Book). English edition: *Exact Thinking in Demented Times. The Vienna Circle and the Epic Quest for the Foundations of Science.* Karl Sigmund, Basic Books, NY 2017. ISBN 978-0-4650-9 [Sigmund (2015)/(2017)].

45. ***When Einstein Walked with Gödel. Excursions to the Edge of Thought***. Jim Holt, Farrar, Straus & Giroux, 2018. ISBN 9-780-3-7414-60-2 [A loose history of time and mathematics, one essay on Einstein/Gödel.] [Holt (2018)].

46. ***The Order of Time***. Carlo Rovelli, Penguin UK, 2018. ISBN: 978-0-24129-253-2 [Rovelli (2018)].

47. ***Gödel, Tarski, and the Lure of Natural Language***. Juliette C. Kennedy, Cambridge university Press, 2020. ISBN: 978-0-51199-839-3 [Kennedy (2020)].

48. ***Journey to the Edge of Reason: The Life of Kurt Gödel***. Stephen Budiansky, W.W. Norton/Oxford Univ. Press/Ullstein, 2021. ISBN 978-0-19-886633-6 [Claims to be the first serious biography of Gödel, giving new insights through the correspondence.] [Budiansky (2021)].

49. ***Gödel Without (Too Many) Tears***. Peter Smith, *Logic Matters*, Cambridge 2021, ISBN 978-1-916-90630-3 [A brief exposition of the Incompleteness theorems [Smith (2021)]; cf. also [Smith (2020)] (Item 32, Appendix B)].

50. ***Your Wit is my Command: Building AIs with a Sense of Humor***. Tony Veale, MIT Press, Cambridge, MA, 2021. ISBN: 978-0-26204-4599-5 [A book on computation and artificial intelligence which discusses Gödel numbering] [Veale (2021)]

Appendix C: Publications by the CBPF Group on Gödel's Universe 1978–2018

1. *The stability of a rotating universe*. M. Novello and Marcelo J. Rebouças, *The Astrophysical Journal* **225** (3), 719–724 (1978).
2. *Rotating universe with successive causal and noncausal regions*. M. Novello and Marcelo J. Rebouças, *The Physical Review D: Particles and Fields* **19**, 10 (1979).
3. *A rotating universe with violation of causality*. M.J. Rebouças *Physics Letters A* **70** (3), 161–163 (1979).
4. *Time-dependent, finite, rotating universes*. M.J. Rebouças and J.A.S. de Lima, *Journal of Mathematical Physics* **22** (11), 2699–2703 (1981).
5. *Geodesic Motion and Confinement in Gödel's Universe*. M. Novello, I.D. Soares and Jayme Tiomno, *The Physical Review D* **27**, 779 (1983).
6. *Gödel-type metric in Einstein-Cartan Spaces*. J. Tiomno, A. Teixeira and J. Duarte, Contributed Papers to the *X. International Conference on General Relativity and Gravitation*, p. 507 (1983).
7. *Homogeneity of Riemannian Space-times of Gödel type*. M.J. Rebouças and J. Tiomno, *The Physical Review D* **28**, 1251 (1983).
8. *A class of Inhomogeneous Gödel-type models*. M. Rebouças and Jayme Tiomno, *Il Nuovo Cimento* **90B**, 204–210 (1985).
9. *Isometries of homogeneous Gödel-type spacetimes*. A.F.F. Teixeira, M.J. Rebouças and J.E. Åman *Physical Review D* **32** (12), 330956 (1985).
10. *Time travel in the homogeneous Som-Raychaudhuri universe*. F.M. Paiva, M.J. Rebouças and A.F.F. Teixeira, *Physics Letters A* **126** (3), 168–170 (1987).

© The Editor(s) (if applicable) and The Author(s), under exclusive
license to Springer Nature Switzerland AG 2022, corrected publication 2023
W. D. Brewer, *Kurt Gödel*, Springer Biographies,
https://doi.org/10.1007/978-3-031-11309-3

11. *Notes on a class of homogeneous space-times.* M.O. Calvão, M.J. Rebouças, A.F.F. Teixeira and W.M. Silva Jr., *Journal of mathematical physics* **29** (5), 1127–1129 (1988)

12. *Geodesics in Gödel-Type Space-times.* I.D. Soares, M.O. Calvão, and J. Tiomno, *General Relativity and Gravitation* **22**, 683 (1990).

13. *Synchronized frames for Gödel's Universe.* M. Novello, N.F. Svaiter, and M.E.X. Guimaraes *General Relativity and Gravitation* **25** (2), 137–164 (1993).

14. *Imparting rotation to a Bianchi type II space-time.* Marcelo J. Rebouças, J.B.S. d'Olival, *Journal of mathematical physics* **27** (1), 417–418 (1996).

15. *Riemann-Cartan spacetimes of Gödel type.* J.E. Åman, J.B. Fonseca-Neto, M.A.H. MacCallum and M.J. Rebouças, *Classical and Quantum Gravity* **15** (4), 1089 (1998).

16. *On a class of Riemann-Cartan space-times of Gödel type.* J.B. Fonseca-Neto and M.J. Rebouças, *General Relativity and Gravitation* **30** (9), 1301–1317 (1998).

17. *Riemannian space-times of Gödel type in five dimensions.* M.J. Rebouças, A.F.F. Teixeira, *Journal of Mathematical Physics* **39** (4), 2180-2192 (1998).

18. *Gödel-type space–times in induced matter gravity theory.* H.L. Carrion, M.J. Rebouças and A.F.F. Teixeira, *Journal of Mathematical Physics* **40** (8), 4011–4027 (1999).

19. *Gödel-type Universe in f(R) Gravity.* M.J. Rebouças and J. Santos, *The Physical Review D, Particles, Fields, Gravitation, and Cosmology* **80**, 063009 (2009).

20. *Gödel-type universes in Palatini f(R) gravity.* J. Santos, M.J. Rebouças and T. Oliveira, *Physical Review D* **81** (12), 123017 (2010).

21. *Gödel-type universes and chronology protection in Hořava–Lifshitz gravity.* JB Fonseca-Neto, AY Petrov, MJ Rebouças, *Physics Letters B* **725** (4–5), 412–418 (2013).

22. *Homogeneous Gödel-type solutions in hybrid metric-Palatini gravity.* J. Santos, M.J. Rebouças, A.F.F. Teixeira, *The European Physical Journal C* **78** (7), 1–10 (2018).

References

Albers, D.J., Alexanderson, G.L., Reid, C.: *International mathematical congresses. An illustrated history 1893–1986*. Revised edition including ICM 1986. Springer, New York (1986). ISBN: 978-1-4684-0301-5

Audureau, E.: *Gödel: from the pure theory of gravitation to Newton's absolute*. In: Crocco, G., Engelen, E.-M. (eds.) *Kurt Gödel, Philosopher-Scientist*. Presses universitaires de Provence, Aix-en-Provence/FR (2016). Print edition 2016, OpenEdition Books version online 2021. ISBN: 978-2-853-99976-2

Barendregt, H.: *Kreisel, lambda calculus, a windmill and a castle*. Festschrift for Kreisel's 70th birthday [Odifreddi (1996)]. Online at the repository of the University of Nijmegen (1996). https://repository.ubn.ru.nl/bitstream/handle/2066/17292/17292.pdf

Barrow, J.D., Tsagas, C.G.: *Dynamics and stability of the Gödel universe*. In: *Classical Quantum Gravity*, vol. 21, pp. 1773–1789 (2004)

Bell, J.L.: *The Axiom of Choice*. The Stanford Encyclopedia of Philosophy. First published 2008 (2015). https://plato.stanford.edu/entries/axiom-choice/

Benacerraf, P., Putnam, H. (eds.): *Philosophy of Mathematics. Selected Readings*. Prentice-Hall Inc., NY (1964). Later editions: Cambridge University Press (1983–1998). ISBN: 978-0-52122-796-4

Bernardete, S.: *Socrates' Second Sailing: On Plato's Republic*. University of Chicago Press (1989). ISBN: 0-226-04242-1

Bernays: *Axiomatische Untersuchung des Aussagen-Kalküls der Principia Mathematica* (Göttingen 1918). Bernays, P.: *Mathematische Zeitschrift* **25**, 305–320 (1926). Based on his *Habilitationsschrift*, Göttingen 1918

Bernays, P.: Review of '*Russell's Mathematical Logic* by Kurt Gödel'. *J. Symb. Logic* **11**(3), 75–79 (1946)

© The Editor(s) (if applicable) and The Author(s), under exclusive license to Springer Nature Switzerland AG 2022, corrected publication 2023
W. D. Brewer, *Kurt Gödel*, Springer Biographies,
https://doi.org/10.1007/978-3-031-11309-3

Bleaney, M.: *Wittgenstein and Frege* (2016). In: Glock, Hyman (eds.) A Companion to Wittgenstein. Wiley (2017). ISBN: 978-1-11930-794-5

Brendle, J., Larson, P., Todorčević, S.: *Rectangular axioms, perfect set properties and decomposition*. The Bulletin (*Académie Serbe des Sciences et des Arts. Classe des Sciences Mathématiques et Naturelles. Sciences Mathématiques*) **33**, 91–130 (2008)

Brewer, W.D., Tolmasquim, A.T.: Jayme Tiomno. *A Life for Science, a Life for Brazil*. Springer Nature, Cham (2020). ISBN: 978-3-030-41010-0

Budiansky, S.: *Journey to the Edge of Reason: The Life of Kurt Gödel*. W.W. Norton/Oxford University Press/Ullstein (2021). ISBN: 978-1-324-00544-5

Bulloff, J., Holyoke, T.C., Hahn, S.W. (eds.): *Foundations of Mathematics*: Symposium Papers Commemorating the Sixtieth Birthday of Kurt Gödel. Springer, New York (1969)

Buss, S.: *On Gödel's theorems on lengths of proofs I*: number of lines and speedups for arithmetic. *J. Symb. Logic* **39**(1994), 737–756 (1994)

Buss, S.: *On Gödel's theorems on lengths of proofs II*. In: *Feasible Mathematics*, pp. 57–90. Birkhauser, Boston (1995)

Call, B.: *The Compactness Theorem and Applications*. Paper for an REU (Research Experience for Undergraduates), University of Chicago, Mathematics Department (2013). https://math.uchicago.edu/~may/REU2013/REUPapers/Call.pdf

Cantor, G.: *Grundlagen einer allgemeinen Mannigfaltigkeitslehre*. Teubner, Leipzig, vol. 2, pp. 165–208 (1883). In Cantor 1932. English translation in Ewald 1996

Cantor, G.: *Beiträge zur Begründung der transfiniten Mengenlehre*. In Cantor 1932, pp. 282–351. English translation in Cantor, G.: Contributions to the Founding of the Theory of Transfinite Numbers, Dover, New York (1955)

Cantor, G.: 1932: *Gesammelte Abhandlungen mathematischen und philosophischen Inhalts*, Georg Cantor. Ernst Zermelo (ed.), Springer-Verlag, Berlin (1932). Reprinted by Olms Verlag, Hildesheim, 1966

Carnap, R.: *Der logische Aufbau der Welt ('Aufbau')*. Felix Meiner Verlag (1928). Reprinted 1966, 1999. English translation by Rolf A. George, 1967

Carnap, R.: *Logische Syntax der Sprache*. Springer, Vienna, 1934/37 (1934). Translated by Smeaton, A.: The Logical Syntax of Language. Routledge, London (1937)

Carnap: *The philosophy of Rudolf Carnap*. In: Schilpp, P.A. (ed.) Library of Living Philosophers, vol. XI. Open Court Publishing, Peru IL (1963). Re-issued in 1997

Casti, J.L., DePauli-Schimanovich, W.: *Gödel. A Life of Logic*. Perseus Publishing (2000). Item 17 in Appendix B. ISBN 0-738-2-0518-4

Cellucci, C. (ed.): *La filosofia della mathematica*. Laterza, Bari (1967)

Chaitin, G.J.: *Thinking about Gödel and Turing*. World Scientific, Singapore (2007). ISBN 9-812-7-0897-9

Chandrasekhar, S., Wright, J.P.: *The geodesics in Gödel's Universe. Proc. Natl. Acad. Sci.* (USA) (PNAS) **47**, 341–347 (1961)

Chang, C.C., Keisler, H.J.: *Model Theory*. Elsevier Science, Amsterdam (1973, 1992). ISBN 0-444-88054-2.

Christian, C.: *Leben und Wirken Kurt Gödels. Monatshefte für Mathematik* **89**(1980), 261–273 (1980)

Church, A.: *On the law of excluded middle. Bull. Am. Math. Soc.* **34**(1928), 75–78 (1928)

Church, A.: *An unsolvable problem of elementary number theory. Am. J. Math.* **58**(2), 345–363 (1936)

Church, A.: Review of "*on computable numbers, with an application to the Entscheidungsproblem*", by A.M. Turing. *J. Symb. Logic* **2**(1), 42–43 (1937)

Cohen, P.J.: *The independence of the continuum hypothesis, part I. Proc. U.S. Natl. Acad. Sci.* **50**(6), 1143–1148 (1963); *The independence of the continuum hypothesis, part II. Ibid.* **51**(1), 105–110 (1964)

Cohen, P.J.: *Set theory and the continuum hypothesis*. W.A. Benjamin, New York (1966). Renewed 1994, republished 2008 by Dover Publications, Mineola, NY. ISBN 13: 978-0-486-46921-8.

Conradi, P.: *Iris Murdoch: A Life*, pp. 264–265. Harper-Collins (2001)

Cook, W.J.: *Zermelo-Fraenkel Set Theory with the Axiom of Choice*. Math handout (2010). https://billcookmath.com/courses/math2510-spring2010/ZFC.pdf

Copeland, B.J., Posy, C.J., Shagrir, O. (eds.): *Computability: Turing, Gödel, Church and Beyond*. MIT Press, Cambridge MA (2013). ISBN 978-0-262-0-1899-3. See also Appendix B, Item 39 (CPS)

Crocco, G.: *Gödel, Leibniz and "Russell's mathematical logic"*. In: Kromer, R., Chin-Drian, Y. (eds.) *New Essays in Leibniz Reception. In Science and Philosophy of Science 1800–2000*, pp. 217–256. Birkhäuser-Springer Science (2013). ISBN: 978-3-034-60503-8

Crocco, G., Engelen, E.-M. (eds.): *Kurt Gödel, Philosopher-Scientist*. Presses universitaires de Provence, Aix-en-Provence/FR (2016). Print edition 2016, OpenEdition Books version online 2021. ISBN: 978-2-853-99976-2

Crossley, J.N. (ed.): *Reminiscences of logicians*. In: *Algebra and Logic Proceedings*, pp. 1–57 (1974). Lecture Notes in Mathematics, vol. 450, Springer, Heidelberg (1975). ISBN: 3-540-07152-0

Crossley, J.N., Ash, C.J., Brickhill, C.J., Stillwell, J.C., Williams, N.H.: *What is Mathematical Logic?* Oxford University Press (1972). Reprinted by Dover Publications, N.Y., 1990 and 2012. ISBN-13: 978-0-486-26404-2

Cvrcek, T., Zajicek, M.: *Cliometrica* **13**(3)(2), 367–403 (2009) (t.cvrcek@ucl.ac.uk and zajicek100@gmail.com)

Czech, H.: Hans Asperger, *National socialism, and "race hygiene" in Nazi-era Vienna. Mol Autism* **9**, 43pp. (2018)

Davis, M. (ed.): *The Undecidable*. Basic papers on undecidable propositions, unsolvable problems and computable functions. Raven Press, Hewlitt NY (1965). ISBN: 0-486-43228-9

Davis, M.: *Why Gödel didn't have Church's thesis. Inf. Control* **54**(1982), 3–24 (1982)

Davis, M.: *How subtle is Gödel's theorem? More on Roger Penrose. Behav. Brain Sci.* **16**(1993), 611–612 (1993)

Davis, M.: *The incompleteness theorem. Not. AMS* **53**(4), 414–418 (2006). http://www.ams.org/notices/200604/200604-toc.html

Dawson, J.W., Jr.: *The published work of Kurt Gödel: an annotated bibliography. Notre Dame J. Form. Logic* **24**, 255–284 (1983)

Dawson, J.W., Jr.: *Discussion on the foundation of mathematics. Hist. Philos. Logic* **5**, 111–129 (1984a)

Dawson, J.W., Jr.: Addenda and corrigenda to *'The published work of Kurt Gödel'. Notre Dame J. Form. Logic* **25**(3), 293–287 (1984b)

Dawson, J.W., Jr.: *The reception of Gödel's incompleteness theorems.* In: PSA: Proceedings of the Biennial Meeting of the Philosophy of Science Association, Volume Two: Symposia and Invited Papers, pp. 253–271 (1984c)

Dawson, J.W., Jr.: *Completing the Gödel-Zermelo correspondence. Hist. Math.* **12**(1), 66–70 (1985)

Dawson, J.W., Jr.: *Logical Dilemmas: The Life and Work of Kurt Gödel.* Taylor & Francis (1997). Reprint CRC Press (2005). ISBN 1-568-8-1256-6. [German edition (ISBN-13: 978-067-1-73335-3, Translation J. Kellner, Springer (1999)]

Dawson, J.W., Jr.: *Gödel and the limits of logic. Sci. Am.* **280**(6), 76–81 (1999)

Dawson, J.W., Jr.: *Max Dehn, Kurt Gödel, and the Trans-Siberian escape route. Not. Am. Math. Soc.* **49**(9), 1068–1075 (2002)

Dawson, J.W., Jr.: *In quest of Kurt Gödel: Reflections of a Biographer. Not. AMS* **53**(4), 444–447 (2006). http://www.ams.org/notices/200604/200604-toc.html

Dawson: Interview with Linda Arntzenius for the IAS Princeton's *Oral History Project*, May 2013 (2013). https://albert.ias.edu/bitstream/handle/20.500.12111/1020/Dawson_John_OH_20130530_final.pdf

Dennett, D., Lambert, K.: Kleene. In: The Philosophical Lexicon, 7th edn, p. 5. American Philosophical Association, Newark, DE (1978)

DePauli-Schimanovich: See Schimanovich

DePauli-Schimanovich, W.: *Kurt Gödel und die mathematische Logik.* University of Vienna (2005) [Volume 5 of *Europolis*. Part of an extensive scholarly study of Gödel and his works, including 4 other books to which the author (from the Vienna "Gödel-Gesellschaft") contributed. ISBN 3-854 8 7815X

DePauli-Schimanovich, W., Weibel, P.: *Kurt Gödel: ein mathematischer Mythos.* Verlag Hölder-Pichler-Tempsky, Vienna (1997). Item 13 in Appendix B. ISBN 3-209-00865-5

Doxiadis, A.K., Papadimitriou, C.H.: *Logicomix: An Epic Search for Truth.* Bloomsbury (2009). ISBN: 978-3-86497-004-7

Edmonds, D.: *The Murder of Professor Schlick. The Rise and Fall of the Vienna Circle.* Princeton University Press, Princeton, NJ/USA (2020). ISBN: 978-0-69116-490-8

Eilenberger, W.: *Die Zeit der Zauberer. Das Große Jahrzehnt der Philosophie 1919–1929.* Klett-Cotta, Stuttgart (2018). ISBN 978-3-608-96451-6

Enderton, H.B.: *A Mathematical Introduction to Logic.* Harcourt/Academic Press, Burlington/MA (1972, 2001). ISBN 0-12-238452-0

Engelen, E.-M. (ed.): *Kurt Gödel: Philosophische Notizbücher*, vols. 1 and 2. Verlag Walter DeGruyter, Berlin, New York (2019, 2020). ISBN: 978-3-11067-409-5

Engelen, E.-M.: Interview with Ulf von Rauchhaupt, Frankfurter Allgemeine Zeitung (FAZ), 12.01.2020. See also the review of the *Philosopische Notizbücher* (Engelin (ed.) (2019, 2020a). FAZ, 16.06.2021, and Appendix A

Engelen, E.-M.: *Rudolf Carnap und Kurt Gödel: Die beiderseitige Bezugnahme in ihren philosophischen Selbstzeugnissen* (Chap. 11) In: Damböck, C., Wolters, G. (eds.) *Der junge Carnap in historischem Kontext: 1918–1935'/'Young Carnap in an Historical Context: 1918–1935'*. Springer, Cham/CH (2021). ISBN 978-3-030-58250-0

Ewald, W.B.: *From Kant to Hilbert: A Source Book in the Foundations of Mathematics*, 2 vols. Oxford University Press, Oxford (1996)

Ewald, W.B.: *The emergence of first order logic*. Stanford Encyclopedia of Philosophy (2018). https://plato.stanford.edu/entries/logic-firstorder-emergence/

Fasora, L. (ed.): *Der Mährische Ausgleich von 1905. Möglichkeiten und Grenzen für einen nationalen Ausgleich in Mitteleuropa*. Matice moravská, Brno (2006). ISBN 80-86488-36-5

Feferman, S.: *Arithmetization of metamathematics in a general setting. Fundam. Math.* **49**(1960), 35–92 (1960)

Feferman, S.: *Penrose's Gödelian argument: a review of shadows of the mind* by Roger Penrose. *Psyche* **2**(7), 21–32 (1995)

Feferman, S.: *My Route to arithmetization. Theoria* **63**, 168–181 (1997)

Feferman, S.: *In the light of logic*. Oxford University Press, NY/Oxford (1998). ISBN 0-19-5-08030-0. See Appendix B, Item 15

Feferman, S.: *The impact of the incompleteness theorems on mathematics. Not. AMS* **53**(4), 434–439 (2006). http://www.ams.org/notices/200604/200604-toc.html

Feferman, S., Dawson, J.W., Jr. et al. (eds.): *Kurt Gödel: Collected Works*, Five vols. Oxford University Press, Clarendon Press (1986–2014)

Ferreira, F.: '*Spector's proof of the consistency of analysis*'. In: Kahle, R., Rathjen, M. (eds.) *Gentzen's Centenary*. Springer, Cham. ISBN: 978-3-319-10102-6 (2012)

Findlay, J.M.: *Godelian sentences: a non-numerical approach. Mind* **51**(1942), 259–265 (1942)

Fisette, D.: *Phenomenology and phenomenalism*: Ernst Mach and the genesis of Husserl's phenomenology. *Axiomathes* **22**, 53–74 (2012). A longer version (in Portuguese) was published in *Studia Scientiae* (São Paulo) 7(4), 535–576 (2009). Also available online

Flexner, A.: *I Remember*: The Autobiography of Abraham Flexner. Simon and Schuster, New York (1960). ISBN: 978-1-40478-133-7

Floyd, J., Kanamori, A.: *How Gödel transformed set theory. Not. AMS* **53**(4), 419–427 (2006). http://www.ams.org/notices/200604/200604-toc.html

Føllesdal, D.: Introductory note to Gödel (*1961/?). Collected Works, Volume III **1995**, 364–373 (1995)

Fölsing, A.: *Albert Einstein: Eine Biographie*. Suhrkamp Verlag (1993) (Also: vol. 7. *Cambridge Studies in Advanced Mathematics*; vol. 2490, Suhrkamp Taschenbuch). ISBN: 978-3-518-40489-8

Fraenkel, A.A.: *Zu den Grundlagen der Cantor-Zermelo'schen Mengenlehre*. Math. Ann. **86**, 230–237 (1922)

Fraenkel, A.A., Bar-Hillel, Y., Levy, A., et al.: *Foundations of Set Theory*, 2nd edn. North-Holland, Amsterdam (1973). ISBN: 0-7204-2270-1

Franzén, T.: *Gödel's Theorem*. An Incomplete Guide to its Use and Abuse. CRC Press (2005). ISBN 978-1-56881-238-0

Franzén, T.: *The popular impact of Gödel's incompleteness theorem. Not. AMS* **53**(4), 440–443 (2006). http://www.ams.org/notices/200604/200604-toc.html

Frege, G.: *Über Sinn und Bedeutung. Zeitschrift für Philosophie und philosophische Kritik* **100**, 25–50 (1892a). English translation (as 'On Sense and Reference') by M. Black appeared in Geach and Black (eds. and transl.) (1980), pp. 56–78

Frege, G.: *Über Begriff und Gegenstand. Vierteljahresschrift für wissenschaftliche Philosophie* **16**, 192–205. Translated (as 'Concept and Object') by P. Geach in Geach & Black (eds.) (1980)

Fuhrmann, A.: *Existenz und Notwendigkeit. Kurt Gödels axiomatische Theologie*. In: Spohn, W. (ed.) *Logik in der Philosophie*, 396pp. Synchron, Heidelberg (2005). See the review https://www.information-philosophie.de/?a=1&t=4435&n=2&y=1&c=50

Gamov, G.: *Rotating universe? Nature* **158**(1946), 549 (1946)

Geach, P., Black, M. (eds.): *Translations from the philosophical writings of Gottlob Frege*, Third edn. with an index by E. D. Klemke. Basil Blackwell, Oxford, and Rowman & Littlefield, Totowa, N.J. (1980)

Geier, M.: *Wittgenstein und Heidegger—Die letzten Philosophen*. Rowohlt, Hamburg (2017). ISBN 978-3-498-02528-1

Gilder, L.: *The Age of Entanglement*. When Quantum Physics was Reborn. Alfred A. Knopf, New York (2008). ISBN: 978-1-4000-4417-7

Glock, H.-J., Hyman, J. (eds.): *A Companion to Wittgenstein*. John Wiley and Sons, Ltd., London (2017). ISBN 978-1-118-64116-3

Gödel, K.: *Uber die Vollstandigkeit des Logikkalküls*. Doctoral thesis. Submitted to the Philosophical Faculty of the University of Vienna in September (1929). Unpublished during Gödel's lifetime. Reconstructed from the copy in the University of Vienna archives, in Gödel's *Collected Works*, vol I, pp. 60–101 (including English translation by Stefan Bauer-Mengelberg & Jean van Heijenoort)

Gödel, K.: *Die Vollständigkeit der Axiome des logischen Funktionenkalküls. Monatshefte für Math. und Phys.* **37**, 349–360 (1930). See also Appendix A

Gödel, K.: *Über formal unentscheidbare Sätze der Principia Mathematica und verwandter Systeme I'. Monatshefte für Math. Phys.* **38**(1931), 173–198 (1931)

Gödel, K.: *The present situation in the foundations of mathematics*. In: Collected Works, vol. III [Gödel *1933o] (1933). Oxford University Press, pp. 45–53 (1995)

Gödel, K.: *On undecidable propositions of formal mathematical systems*. Kurt Gödel, his 'Princeton lectures'. Published in Davis (ed.) (1965), pp. 39–74.

Gödel, K.: *Über die Länge von Beweisen. Ergebnisse eines Mathematischen Kolloquiums* **1936**, 23–24 (1936)

Gödel, K.: *The consistency of the generalized continuum-hypothesis. Proc. U.S. Natl. Acad. Sci.* **24**, 556–557 (1938)

Gödel, K.: *Consistency-proof for the generalized continuum-hypothesis. Bull. Am. Math. Soc.* **45**(1939), 93 (1939)

Gödel, K.: *Consistency-proof for the generalized continuum hypothesis. Proc. U.S. Natl. Acad. Sci.* **25**(4), 220–224 (1939a). Gödel [1939a] in the *Collected Works*, vol. II

Gödel, K.: Gödel's lectures on logic, Notre Dame (1939b). *Bull. Symb. Logic* **22**(4), 469–481 (Adžić, M., Došen, K. (eds.)) (2016)

Gödel, K.: *The consistency of the axiom of choice and the generalized continuum hypothesis with the axioms of set theory. Ann. Math. Stud.* **3**. Princeton University Press, Princeton (1940). Based on lecture notes taken by George W. Brown from Gödel's 1938 lectures at the IAS. Gödel [1940] in the *Collected Works*, vol. II

Gödel, K.: *Ontological proof* (1941). In: *Collected Works* Volume III: Unpublished Essays and Letters. Oxford University Press (1970), and Gödel, K.: Appendix A. Notes in Kurt Godel's Hand. In: Sobel, J.H. (ed.) *Logic and Theism*: Arguments for and Against Beliefs in God, pp. 144–145. Cambridge University Press (2004). https://www.spiegel.de/wissenschaft/mensch/formel-von-kurt-goedel-mathematiker-bestaetigen-gottesbeweis-a-920455.html

Gödel, K.: *Russell's Mathematical Logic* (Schilpp (ed.)) (1944) (q.v.)

Gödel, K.: Letters from Kurt Godel in Princeton to his mother Marianne in Vienna, February 1946–July 1966 (1946–66). See also Appendix A. https://www.digital.wienbibliothek.at/wbr/nav/classification/2559756

Gödel, K.: *What is Cantor's continuum problem? Am. Math. Mon.* **54**(9), 515–525 (1947). Jstor at https://www.jstor.org/stable/pdf/2304666.pdf?refreqid=exc elsior%3Af68a21818e1a78cad246bcaf20c79269

Gödel, K.: *A remark about the relationship between relativity theory and idealistic philosophy.* In: Schilpp, P.A. (ed.) *Albert Einstein. Philosopher—Scientist*, vol. VII, pp. 557–562. The Library of Living Philosophers) La Salle, Ill (Reprint of the first edition from 1949a)

Gödel, K.: *An example of a new type of cosmological solutions of Einstein's field equations of gravitation. Rev. Mod. Phys.* **21**(1949), 555–562 (1949b)

Gödel, K.: Lecture on rotating universes. A lecture held at the IAS on May 7th, 1949, in which he presented his new results on GR to his colleagues. Reconstructed in the *Collected Works*, vol. III, pp. 261 ff. (1949c). Referenced there as [*Gödel *1949b*]. Introduction by David B. Malament

Gödel, K.: *Some basic theorems on the foundations of mathematics and their implications.* In: *Collected Works*, vol. III, pp. 304–323 (1951) [Gödel *1951]

Gödel, K.: *Rotating universes in general relativity theory.* In: Proceedings of the 11th International Congress of Mathematicians, vol. 1, pp. 175–181. Cambridge, MA (1952)

Gödel, K.: From the German translation of Gödel's article in the Einstein Festschrift. He included some important 'additional remarks' to two footnotes in that edition, not in the original English edition of 1949. His article has the German title '*Eine Bemerkung über die Beziehungen zwischen der Relativitätstheorie und der indealistischen Philosophie*'. In: Schlipp, P.A. (ed.) *Albert Einstein als Philosoph und Naturforscher*. Kohlhammer, pp. 406–412 (1956). ISBN: 978-3-52808-538-4

Gödel, K.: *Über eine bisher noch nicht benützte Erweiterung des finiten Standpunktes*. *Dialectica* **12**(3–4), 280–287 (1958)

Gödel, R.: Memoir of the Gödel family. In: Weingartner, P., Schmetterer, L. (eds.) *Gödel remembered*, Salzburg 10–12 July 1983 (Conference Proceedings). Österreichische Akademie der Wissenschaften, Institut für Wissenschaftstheorie (Salzburg, Austria). Bibliopolis, Naples (1987)

Gold, T., Bondi, H. (eds.): *The Nature of Time*. Conference Proceedings of a meeting held at Cornell University, 30th May–1st June, 1963. Cornell University Press, Ithaca, NY (1967). See the paper '*Geodesics in Godel's universe*' by S. Chandrasekhar

Goldstein: *Incompleteness: The Proof and Paradox of Kurt Gödel* (Great Discoveries). Rebecca Goldstein (2005): Atlas Books (2005), W.W. Norton (2006). ISBN 978-0-7394-5744-3

Graf-Grossmann, C.: *Marcel Grossmann. For the love of mathematics*. Scientific-historical Epilogue by Tilman Sauer, translated by W.D. Brewer. Springer Biographies, Springer Nature, Cham, CH (2018). ISBN: 978-3-319-90076-6

Graham, E.: *Adventures in Fine Hall*. The weirdness of math's golden age. Princeton Alumni Weekly (PAW), 10th January 2018. https://paw.princeton.edu/article/adventures-fine-hall (consulted 30 January 2022)

Grannec, Y.: *La Déesse des Petites Victoires*. Èditions Anne Carrière, Paris (2012). ISBN 978-2-84337-666-5

Grattan-Guinness, I.: *In Memoriam Kurt Gödel*: His 1931 Correspondence with Zermelo on his Incompleteness Theorem. *Hist. Math.* **6**, 294–304 (1979)

Gray, R.: *Georg Cantor and transcendental numbers*. *Am. Math. Mon.* **101**(9), 819–832 (1994). https://doi.org/10.2307/2975129, JSTOR 2975129

Greiser, D.: *Eine Liebe in Wien* ('A Love in Vienna'), pp. 170–177. Verlag Niederösterreichisches Pressehaus, St. Pölten & Vienna (1989). ISBN: 978-3-85326-881-0

Grzegorczyk, A.: *On the concept of categoricity*. *Stud. Log. T* **13**(1962), 39–66 (1962)

Guerra-Pujol, F.E.: *Gödel's Loophole*. *Cap. Univ. Law Rev.* **41**, 637–673 (2013)

Heer, F.A.: *The Holy Roman Empire*. Praeger, New York (1967). ISBN 978-0-297-17672-5 (Original in German; English translation 1968)

Henkin, L.: *The completeness of the first-order functional calculus*. *J. Symb. Logic* **14**(3), 159–166 (1949)

Henkin, L.: *The discovery of my completeness proofs*. *Bull. Symb. Logic* **2**(2), 127–158 (1996)

Hentschel, K.: Review of Palle Yourgrau's (2005). *Phys. Today* **58**(12) (2005). https://physicstoday.scitation.org/doi/10.1063/1.2169448

Hilbert, D.: *Über die Theorie der algebraischen Formen. Math. Ann.* **36**(1890), 473–534 (1890)

Hilbert, D.: *Über den Zahlbegriff. Jahresber. Deutsch. Math.-Verein.* **8**(1900), 180–183 (1900)

Hilbert, D.: *Über das Unendliche. Math. Ann.* **95**(1), 161–190 (1926). See page 170 for 'Cantor's paradise'. An English translation can be found in van Heijenoort, pp. 367–392 (1967)

Hilbert, D., Ackermann, W.: *Grundzüge der theoretischen Logik (HA)*. Springer, Heidelberg/Berlin (1928). Reissued 1959. English translation: *Principles of Mathematical Logic*, Hilbert & Ackermann, translated by L.M. Hammond, G.G. Leckie, F. Steinhardt. AMS Chelsea Publishing, Providence RI (1999). ISBN 978-0-8218-2024-7

Hilbert, D., Bernays, P.: The second volume of '*Grundlagen der Mathematik*', a two-volume work on the fundamentals of mathematics; the first volume appeared in 1934. A detailed proof of Gödel's Second Incompleteness Theorem was given (presumably by Bernays) in the second volume: D. Hilbert and P. Bernays (new edition, 1970), '*Grundlagen der Mathematik II*', in the series '*Die Grundlehren der mathematischen Wissenschaften*', vol. 50, 2nd edn. Springer, Berlin, New York (1970). ISBN 978-3-64286-897-9

Hindlyckc, C.: *The relative consistency of the axiom of choice and the generalized continuum hypothesis with the Zermelo-Fraenkel axioms*: the constructible sets L. University of Uppsala, U.U.D.M. Project Report 2017:33 (2017). https://uu.diva-portal.org/smash/get/diva2:1138999/FULLTEXT01.pdf

Hintikka, J. (ed.): *The Philosophy of Mathematics*. Oxford University Press (1969)

Hintikka, J.: *On Gödel* (1999). Wadsworth, Belmont/CA (2000). ISBN 0-534-57595-1. See Item 16 in Appendix B

Hintikka, J.: *What platonism?* Reflections on the thought of Kurt Gödel. In: *Revue internationale de philosophie* **234**, 535–552 (2005/4)

Hoffmann, D.W.: *Die Gödel'schen Unvollständigkeitssätze: Eine geführte Reise*. Springer Spektrum (2013/2017). ISBN 3-827-4-3000-3

Hofstadter, D.R.: *Gödel, Escher, Bach. An Eternal Golden Braid*. Penguin Philosophy Series (1979) (reprinted 1980). ISBN: 978-0-465-02685-2. GEB was re-issued in 1999. See Item 3 in Appendix B

Hofstadter, D.R.: *Time Magazine*, Monday, 29 March 1999 (1999). See also Hofstadter's Foreword to Nagel & Newman (2001) and to Sigmund (2006)

Hofstadter, D.R.: *I am a Strange Loop*. Basic Books, NY (2007). ISBN: 978-0-465-03078-1

Holt, J.: *When Einstein Walked with Gödel*. Excursions to the Edge of Thought. Farrar, Straus & Giroux (2018). ISBN 9-780-3-7414-60-2

Huber-Dyson, V.: Review of '*Gödel, Escher, Bach: An Eternal Golden Braid*' by D. R. Hofstadter. *Can. J. Philos.* **11**(4), 775–792 (1981)

Huber-Dyson, V.: *Gödel's theorems; a workbook on formalization.* Teubner-Texte zur Mathematik, vol. 122. Teubner Verlagsgesellschaft, Stuttgart-Leipzig (1991)

Huber-Dyson, V.: Interview in *The Edge*, entitled 'Gödel and the Nature of Mathematical Truth', on 25.07.2005 (2005). https://www.edge.org/3rd_culture/vhd05/vhd05_index.html

Infeld, L., Schild, A.: *On the motion of test particles in general relativity. Rev. Mod. Phys.* **21**, 408 (1949)

Jourdain, P.E.B.: *On transfinite cardinal numbers of the exponential form. Philos. Mag. Ser.* 6 **9**, 42–56 (1905)

Jung, T.: *"Don't you think it is very exciting?"*—Ein Beitrag zur Untersuchung des Hintergrunds der Entstehung des rotierenden Gödel-Universums. *Berichte zur Wissenschafts-Geschichte* **29**(2006), 325–340 (2006)

Kahle, R.: *Die philosophische Bedeutung des Gödel-Universums* ('The Philosophical Significance of the Gödel Universe'). Essay (2021). See the website of the *Kurt-Gödel-Freundeskreis Berlin.* https://kurtgoedel.de/

Kanamori, A.: *Gödel and set theory. Bull. Symb. Logic* **13**(2), 153–188 (2007)

Kennedy, H.C.: *What Russell learned from Peano. Notre Dame J. Form. Logic* **XIV**(3), 367–372 (1973)

Kennedy, H.C.: *Giuseppe Peano.* Birkhäuser Verlag, Basel (1974). (Beiheft 14 of *Elemente der Mathematik*)

Kennedy, H.C.: *Peano: life and works of Giuseppe Peano.* In: *Studies in the History of Modern Science*, vol. IV. Kluwer, Boston/Dordrecht (1980). ISBN 978-90-277-1068-0

Kennedy, J.C.: Review of *Incompleteness* by Rebecca Goldstein. *Not. AMS* **53**(4), 448–455 (2006). http://www.ams.org/notices/200604/200604-toc.html

Kennedy, J.C.: *Gödel's thesis. An appreciation.* In: Baaz, M., et al. (eds.) *Kurt Gödel and the Foundations of Mathematics.* Cambridge University Press, published online 2011. ISBN: 978-0-51197-423-6

Kennedy, J.C.: Review of Tieszen (2011), 4th October 2013. *Notre Dame Philos. Rev.* (2013). https://ndpr.nd.edu/reviews/after-gdel-platonism-and-ration alism-in-mathematics-and-logic/

Kennedy, J.C. (ed.): *Interpreting Gödel.* Cambridge University Press, Cambridge, UK (2014). ISBN: 978-0-511-75630-6

Kennedy, J.C.: *Turing, Gödel and the "Bright Abyss".* Springer Monographs in Mathematics (2017). https://doi.org/10.1007/978-3-319-53280-6_3

Kennedy, J.C.: *Gödel, Tarski, and the Lure of Natural Language.* Cambridge University Press (2020). ISBN: 978-0-51199-839-3

Kennedy, J.C.: *Gödel and the integrated self*, or: on the philosopher's second sailing. *Theoria* **87**(2021), 874–884 (2021)

Kennedy, J., Kossak, R. (eds.): *Set Theory, Arithmetic, and Foundations of Mathematics*: Theorems, Philosophies. Lecture Notes in Logic, vol. 36. Cambridge University Press (2011). ISBN 978-1-139-50481-2

Kiefer, C.: *Was bedeutet es für unser Weltbild, wenn wir mit Gödel die Nichtexistenz der Zeit annehmen?* (What Does it Mean for our Worldview, if we assume— with Gödel—the Nonexistence of Time?). Essay (2021). See the website of the *Kurt-Gödel-Freundeskreis Berlin*. https://kurtgoedel.de/

Kleene, S.C.: *Introduction to Metamathematics*. Ishi Press International, NY (2009). First printing in Amsterdam (1952). ISBN 978-0-923891-57-2

Kleene, S.C., Post, E.L.: *The upper semi-lattice of degrees of recursive unsolvability*. *Ann. Math.* **59**, 379–407 (1954)

Kleene, S.C.: *The work of Kurt Gödel*. *J. Symb. Logic* **41**(4), 761–778 (1976)

Kleene, S.C.: *Origins of recursive function theory*. *Ann. Hist. Comput.* **3**, 52–67 (1981)

Kleene, S.C., Rosser, J.B.: Interview with Stephen C. Kleene and J. Barkley Rosser about their time in Princeton, conducted by William Aspray of the Charles Babbage Institute, in Madison, WI, 26th April 1984. Studying Logic. https://studyinglogic.tumblr.com/post/143315364035/interview-with-kleene-and-rosser. Part of '*The Princeton Mathematics Community in the 1930s*', Transcript Number 23 (PMC23). Copyright The Trustees of Princeton University (1985)

Köhler, E., et al. (eds.): *Kurt Gödel. Wahrheit und Beweisbarkeit*. Documents and historical analyses, vol. 1. Öbv Et Hpt, Vienna (2002). ISBN 3-209-03834-1

Koons, R.C.: *Sobel on Gödel's Ontological Proof* (2005). http://www.phil.pku.edu.cn/documentsl20050317075553Sobel.pdf

Kreisel, G.: *Kurt Gödel*. In: Biographical Memoirs of Fellows of the Royal Society, vol. 26, pp. 148–224 (1981)

Kripke, S.A.: *Gödel's Theorem and Direct Self-Reference*, 7 June 2021. arXiv:2010.11979v2[math.LO], https://arxiv.org/pdf/2010.11979.pdf

Lanczos, K.: *Über eine stationäre Kosmologie im Sinne der Einstein'schen Gravitationstheorie*. *Z. für Phys.* **21**(1924), 73–110 (1924)

Leonard: *Ethics and the Excluded Middle*. Karl Menger and Social Science in Interwar Vienna (1998). Leonard, R.J.: *Isis* **89**, 1–26 (1998)

Leven, T.: *The role of intuition in Gödel's and Robinson's points of view*. *Axiomathes* **29**(2019), 441–461 (2019)

Levine, H.: *In Search of Sugihara*: The Elusive Japanese Diplomat Who Risked his Life to Rescue 10,000 Jews From the Holocaust. Free Press (1996). ISBN 978-0684832517

Liddle, A.: *The cosmological constant and its interpretation*. In: Murdin, P. (ed.) Encyclopedia of Astronomy and Astrophysics. IOP Publishing Ltd., London (2006). ISBN 0-333-75088-8

Livio, M.: *Is God a Mathematician?*. Simon and Shuster (2011). ISBN 1-416-5-9443-4

Liouville, J.: *Sur des classes très étendues de quantités dont la valeur n'est ni algébrique ni même réductible à des irrationnelles algébriques. J. Math. Pures et Appl.* **16**(1), 133–142 (1851) (There is some evidence that Liouville actually originated the concept of transcendental numbers in 1844, and published it in the *Comptes rendus* then)

Löwenheim, L.: *Über Möglichkeiten im Relativkalkül. Math. Ann.* **76**, 447–470 (1915). English translation in van Heijenoort 1967

Lucas, J.R.: *Minds, Machines, and Gödel. Philosophy*, vol. XXXVI, pp. 112–127 (1961). Reprinted in books published in 1963 and 1964.

Lyons, V., Fitzgerald, M.: *Kurt Gödel (1906–1978). The mathematical genius who had Asperger syndrome. Autism—Asperger's Digest*, pp. 46–47 (2004)

Lyons, V., Fitzgerald, M.: *Asperger Syndrome: A Gift or a Curse?* (See Chap. VIII on Kurt Gödel). Nova Biomedical Books, NY (2005). ISBN 1-594-5-4387-9

Macintyre, A.J.: *Abraham Robinson, 1918–1974.* An obituary notice by Angus J. Macintyre, in: the *Bull. Am. Math. Soc.* **83**(4), 646–666 (1977)

Majer, U.: *Geometry, Intuition and Experience: From Kant to Husserl. Erkenntnis* **42**(2), 261–285 (1995)

Malament 1984: *"Time Travel" in the Gödel Universe*, David B. Malament, *Proceedings of the Biennial Meeting of the Philosophy of Science Association* (1984), Volume Two: Symposia and Invited Papers, pp. 91–100.

Mal'cev, A.I.: *The Metamathematics of Algebraic Systems*, Collected Papers: 1936–1967. North-Holland Pub. Co., Amsterdam (1971). ISBN 0-7204-2266-3 [Note that the name of this author is often transliterated as 'Maltsev' rather than 'Mal'cev']

Mar, G.R.: *Hao Wang's logical journey. J. Chin. Philos.* **42**(51) (2017)

Marino, E.C.: *Jorge André Swieca: Uma figura ímpar na física brasileira. Revista Brasileira de Ensino de Física* **37**(3) (2015)

McGuinness, B. (ed.): *Collected Papers on Mathematics, Logic, and Philosophy by Gottlob Frege.* Translated by Max Black ... [et al.]. Blackwell, Oxford/New York (1984). ISBN: 063-1-12728-3

McTaggart, J.M.E.: *The unreality of time. Mind* **17**(68), 457–474 (1908)

Menger, K.: *Reminiscences of the Vienna Circle and the Mathematical Colloquium*, Karl Menger, Kluwer (1994). Reprinted 2013: Vienna Circle Collection: Series (Mulder, H.L., Cohen, R.S., McGuinness, B., Haller, R. (eds.)). Springer, Netherlands (2013). Kindle-Version, ISBN 978-94-011-1102-7

Misner, C.W., Thorne, K.S., Wheeler, J.A.: *Gravitation.* Princeton University Press (1973). Reissued 2017. ISBN: 978-1-400-88909-9

Moore, G.H.: *Zermelo's Axiom of Choice.* Its Origins, Development & Influence. Dover Publications, Mineola, NY (1982). ISBN: 978-0-486-48841-7

Morgenstern: Diaries, Oskar Morgenstern Papers, archived in the Rare Book and Manuscript Library. Duke University, Durham, N.C. http://gams.uni-graz.at/con text:ome

Mostowski, A.: *Sentences Undecidable in Formalized arithmetic*: An Exposition of the Theory of Kurt Gödel. Andrzej Mostowski, with Kurt Gödel. North-Holland Publishing Co. (1952)

Müller, D.: *Kurt Gödels Brünner Verwandte* (Kurt Gödel's Relatives in Brünn). *J. Phys. Conf. Ser.* **82**, Article 012001 (2007)

Myhill, J.: *Some Philosophical Implications of Mathematical Logic. Rev. Metaphys.* **6**(1952), 165–198 (1952)

Nagel, E., Newman, J.R.: *Gödel's proof. Sci. Ame.* **194**(6), 71–90 (1956)

Nagel, E., Newman, J.R.: *Gödel's Proof*. NYU Press (2001) (Originally published in 1958. Reissued 2001, 2005, 2008). ISBN 978-0-814-75816-8

Natário, J.: *Optimal time travel in the Gödel universe*. arXiv:1105.6197v2 [gr-qc] 14th October 2011

Neurath, O.: *Wissenschaftliche Weltauffassung: der Wiener Kreis*. In: Neurath, M., Cohen, R.S. (eds.) *Empiricism and Sociology*, Vienna Circle Collection, vol. 1. Springer, Dordrecht (1973). ISBN 978-90-277-0259-3

O'Connor, J.J., Robertson, E.F.: *Abraham Robinson*. In: MacTutor Biographies (2000). https://mathshistory.st-andrews.ac.uk/Biographies/Robinson/

Odifreddi, P. (ed.): *Kreiseliana: About and Around Georg Kreisel*. Taylor & Francis (1996). A Festschrift for Kreisel's 70th birthday. ISBN: 978-1-568-81061-4

Odifreddi: *Gödel's mathematics of philosophy*. In: Baaz, M., Papadimitriou, C.H., Putnam, H.W., Scott, D.S., Harper, C.L., Jr. (eds.) *Kurt Gödel and the Foundations of Mathematics*. Cambridge University Press (2011). ISBN: 978-0-51197-4438 6

Oppenheimer, J.R., Snyder, H.: *On continued gravitational contraction. Phys. Rev.* **56**, 455–459 (1939)

Pais, A.: *Subtle is the Lord*. Oxford University Press, Oxford and New York (1982). Re-issued 2005. ISBN: 978-0-19-280672-7

Parsons, C.: *Platonism and mathematical intuition in Kurt Gödel's Thought. Bull. Symb. Logic* **1**(1), 44–74 (1995)

Parsons, C.: *In memoriam: Hao Wang. Bull. Symb. Logic* **2**(1), 108–109 (1996a)

Parsons, C.: *Hao Wang as philosopher*. In: Hájek, P. (ed.) Lecture Notes in Logic, pp. 64–80 (1996b)

Pfarr 1981: '*Time Travel in Gödel's Space*', Joachim Pfarr (1980), *General Relativity and Gravitation*, Vol. 13, No. 11 (1981), pp. 1073–1091. See page 1089.

Post, E.L.: *Introduction to a general theory of elementary propositions. Am. J. Math.* **43**(3), 163–185 (1921)

Post, E.L.: *Finite combinatory processes—formulation 1. J. Symb. Logic* **1**(3), 103–105 (1936)

Post, E.L.: *Recursively enumerable sets of positive integers and their decision problems. Bull. Am. Math. Soc.* **50**(5), 284–316 (1944)

Pruss, A.R.: *A Gödelian ontological argument improved. Relig. Stud.* **45**(3), 347–353 (2009)

Quine, W.V.O.: *Completeness of the propositional calculus. J. Symb. Logic* **3**(1938), 37–40 (1938)

Quine, W.V.O.: *Paradox. Sci. Am.* **206**(4) (1962). Reprinted as *"The Ways of Paradox"*: The Ways of Paradox and Other Essays, pp. 1–21. Harvard University Press, Cambridge (1966)

Quine, W.V.O.: *Paradoxes.* In: *Quiddities*: An Intermittently Philosophical Dictionary, pp. 145–149. Harvard University Press, Cambridge MA (1987). ISBN 0-674-74352-0

Raatikainen, P.: *Gödel's incompleteness theorems.* Stanford Encyclopedia of Philosophy. https://plato.stanford.edu/entries/goedel-incompleteness/#:~:text=The%20first%20incompleteness%20theorem%20states,disproved%20in%20%5C(F%5C)

Rindler, W.: *Relativity: Special, General and Cosmological.* Oxford University Press (2001). Reissued 2006. ISBN: 978-0-198-50835-9

Rindler, W.: *Gödel, Einstein, Mach, Gamow, and Lanczos*: Gödel's remarkable excursion into Cosmology. *Am. J. Phys.* **77**(2009), 498–510 (2009)

Rindler, W.: *Gödel, Einstein, Mach, Gamow, and Lanczos*: Gödel's remarkable excursion into cosmology. In: Baaz, M., et al. (eds.) *Kurt Gödel and the Foundations of Mathematics*, pp.185–212. Cambrige University Press, Cambridge UK (2011). See Item 35 in Appendix B

Roberts, S.: *Genius at Play*: The Curious Mind of John Horton Conway. Bloomsbury Publishing, USA (2015). ISBN978-1-620-40594-9

Robinson, R.M.: *An essentially undecidable axiom system.* In: *Proceedings of the International Congress of Mathematics*, pp. 729–730 (1950)

Robinson, A.: *Concerning progress in the philosophy of mathematics.* In: *Bristol Logic Colloquium* (1973), and published in its Proceedings (eds. H.E. Rose & J.C. Shepherdson), pp. 41–52. North-Holland, Amsterdam (1975)

Rosser, J.B.: *Extensions of some theorems of Gödel and Church. J. Symb. Logic* **1**, 87–91 (1936)

Rosser, J.B.: *An informal exposition of proofs of Gödel's theorems and Church's theorem. J. Symb. Logic* **4**(2), 53–60 (1939)

Rovelli, C.: *The Order of Time.* Penguin UK (2018). ISBN: 978-0-24129-253-0

Russell, B.: *A Critical Exposition of the Philosophy of Leibniz.* Cambridge University Press (1900). Reprinted 1937 and 2013: ISBN 978-1-107-68016-6

Russell, B.: *The Principles of Mathematics*, 2nd edn. Reprint, W. W. Norton & Company, New York (1996) (First published in 1903)

Russell, B.: *Introduction to Mathematical Philosophy* (1919), 2nd edn. George Allen and Unwin, London (1920)

Russell, B.: *Autobiography*, Three vols. (1967, 1968, 1969). Allen & Unwin (1967–69). Reissued by Routledge, London/New York (1998). ISBN 0-415-22862-X

Russell, B., Whitehead, A.N.: *Principia Mathematica (PM)*, Three vols. Cambridge University Press, Cambridge/UK (1910, 1912, 1913)

Sacks, G.: *Reflections on Gödel.* Lecture given at the University of Pennsylvania, 11th April 2007 (2007). Available on YouTube as an audio file. A brief summary is given at https://www.math.upenn.edu/events/reflections-goedel

Schilpp, P.A. (ed.): *The Philosophy of Bertrand Russell*. Library of Living Philosophers, vol. V. Northwestern University Press, Evanston/Chicago (1944)

Schilpp, P.A. (ed.): *Albert Einstein, Philosopher-Scientist*. Library of Living Philosophers Northwestern University Press, Evanston (1949)

Schimanovich-Galidescu, M.-E.: Gödel's Letters to his Mother, 1946–1966 (Köhler et al. (eds.)). Princeton–Vienna (2002)

Schindler, R.: *Kurt Gödel (1906–1978)*. *DMV-Mitteilungen* **14–1**, 42–45 (2006)

Schmidthuber, J.: *Als Kurt Gödel die Grenzen des Berechenbaren entdeckte* ('When Kurt Gödel discovered the limits of the computable'). *Frankfurter Allgemeine Zeitung* (FAZ), 16th June 2021. See also Schmidthuber's blog at https://people. idsia.ch/~juergen/goedel-1931-begruender-theoretische-informatik-KI.html

Schroer, B.: *Jorge A. Swieca's contributions to quantum field theory in the 60's and 70's and their relevance in present research*, May 2010. arXiv:0712.0371v3 [physics.hist-ph] (2010). Published in *Eur. Phys. J. H* **35**(1), 53–88 (2010)

Schwarzschild, K.: *Über das Gravitationsfeld eines Massenpunktes nach der Einstein'schen Theorie*. Reimer, Berlin (1916); Sitzungsberichte der Königlich-Preussischen Akademie der Wissenschaften, pp.189 ff

Scott, J.H.: Cited in Sobel (1987), *Gödel's ontological proof*. In: Thomson, J.J. (ed.) *On Being and Saying*: Essays for Richard Cartwright, pp. 241–261. MIT Press, Cambridge/MA and London (1987). ISBN 978-0-2622-0063-9

Sieg, W.: *Hilbert's programs: 1917–1922*. *Bull. Symb. Logic* **5**(1), 1–44 (1999)

Sieg, W.: *Only two letters: the correspondence between Herbrand and Gödel*. *Bull. Symb. Logic* **11**(2), 172–184 (2005)

Sieg, W.: *Gödel on computability*. *Philosophia Mathematica* **14**(2), 189–207 (2006)

Sigmund, K.: *Pictures at an exhibition*. *Not. AMS* **53**(4), 428–432 (2006). http:// www.ams.org/notices/200604/200604-toc.html

Sigmund, K.: *Sie nannten sich der Wiener Kreis. Exaktes Denken am Rand des Untergangs*. Springer Spektrum (2015). ISBN 978-3-658-08534-6 (Print); ISBN 978-3-658-08535-3 (e-Book). English edition: Exact Thinking in Demented Times. *The Vienna Circle and the Epic Quest for the Foundations of Science*. Basic Books, NY (2017). ISBN 978-0-46509

Silberman, S.: *NeuroTribes: The Legacy of Autism and How to Think Smarter about People who Think Differently*, 544pp. Allan and Unwin (2015). ISBN: 978-1-760-11363-6

Skolem, T.A.: *Einige Bemerkungen zur axiomatischen Begründung der Mengenlehre*. In: *Matematikerkongressen in Helsingfors 4–7 Juli 1922, Den femte skandinaviske matematiker-kongressen. Proceedings: Redogörelse*, pp. 217–232 (1923)

Skolem, T.A.: *Selected Works in Logic* (Fenstad, J.E. (ed.)). Scandinavian University Books, Oslo (1970) [It contains 22 articles in German, 26 in English, 2 in French, 1 English translation of an article originally published in Norwegian, and a complete bibliography]

Smith, P.: *An Introduction to Gödel's Theorems* (2005). Cambridge University: Preliminary version of the book listed as Item 32 in Appendix B (2007 and 2020). See www.godelbook.net.

Smith, P.: *An Introduction to Gödel's Theorems.* Cambridge University Press (2007, 2020)

Smith, P.: *Gödel without (too many) tears.* In: Logic Matters, Cambridge (2021)

Smullyan, R.M.: *Gödel's Incompleteness Theorems.* Oxford University Press, New York (1992). See Item 11 in Appendix B. ISBN 0-19-5046-72-2

Sobel, J.H.: *Logic and Theism:* Arguments For and Against Belief in God. Cambridge University Press, Cambridge, UK (2004). ISBN: 978-1-13944-998-4

Spector, C.: *Provably recursive functionals of analysis: a consistency proof of analysis by an extension of principles in current intuitionistic mathematics.* In: Dekker, F.D.E. (ed.) *Recursive Function Theory:* Proceedings of Symposia in Pure Mathematics, vol. 5, pp. 1–27. American Mathematical Society, Providence, Rhode Island (1962)

Stachel, P.: *Das österreichische Bildungssystem zwischen 1749 und 1918.* In: Acham, K. (ed.) *Geschichte der österreichischen Humanwissenschaften. Historical context, sociological results and methodological prerequisites,* vol. 1, pp. 115–146. Passagen-Verlag, Vienna (1999)

Stachel, J.: Review of '*A World without Time: The Forgotten Legacy of Gödel and Einstein*', by Palle Yourgrau. *Not. Am. Math. Soc.* **54**(7), 861–868 (2007)

Stein, H.: *On the paradoxical time-structures of Gödel. Philos. Sci.* **37**(1970), 589–601 (1970)

Stillwell, J.: *Emil Post and his anticipation of Gödel and Turing. Math. Mag.* **77**(1), 3–14 (2004)

Stölzner, M., Uebel, T.: '*Wiener Kreis*', Texts on the scientific worldview of Rudolf Carnap, Otto Neurath, Moritz Schlick, Philipp Frank, Hans Hahn, Karl Menger, Edgar Zilsel and Gustav Bergmann. In: Stöltzner, M., Uebel, T. (eds.) Philosophische Bibliothek (2009). ISBN 978-3-7873-2109-4

Strauss, E.G.: *Reminiscences.* In: Holton, G., Elkana, Y. (eds.) Albert Einstein: Historical and Cultural Perspectives. The Centennial Symposium in Jerusalem, p. 422. Princeton University Press, Princeton (1982). Cited in Feferman (1998)

Strogatz, S.: *Einstein's First Proof.* The New Yorker, 19th November 2015 (2015)

Sugihara, Y.: *Visas for Life.* Translated by Hiroki Sugihara. San Francisco, Edu-Comm Plus (1993). ISBN 10: 0-964-96740-5

Synge, J.L.: *Relativity: The General Theory.* North Holland-Interscience Publishers, New York (1960). ISBN 978-0-4441-0279-9

Takeuti, G.: *Suugaku ni tsuite* ('About Mathematics'). *Kagakukisoron Kenkyu* (Ann. Jpn. Assoc. Philos. Sci.) **4**, 170–174 (1972)

Tarski, A.: *The Concept of Truth in Formalized Languages,* first published in Polish in 1933, then as a 'preprint' in German in 1935, then as '*Der Wahrheitsbegriff in den formalisierten Sprachen*' in Studia Philosophica, vol. I, pp. 261–405 (1936), and finally in English in Tarski's book (1956)

Tarski, A.: *Logic, Semantics, Metamathematics.* Clarendon Press, Oxford (1956)

Taußky-Todd, O.: *Remembrances of Kurt Gödel.* In: Weingartner & Schmetterer (1987). A reprint of the article as a pdf: https://calteches.library.caltech.edu/605/2/Todd.pdf

Tieszen, R.L.: *After Gödel: Platonism and Rationalism in Mathematics and Logic*. Oxford University Press (2011). ISBN: 978-0-1996-0620-7

Tieszen, R.L.: *Simply Gödel*. Simplycharly, NYC (2017). ISBN 978-1-943657-15-5

Todorov, I.T.: *Einstein and Hilbert: The Creation of General Relativity*. Institut für Theoretische Physik der Universität Goettingen. arXiv:physics/0504179v1, 25 April 2005. Written version of a lecture given at the ICTP, Trieste on 9th December 1992 and updated in a lecture at the International University Bremen on 15th March 2005.

Toledo, S.: *Conversations with Gödel, 1972–75*. Kennedy & Kossak, pp. 200–207 (2011)

Tully, R.: Website entitled '*Stanley Tennenbaum American Original*', Estate of Rob Tully © 2022 (2015). https://stanleytennenbaumamericanoriginal.com/. Accessed 6 Mar 2022

Turing, A.M.: *On computable numbers, with application to the "Entscheidungsproblem"*. Proc. Lond. Math. Soc. **42**, 230–265 (1937) (Submitted on 28th May 1936)

Turing, A.M.: *Systems of logic based on ordinals*. Proc. Lond. Math. Soc. (Series 2) **45**, 161–228 (1939)

Turing, A.M.: *Computing machinery and intelligence*. Mind. A Q. Rev. Psychol. Philos. **LIX**(236), 433–460 (1950)

Turing, A.M.: *Solvable and unsolvable problems*. Science News **31**, 7–23 (1954)

Uebel, T.: *"Logical Positivism"—"Logical Empiricism": What's in a Name?* Perspect. Sci. **21**(1), 58–99 (2013)

Urquhart, A.: *Russell and Gödel*. Bull. Symb. Logic **22**(4), 504–520 (2016)

van Atten, M., Kennedy, J.: *On the philosophical development of Kurt Gödel*. Bull. Symb. Logic **9**(4), 425–476 (2003)

van Atten, M., Kennedy, J.: *Gödel's logic*. In: Gabbay, D.M., Woods, J. (eds.) *Handbook of the History of Logic*, vol. 5. Elsevier B.V., Amsterdam (2009). ISBN: 978-0-444-51620-6

van Atten, M. (ed.): *Essays on Gödel's Reception of Leibniz, Husserl, and Brouwer*. Springer, Cham (2015). ISBN: 978-3-319-10030-2. See van Atten's essay on '*Gödel, Mathematics, and Possible Worlds*', pp. 147–155

van Heijenoort, J. (ed.): *From Frege to Gödel: A Sourcebook in Mathematical Logic, 1879–1931*. Harvard University Press, Cambridge MA (1967)

van Stockum, J.: *The gravitational field of a distribution of particles rotating about an axis of symmetry*. Proc. R. Soc. Edinb. **57**, 135 (1937)

Veale, T.: *Your Wit is my Command: Building AIs with a Sense of Humor*. MIT Press, Cambridge, MA (2021). ISBN: 978-0-26204-4599-5

Veblen, O.: *A system of axioms for geometry*. Trans. Am. Math. Soc. **5**(3), 343–384 (1904)

von Doderer, H.: *Die Strudlhofstiege: oder, Melzer und die Tiefe der Jahre*. Biederstein Verlag (1951). ISBN: 978-3-764-20050-3

von Neumann, J.: *Eine Axiomatisierung der Mengenlehre*. J. für die reine und angewandte Math. **154**(1925), 219–240 (1925)

Waismann, F., Wittgenstein, L. (eds.): *Ludwig Wittgenstein und der Wiener Kreis. Gespräche, aufgezeichnet von Friedrich Waismann.* Brian F. McGuinness, Suhrkamp (1984). ISBN: 978-3518281031

Wang, H.: *From Mathematics to Philosophy.* Routledge (1974). ISBN 978-1-1138-68773-8

Wang, H.: *Some facts about Kurt Gödel. J. Symb. Logic* **46**(3), 653–659 (1981)

Wang, H.: Reflections on Kurt Gödel. MIT Press (1987). ISBN 978 0 26223 127-5

Wang, H.: *Time in philosophy and physics: from Kant and Einstein to Gödel. Synthese* **102**(1995), 215–234 (1995)

Wang, H.: *A logical journey: from Gödel to philosophy.* Bradford Books. MIT Press (1997). ISBN 0-262 2 61251

Weingartner, P., Schmetterer, L. (eds.): *Gödel remembered.* In: Proceedings of a Symposium held in Salzburg, July 1983. Bibliopolis, Naples (1987). ISBN 8-870-88141-5. Contains articles by Rudolf Gödel, Olga Taußky-Todd, Ernest Zermelo, Stephen C. Kleene, and Georg Kreisel

Wheeler, J.A., Ford, K.: *Geons, Black Holes, and Quantum Foam.* W.W. Norton & Co., NY (1998). ISBN: 0-393-04642-7

Whitehead, A.N., Russell, B.: *Principia Mathematica (PM)*, three vols (1910, 1912 and 1913), reprinted 1927, 1963 and 2004. Cambridge University Press, Cambridge U.K. (2004). ISBN: 0-521-06791-X

Wittgenstein, L.: *Tractatus Logico-Philosophicus* (TLP), C. K. Ogden (transl.), London: Routledge & Kegan Paul. Originally published as "*Logisch-Philosophische Abhandlung*", in *Annalen der Naturphilosophie*, XIV (3/4), 1921 (1922)

Wittgenstein, L.: *Philosophical Investigations*, G.E.M. Anscombe and R. Rhees (eds.), G.E.M. Anscombe (transl.), Oxford: Blackwell (1953)

Wittgenstein, L.: *Remarks on the Foundations of Mathematics*, G. H. von Wright, R. Rhees and G. E. M. Anscombe (eds.), G. E. M. Anscombe (transl.), Oxford: Blackwell, revised edition 1978 (1956)

Yourgrau, P.: *A World Without Time: The Forgotten Legacy of Gödel and Einstein*, 224 pp. Basic Books, NY (2005). ISBN-13: 978-0-46509-293-2

Yourgrau, P.: *Does Reason Have Limits?* Review of Budiansky (2021). https://kurtgo edel.de/review-essay-by-palle-yourgrau/

Zach, R.: *Completeness before Post: Bernays, Hilbert, and the development of propositional logic. Bull. Symb. Logic* **5**(1), 331–366 (1999)

Závodnik, J.: *Die Geschichte des Zusammenlebens der Tschechen und Deutschen in Brünn* ('History of coexistence between the Czechs and the Germans in Brno'). Masters thesis, Brno (2008). https://is.muni.cz/th/njo0x/oponentsky_posudek_Zavodnik.doc

Zermelo, E.: *Untersuchungen über die Grundlagen der Mengenlehre*. I. Math. Ann. **65**(2), 261–281 (1908). Digital version from the University of Göttingen. http://resolver.sub.uni-goettingen.de/purl?GDZPPN002262002

Zuckmayer, C.: *Als wär's ein Stück von mir*. Fischer-Verlag, Frankfurt am Main (1966). ISBN 3-436-01035-9 (1969 edition). English translation: 'A Part of Myself'. Carl Zuckmayer, transl. Richard Winston, Clara Winston, Secker & Warburg, London (1970). ISBN-13: 978-0-43659-261-4

Index

© The Editor(s) (if applicable) and The Author(s), under exclusive
license to Springer Nature Switzerland AG 2022, corrected publication 2023
W. D. Brewer, *Kurt Gödel*, Springer Biographies,
https://doi.org/10.1007/978-3-031-11309-3

Carnap, Rudolf, 47, 48ff, 101,
    128n4, 133, 140, 141,
    168n1, 176, 186, 200,
    376ff, 379, 433
–*Logical structure of the World,*
    The, 50
Categoricity, 109, 122
Catholic University of Rio de Janeiro
    (PUC/ RJ), 292
Chaitin, Gregory J., ix, 282, 435
–complexity, 282
Chandrasekhar, Subrahmanyan, 347
–& Wright, 426
–& Wright, 348, 361n34
Chang, Chen Chung, 112
Charlemagne, 2, 4
Chernowitz (*Chernivitsi*), 12, 46
Choice function, 236
Christoffel, Elwin Bruno, 323
Church, Alonzo, 99, 126, 154, 222,
    264, 268ff, 283n11, 358,
    362n51, 366, 380, 383,
    419
–λ-calculus, 222, 265, 269, 383
–Church-Turing thesis, 222, 265,
    366
–hypothesis, 270
–logistic method, 270
City College/University (CUNY),
    NY, 125, 402
Cockcroft, John, 334
Cohen, Jeffrey M., 428
Cohen, Paul J., 97, 139, 363, 367,
    396ff, 398ff, 405n6, 417
Columbia University (NY), 126,
    134, 298
Compactness theorem, 116ff
Completeness, 101ff, 137
–of first-order logic, 101ff
Computability, vi, 222, 260, 264ff,
    303, 378
Computers, 87, 149, 218, 260,
    263ff, 298, 342, 381, 409
–and brains, 278ff

–complexity theory, 234
–music, 382
–simulation of thought cf.
    Penrose, neural networks,
    280
Concepts, 302
Conradi, Peter, 369, 386n8
Consistency, 105ff, 133, 154, 190,
    226, 236, 243, 244, 252,
    397
–ω, 154
–simple, 154
Constructability, 303
Constructible
–universe of, Gödel's, 243, 366
Constructible sets, 179, 237, 260
–universe of, Gödel's, 237
–universe of, von Neumann's,
    237
Construction, in geometry, 77
Contingent properties, 353
Continuum, 97, 235
Continuum hypothesis (CH), 139,
    169n13, 179, 189, 197,
    234ff, 242ff, 300, 307, 363,
    367, 396, 397, 399, 408,
    410ff
Copeland, B. Jack, 270, 454
–Copeland, Posy & Shagrir (CPS,
    2013), 270, 275, 282n3
Coriolis force, 337
Correspondence lemma, 146, 147
Cosmological constant, 329ff, 338ff
Cosmology, x, 303, 330ff, 363, 427
–stability, cosmos, 342
Crick, Francis, 369
Crocco, Gabriella, 302, 313n5, 333,
    416
Czechoslovakia, 35, 71, 192, 229,
    253, 385

D

Darwin, Charles, 439